Sensors and Circuits

SENSORS, TRANSDUCERS, and SUPPORTING CIRCUITS for ELECTRONIC INSTRUMENTATION, MEASUREMENT and CONTROL

Joseph J. Carr

PTR Prentice Hall, Englewood Cliffs, New Jersey 07632

Library of Congress Cataloging-in-Publication Data

Carr, Joseph J.

Sensors and circuits : sensors, transducers, and supporting circuits for electronic instrumentation, measurement, and control / Joseph Carr.

p. cm.

Includes bibliographical references and index

ISBN 0-13-805631-5

1. Transducers. 2. Detectors. 3. Biosensors. I. Title.

TK7872.T6C375 1993

681′.2—dc20

92-14958
CIP

Editorial Assistant: Diane Spina
Prepress and manufacturing buyer: Mary McCartney
Cover design: Wanda Lubelska
Copyeditor: Barbara Liguori
Proofreader: Kristen Cassereau, Carol A. Flynn, and Steven Fine

The publisher offers discounts on this book when ordered in bulk quantities. For more information, contact:

Corporate Sales Department
PTR Prentice Hall
113 Sylvan Avenue
Englewood Cliffs, NJ 07632.

Phone: 201-592-2863
FAX: 201-592-2249

Printed in the United States of America
10 9 8 7 6 5 4 3 2 1

ISBN 0-13-805631-5

Prentice-Hall International (UK) Limited, *London*
Prentice-Hall of Australia Pty. Limited, *Sydney*
Prentice-Hall Canada Inc., *Toronto*
Prentice-Hall Hispanoamericana, S.A., *Mexico*
Prentice-Hall of India Private Limited, *New Delhi*
Prentice-Hall of Japan, Inc., *Tokyo*
Simon & Schuster Asia Pte. Ltd., *Singapore*
Editora Prentice-Hall do Brasil, Ltda., *Rio de Janeiro*

Contents

6 Force and Pressure Sensors 125

7 Vibration and Acceleration Sensors 158

8 Electrodes for Bioelectric Sensing 166

Preface

Sensors and transducers are the eyes and ears of modern measurement instrumentation and control systems. Many types of machines depend on sensors to provide input data about the environment. Sensors represent both one of the oldest segments of the electronic industry and one of the most modern. Never before have so many different high-quality sensors and transducers been available to designers. In an age when the computer and advanced analog circuitry are making measurement and control advanced arts, it is no surprise that a vast array of old and very modern sensors are needed to interface with the physical world. In short, a sensor is a machine's way to "see," "hear," and "touch" the environment.

The goal of this book is to provide a representative overview of sensors, how they work, how they are applied, and what basic electronic circuits are needed to support them. It is intended to be a practical book that will benefit a variety of professional users of sensors and the instruments they support. It is hoped that it will be useful for many years to come.

Acknowledgments

The author wishes to thank the following for their assistance: Dorothy Rosa, editor of *Sensors: The Journal of Machine Perception*; Charles E. McCullough, MSEE, CCE, Director of Biomedical Engineering at Greater Southeastern Community Hospital in Washington, D.C.; and Mr. John M. Brown of Burr-Brown Corporation. Thanks are also extended to Dr. Michael J. Shaffer, director of The Bioelectronics Laboratory, The George Washington University Medical Center and Professor in the Department of Anesthesiology, and Dr. Marvin Eisenberg of the Department of Electrical Engineering, School of Engineering and Applied Science, The George Washington University, Washington, D.C., for their many years of mentoring and friendship.

Joseph J. Carr, MSEE, CCE

Transducers, Sensors, and Signal Processing

1

This book is about electronic sensors, the artificial sense organs of instruments and machines. Widely used in both analog and digital instrumentation systems, sensors provide the interface between electronic circuits and the "real world" where things happen. The world of electronics would remain a curiosity for laboratory scientists—and a very few others—were it not for sensors.

A *sensor*, or *transducer* (the words are roughly equivalent for our purposes), is a device that converts energy derived from a physical phenomenon into an electrical current or voltage, for purposes of measurement, control, or information (as in scientific research).

ELECTRONIC INSTRUMENT SYSTEMS

Figure 1–1 shows a block diagram for a generic electronic instrument. Although this figure is merely an example, it could easily serve for a large class of actual analog and digital instruments presently on the market. The principal stages or functions of an electronic instrument are *input parameter* (*stimulus*); *sensor* or *transducer*; *input functions*; *amplification* and *signal processing*; *output functions*; and *display*, *recording*, or other means for presenting the data.

Physical stimulus. The physical stimulus that is sensed in an instrumentation system may be temperature, light, displacement, fluid or gas flow, electrical resistance, electrical potential, or any of a host of other physical parameters. The particular stimulus is not important in a discussion of generic

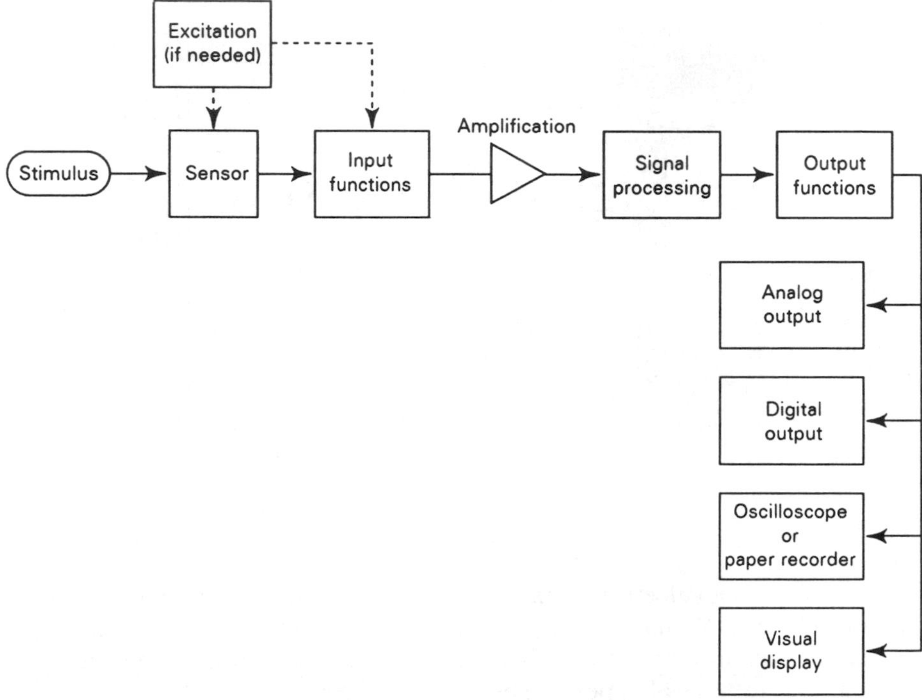

Figure 1–1 A typical sensor-based instrumentation system.

parameters. The particular stimulus is not important in a discussion of generic instruments but becomes important when specific applications are being discussed. Various chapters discuss the applicable sensors for specific forms of physical parameters or stimuli.

Sensors and transducers. The *sensor* or *transducer* is a device that is capable of responding to the applied stimulus and producing an electrical output signal that corresponds to the value of the applied stimulus.

Examples of typical sensors are shown in Fig. 1–2. The device shown in Fig. 1–2(a) is a Grass FT-3 force-displacement sensor. It consists of a Wheatstone bridge strain gage to measure minute forces (in units of gram·force) or minute deflections from the rest position. The devices in Fig. 1–2(b) are photoplethysmograph (PPG) blood oxygen monitors. These sensors are used to measure the blood O_2 level without physically invading the body. They utilize a property of blood, namely, the *isosbestic point*, the wavelength (color) at which the absorptivity of oxygenated and deoxygenated blood is the same (a wavelength of about 800 nm), as the transducible reference point.

There is often some ambiguity in the use of the words *sensor* and *transducer*, and in many cases they are properly used interchangeably. A transducer is a device that converts energy from one form to another (e.g., pressure

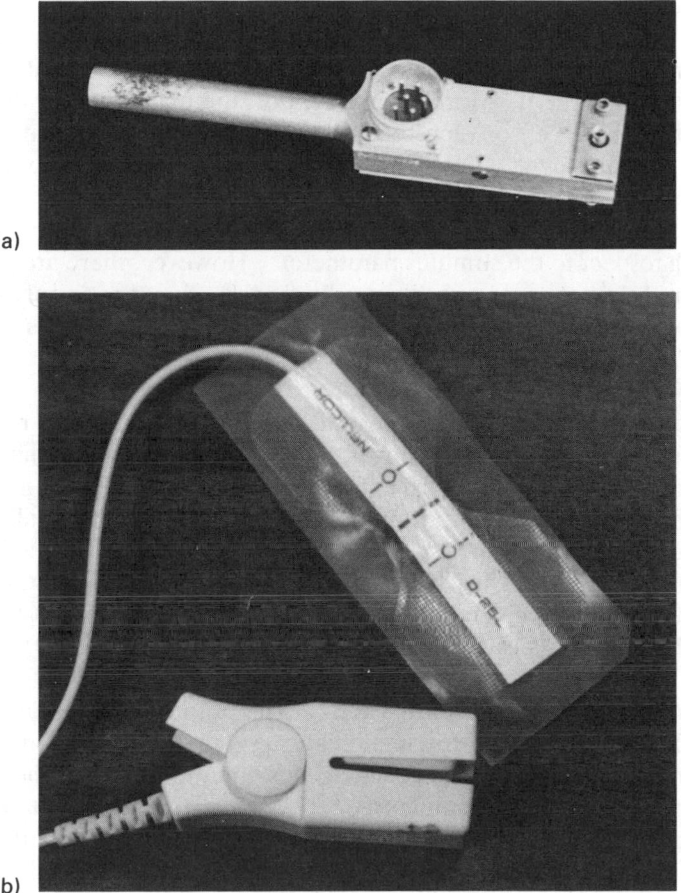

Figure 1–2 (a) Force-displacement sensor; (b) Pulse oximeter sensors used to noninvasively measure blood oxygen level.

to an electrical potential), whereas a sensor may or may not make some sort of conversion, at least in an obvious way (e.g., a biomedical electrode). Thus, an electrode used in medical electrocardiograph (ECG) recording is a sensor but not a transducer, but a pressure transducer is both a sensor and a transducer. A loudspeaker transduces an electrical audio frequency signal into a mechanical acoustical vibration. It is an *output transducer* and not a sensor at all.*

Some physical parameter or stimulus (e.g., temperature, flow, pressure, or displacement) affects the output state of a sensor; the sensor is a device

*For those who know more than is necessary for this discussion, it is recognized that certain loudspeakers, notably those of the *dynamic PMMC* design, can also serve as microphones—so they are both sensors and output devices.

that produces an output signal that is proportional to the applied parameter. Thus, the output of the sensor will be either a voltage or current that represents the parameter being measured (e.g., a temperature sensor that outputs a voltage V of 10 mV/K). More often than not, the magnitude of the voltage or current from the sensor represents the magnitude of the parameter at the instant of measurement. Over time, this voltage or current represents the time-history of the parameter.

The most desirable sensors have an output signal characteristic that is linear with respect to the stimulus parameter. However, there are also many useful transducers that are either quasi-linear (i.e., linear over only a portion of their total range) or even nonlinear. Such transducers are often used over a limited range, or they must be artificially linearized.

Input functions. The purpose of the input circuit is to receive the signal from the transducer and convert it into whatever form (usually a voltage) that is required by the circuits to follow. In this section of the instrument, interfacing becomes very important. The input functions usually include amplification, but they can also include an ac or dc excitation voltage (especially in the case of Wheatstone bridge sensors), dc level shifting, or isolation of the input circuit from the remainder of the instrument (common in medical instruments because of patient safety considerations).

Signal processing/amplification. The output signal from most sensors is not usually suitable for immediate display. Rather, some form of signal conditioning is usually needed. This conditioning may be only amplification, but it may also include frequency-selective filtering; mathematical operations such as differentiation, integration, "logging" or "antilogging"; or simple dc level translation. In other cases the signal processing uses the analog circuit, in effect, as a fixed-program, dedicated analog computer to solve for a mathematical expression. Some of these functions are most reasonably assigned to analog circuits, whereas others are most reasonably assigned to either digital circuits or computer software. In each case the designer must decide the proper choice for the problem at hand.

Output functions. The output of the instrument must often be processed in some manner before it can be displayed. The output functions may include power amplification (as in the case of a control system motor driver), digitization for input to a computer, or voltage scaling for ease of reading by a human operator.

Output and display. Finally, for an instrument to be useful there must be a display, data storage, or a control function. Various forms of output and display function are available, depending on the need. An analog output may be necessary, especially if the signal must be passed to some other instrument for further processing or application. Alternatively, the signal may be digitized (or, indeed, originally generated as a digital signal) and applied to some form of digital display. A y-time or x-y oscillographic display can be interfaced with either a strip-chart recorder or a cathode ray oscilloscope. Various forms of visual (sometimes called alphanumeric) display devices are used, especially when the output information is in the form of a numerical value (e.g., 100 torr of pressure).

The visual display or recording device may be a dc meter movement (Fig. 1–3), an oscilloscope (Fig. 1–4), a strip-chart recorder (Fig. 1–5), a digital printer (Fig. 1–6), a video terminal, or even a simple GO–NO GO lamp. In some systems the signal has to be digitized before it can be stored as data in a digital computer system. In still other systems the output signal is simply stored in a mass storage device for later use.

DIRECT, INDIRECT, AND INFERENTIAL MEASUREMENTS

One way to categorize instrumentation systems is according to how a measurement is derived. It is reasonable to speak of three different categories: direct measurements, indirect measurements, and inferential measurements (although some authorities generally lump together the latter two categories).

The *direct measurement* is, as the term implies, a measurement that is made of the parameter itself without the need for interpretation, calculation, or any form of interpolation. The force-displacement sensor in the preceding discussion is an instrument that directly measures the applied force. Another example of an instrument that performs the direct type of measurement is a pressure gauge

Figure 1–3 Zero-center analog meter.

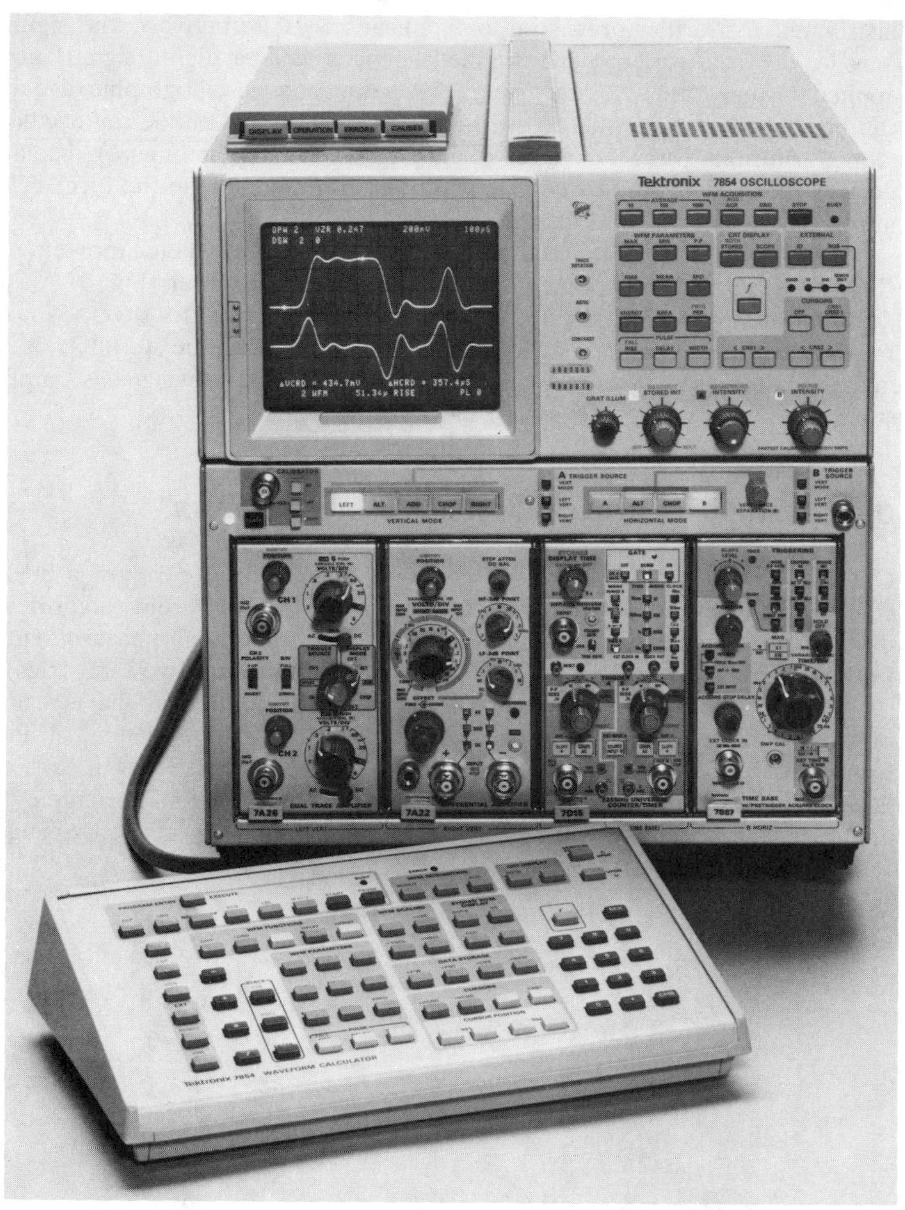

Figure 1–4 Modern oscilloscope uses plug-in modules to enhance versatility.

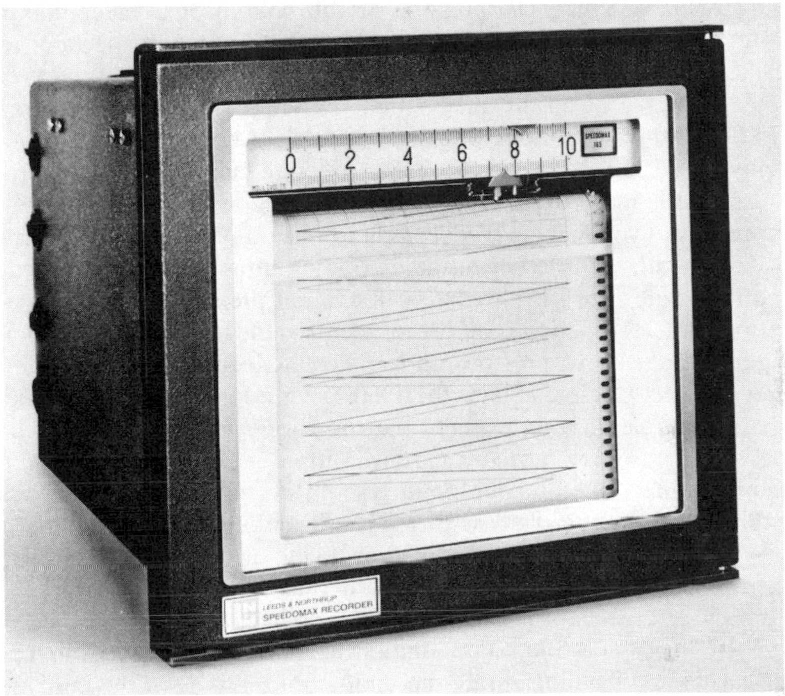

Figure 1–5 Analog strip-chart recorder used to make a permanent paper record of data.

Figure 1–6 Small digital printer.

in a hydraulic system. The pressure at the bottom of a water tank can be measured directly by a gauge connected to the bottom of the tank (and by Pascal's principle, in a closed system, everywhere else in the tank as well).

An example of an instrument that performs an *indirect measurement* is the PPG blood oxygen meter discussed earlier. Another common example is the sphygmomanometer, which is used to measure human blood pressure. The measurement is made by occluding the underlying arteries against the bone of the upper arm by inflating a bladder called a blood pressure cuff. Certain key *Korotkoff sounds* (detected by a stethoscope or ultrasonic detector) occur when the cuff pressure drops below the peak arterial pressure (systolic) value, and these sounds cease when the cuff pressure equals the minimum (diastolic) value. Sphygmomanometry was discovered by Nicolas Korotkoff in 1905, but its use did not become widespread until the 1930s after a long validation period.

An example of an instrument that performs an *inferential measurement* is the system used to measure cardiac output. The output of the heart, measured in liters of blood per minute (l/min), is not easily measured directly or indirectly, except in the case of a patient undergoing open heart surgery. However, it is possible to thread a thermistor or other temperature sensor through the circulatory system (from an outside site) to a location in the pulmonary artery just outside the output chamber of the right side of the heart. By injecting a cold saline solution into the vena cava (at the input side of the heart) and then measuring the temperature profile at the output side, it is possible to calculate the cardiac output from the temperature data. Thus, an inferential measurement is one in which there is either a correlation (but not necessarily causal) or a statistical relationship between an easily (or safely) measured parameter and one that is not easily or safely measured or that is impossible to measure. In this type of measurement the validation process is critically important and is bound to be the most controversial issue about the design.

A related inferential measurement is sometimes used in chemical or petroleum pipelines. These are, after all, not too dissimilar to blood vessels. In these applications an optical or radioactive die injectate is used rather than cold saline solution. A detector downstream measures the concentration time-history as the injectate passes. When integrated, processed, and passed through an equation, the concentration data yield the flow rate. Collectively these methods, in both medical and industrial settings, are called *dye-dilution flow measurements*.

TRANSDUCTION

In studying sensors it is necessary to understand the linked concepts of transduction and transducible property. A *transducible property* is a characteristic of the physical event that is singularly able to represent that parameter and can be transformed into an electrical signal by some device or process. For example, carbon dioxide (CO_2) absorbs electromagnetic radiation of wavelengths of 2.7, 4.3, and 14.7 microns (1 micron = 10^{-6} m). Although water also absorbs 2.7 micron radiation to a small degree, it is possible to make an infrared (IR) "light" sensor that will respond to either 4.3 or 14.7 microns, or to all three wavelengths, to measure CO_2 content of a gas such as air.* Carbon dioxide meters, monitors, and detectors use infrared sensors, as do certain industrial, medical, and general scientific instruments. *Transduction* is the process of converting the transducible property into an electrical signal that can be input to an instrument.

PASSIVE VERSUS ACTIVE SENSORS

Another issue in discussions of sensors is the distinction between passive sensors and active sensors. Unfortunately, competing texts use exactly opposite definitions of these terms. This text adopts the form that is used by most people in the electronic instruments field and is also consistent with usage in other technical areas.

An *active sensor* is one that requires an external ac or dc electrical source to power the device. An example of an active sensor is the resistive strain gage pressure sensor that requires a +7.5 V dc regulated power supply to operate. Without that external excitation potential there is no output from the sensor.

A *passive sensor*, on the other hand, is one that either provides its own energy or derives it from the phenomenon being measured. An example of a passive sensor is the thermocouple (which is often used to measure temperature in research settings).

SOURCES OF SENSOR ERROR

Sensors, like all other devices, suffer from certain errors. To maintain consistency we define an *error* to be the "difference between the measured value and the true value." Although the full range of possible errors is beyond the scope of this book, it is possible to break them into five basic categories: *insertion errors, application errors, characteristic errors, dynamic errors*, and *environmental errors*.

Insertion errors. This class of errors occurs when the act of inserting the sensor into the system changes the parameter being measured. This problem is a general problem with electronic measurements, indeed all measurements. For example, a voltmeter used to measure the voltage in a circuit must have an inherent impedance that is very much larger than the circuit

*IR includes electromagnetic waves with wavelengths longer than those of visible light but shorter than millimeter-length microwaves.

impedance; otherwise circuit loading will occur, and the reading will be in significant error. Possible sources of this form of error include using a transducer that is too large for the system (e.g., pressures), one that is too sluggish for the dynamics of the system, or one that self-heats to the extent that excessive thermal energy is added to the system. Nineteenth-century British physicist Lord Kelvin formulated a "first rule of instrumentation," which stated that "the measuring instrument must not alter the event being measured."

Application errors. These errors are operator-caused (as in the proverbial "cockpit trouble" referred to by airplane mechanics). Again, so many of these errors are possible that we give just a few illustrative examples. Errors seen in temperature measurements include incorrect placement of the probe and erroneous insulation of the probe from the measurement site. The latter problem often occurs in clinical medicine when the sanitary cover over the probe of a digital thermometer is not properly seated (Fig. 1–7). Examples of application errors seen in blood pressure sensor applications include failure to purge the system of air and other gases ("bubbles in the line") and incorrect physical placement of the transducer (above or below the heart line), so that a positive or negative pressure head is erroneously added to the correct reading.

Characteristic errors. This category is most often meant when discussing errors without otherwise qualifying the term. Characteristic errors are those that are inherent in the device itself, that is, the difference between the ideal published characteristic transfer function of the device and the actual characteristic. This form of error may include a dc offset value (e.g., a false pressure head), an incorrect slope, or a slope that is not perfectly linear.

Dynamic errors. Many sensors are characterized and calibrated in a static condition, that is, with an input parameter that is either static or quasi-static in nature. Many sensors are heavily damped, so they will not respond to rapid changes in the input parameter. For example, thermistors tend to require many seconds to respond to a step-function change in temperature.

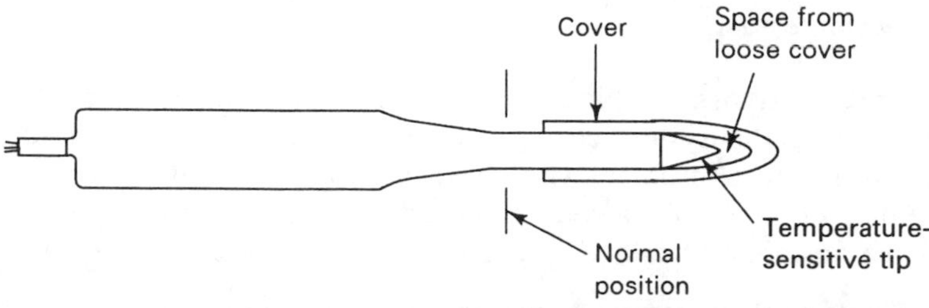

Figure 1–7 Loose cover is an example of an application error.

That is, a thermistor in equilibrium will not immediately jump to the new resistance on an abrupt change in temperature. Rather, the device will change slowly toward the new value. Thus, if an attempt is made to follow a rapidly changing temperature with a sluggish sensor, the output waveform will be distorted because it contains error. Factors to consider with respect to dynamic errors include response time, amplitude distortion, and phase distortion.

Environmental errors. These errors derive from the environment in which the sensor is used. They most often include temperature but may also include vibration, shock, altitude, chemical exposure, or other factors. These factors most often affect the characteristic errors of the sensor, so they are often lumped together with that category in practical application.

SENSOR TERMINOLOGY

Sensors, like other areas of technology, include their own terminology that must be understood before they can be properly applied. The following section defines some of the most common terms.

Sensitivity. The *sensitivity* of the sensor is defined as the slope of the output characteristic curve ($\Delta y/\Delta x$ in Fig. 1–8). More generally, it is the *minimum input of physical parameter that will create a detectable output change*. In some sensors the sensitivity is defined as the input parameter change required to produce a standardized output change. In still others it is defined as an output voltage change for a given change in input parameter. For example, a typical blood pressure transducer may have a sensitivity rating of 10 μV/V/mm Hg; that is, there will be a 10 μV output voltage for each volt of excitation potential and each millimeter of mercury of applied pressure.

Sensitivity error. The *sensitivity error* (shown as a dotted curve in Fig. 1–8) is a departure from the ideal slope of the characteristic curve. For example, the pressure transducer discussed in the previous paragraph may have an actual sensitivity of 7.8 μV/V/mm Hg instead of 10 μV/V/mm Hg.

Range. The *range* of the sensor is the maximum and minimum values of applied parameter that can be measured. For example, a given pressure sensor may have a range of −400 mm Hg to +400 mm Hg; however, it is often the case that the positive and negative ranges are unequal. For example, a certain medical blood pressure transducer is specified to have a minimum (i.e., vacuum) limit of −50 mm Hg (y_{min} in Fig. 1–8) and a maximum (pressure) limit of +450 mm Hg (y_{max} in Fig. 1–8). This specification

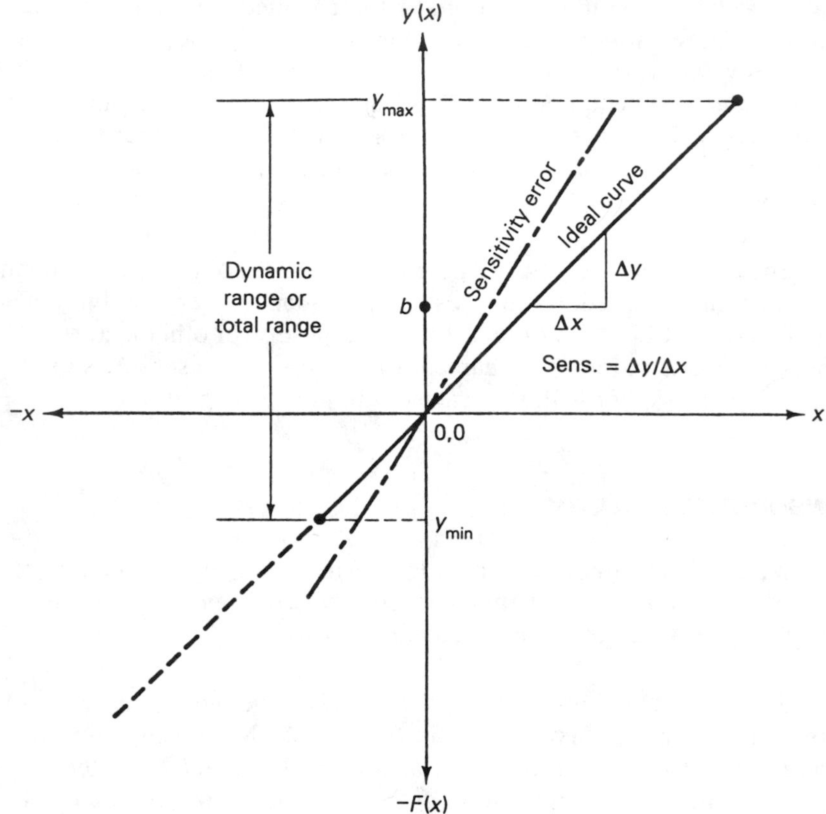

Figure 1–8 Ideal curve and sensitivity error.

is common and reflects the needs of a specific customer (few blood pressures are negative!).

Dynamic range. The *dynamic range* is the total range of the sensor from minimum to maximum. That is, in terms of Fig. 1–8, $R_{dyn} = y_{max} - |-y_{min}|$.

Precision. *Precision* refers to the degree of *reproducibility* of a measurement. In other words, if exactly the same value was measured a number of times, an ideal sensor would output exactly the same value every time. But real sensors output a range of values distributed in some manner related to the actual correct value. For example, suppose a pressure of 150 mm Hg, exactly, is applied to a sensor. Even if the applied pressure never changes, the output values from the sensor will vary considerably. Some subtle problems occur in the matter of precision when the true value and the sensor's mean value are not within a certain distance of each other (e.g., the $1 - \sigma$ range of the statistician's normal distribution curve).

Resolution. *Resolution* is the smallest detectable incremental change of input parameter that can be detected in the output signal. This value can be expressed either as a proportion of the reading (or the full-scale reading) or in absolute terms.

Accuracy. The *accuracy* of the sensor is the maximum difference that will exist between the actual value (which must be measured by a primary or good secondary standard) and the indicated value at the output of the sensor. Again, the accuracy can be expressed either as a percentage of full scale or in absolute terms.

Offset. The *offset error* of a transducer is defined as the output that will exist when it should be zero or, alternatively, the difference between the actual output value and the specified output value under some particular set of conditions. An example of the first situation in terms of Fig. 1–8 would exist if the characteristic curve had the same sensitivity slope as the ideal but crossed the *y*-axis (i.e., output) at *b* instead of zero. An example of the other form of offset is seen in the characteristic curve of a chemist's pH electrode shown in Fig. 1–9. The ideal curve will exist at only one temperature (usually 25°C), while the actual curve will lie between the minimum and maximum temperature limits depending on the temperature of the sample and electrode.

Linearity. The *linearity* of the transducer is an expression of the extent to which the actual measured curve of a sensor departs from the ideal curve. Figure 1–10 shows a somewhat exaggerated relationship between the ideal, or least-squares-fit, line and the actual measured or *calibration* line. (*Note:* in most cases the static curve is used to determine linearity, and this may deviate somewhat from a dynamic linearity.) Linearity is often specified in terms of *percentage of nonlinearity*, which is defined as

$$\text{Nonlinearity } (\%) = \frac{D_{\text{in(max)}}}{\text{IN}_{\text{f.s.}}} \times 100\% \qquad (1\text{–}1)$$

where $D_{\text{in(max)}}$ is the maximum input deviation
$\text{IN}_{\text{f.s.}}$ is the maximum, full-scale input

The static nonlinearity defined by Eq. (1–1) is often subject to environmental factors, including temperature, vibration, acoustical noise level, humidity, and so forth. It is important to know under what conditions the specification is valid.

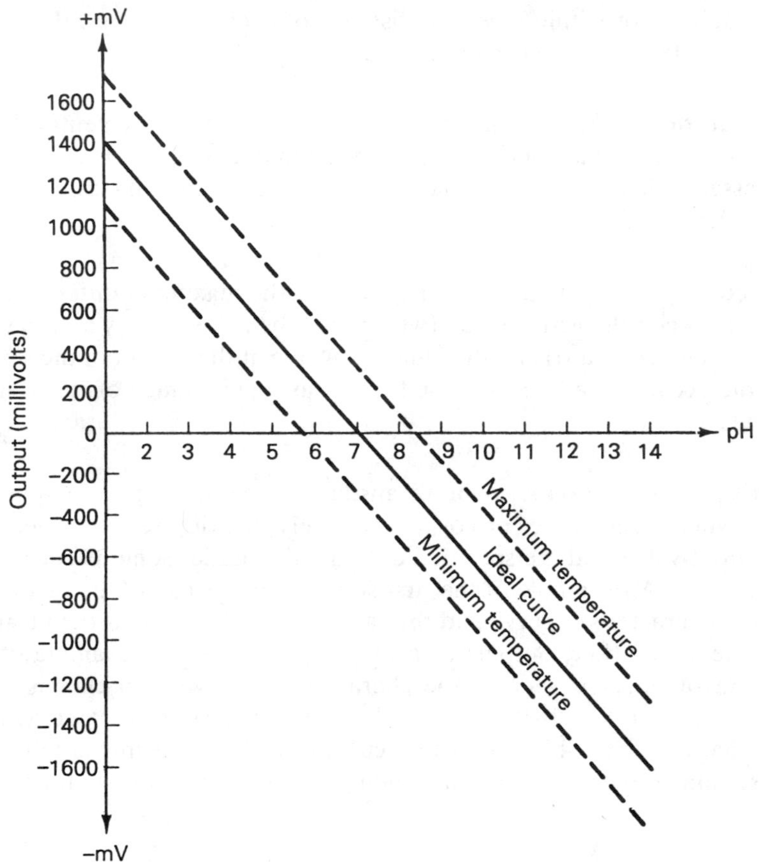

Figure 1–9 Typical pH electrode characteristic curve showing temperature sensitivity.

Monotonicity. Another measure of linearity is monotonicity. A *monotonic curve* is one in which the dependent variable always either increases or decreases as the independent variable increases. In such a curve, $f(x + 1)$ will always be greater than $f(x)$ when x increases from x to $x + 1$, if the slope is positive. Similarly, a monotonic curve with a negative slope will always show $f(x + 1)$ less than $f(x)$ under the same conditions. The opposite never occurs. If a pressure sensor is monotonic, for example, increasing the applied pressure in steps of, say, 10 torr will always yield a slightly more positive output voltage. The output voltage will never be seen to drop when the pressure goes from P to $P + 10$ torr. If a sensor lacks monotonic linearity, then an unexpected result will occasionally be seen.

Do not confuse monotonicity with other forms of nonlinearity in which the rate of increase or decrease varies, as when the output function is S-shaped

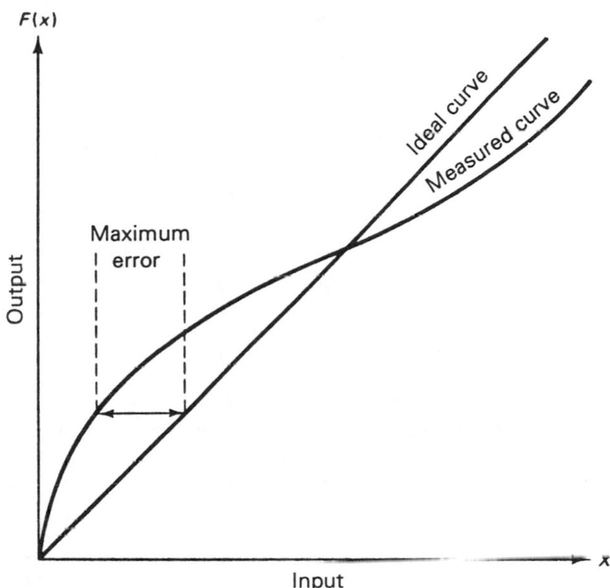

Figure 1–10 Ideal versus measured curve showing linearity error.

or otherwise not a straight line. Some sensors have a quadratic or cubic output function but still qualify as monotonic. Similarly, ordinary random variation, which is present in all sensing and measuring applications, is not necessarily evidence of poor monotonicity.

Hysteresis. A transducer should be capable of following the changes in the input parameter regardless of the direction from which the change occurs; *hysteresis* is the measure of this property. Figure 1–11 shows a typical hysteresis curve. Note that it matters from which *direction* the change is made. Approaching a fixed input value (e.g., point B) from a higher value (e.g., point P) will result in a different output from that obtained by approaching the same value from a lesser value (point Q or zero). Note that input value B can be represented by $F(x)_1$, $F(x)_2$, or $F(x)_3$ depending on the immediate previous value; clearly, there is an error due to hysteresis.

Response time. A sensor does not change output state immediately when an input parameter change occurs. Rather, it will change to the new state over a period of time, called the response time (T_r in Fig. 1–12). The *response time* can be defined as the *time required for a sensor output to change from its previous state to a final settled value within an error tolerance band of the correct new value*. This concept is somewhat different from the notion of the *time constant* T of the system.

The curves in Fig. 1–12 show two types of response time. In Fig.

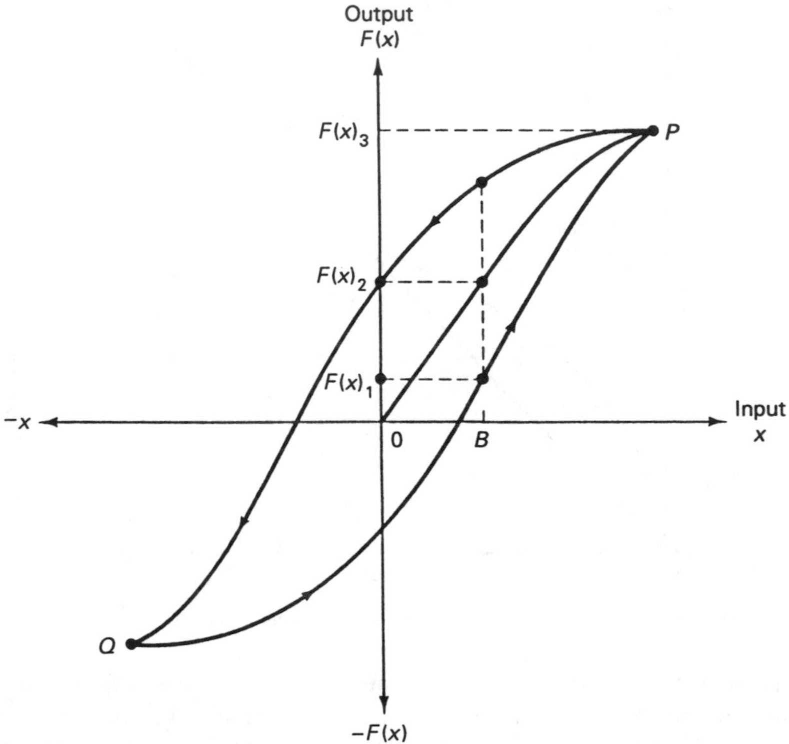

Figure 1–11 Hysteresis curve.

1–12(a) the curve represents the response time following an abrupt positive-going step-function change of the input parameter. The form shown in Fig. 1–12(b) is a decay time (T_d to distinguish from T_r, for they are not always the same) in response to a negative-going step-function change of the input parameter.

Dynamic linearity. The *dynamic linearity* of the sensor is a measure of its ability to follow rapid changes in the input parameter. Amplitude distortion characteristics, phase distortion characteristics, and response time are important in determining dynamic linearity. Given a system of low hysteresis, the amplitude response is represented by the equation

$$F(x) = ax + bx^2 + cx^3 + dx^4 + ex^5 + \ldots + K \tag{1–2}$$

where $F(x)$ is the output signal
 x-terms represent the input parameter and its harmonics
 K is an offset constant (if any).

The harmonics become especially important when the error harmonics generated by the sensor action fall into the same frequency bands as the natural

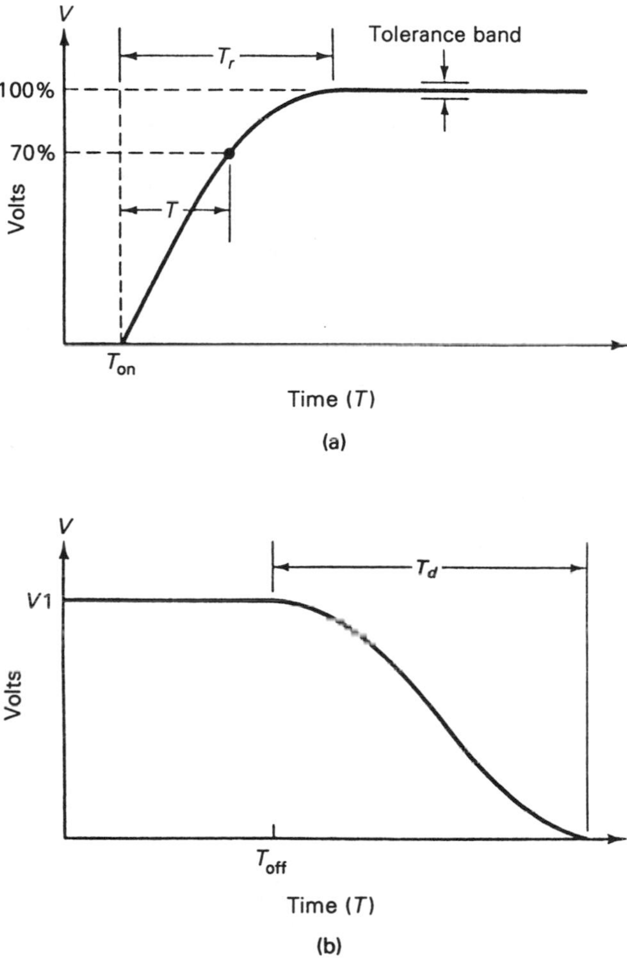

Figure 1–12 (a) Rise-time definition; (b) fall-time definition.

harmonics produced by the dynamic action of the input parameter. All continuous waveforms are represented by a Fourier series of a fundamental sine wave and its harmonics. In the case of any nonsinusoidal waveform (including time-varying changes of a physical parameter), harmonics will be present that can be affected by the action of the sensor.

The nature of the nonlinearity of the calibration curve [(Figs. 1–13(a) and 1–13(b)] reveals information about which harmonics are present. The calibration curve in Fig. 1–13(a) (shown as a dotted line) is *asymmetrical*, so only *odd* harmonic terms exist. If we assume a form for the ideal curve of $F(x) = mx + K$, Eq. (1–2) becomes for the symmetrical case

$$F(x) = ax + bx^2 + cx^4 + \dots + K \tag{1–3}$$

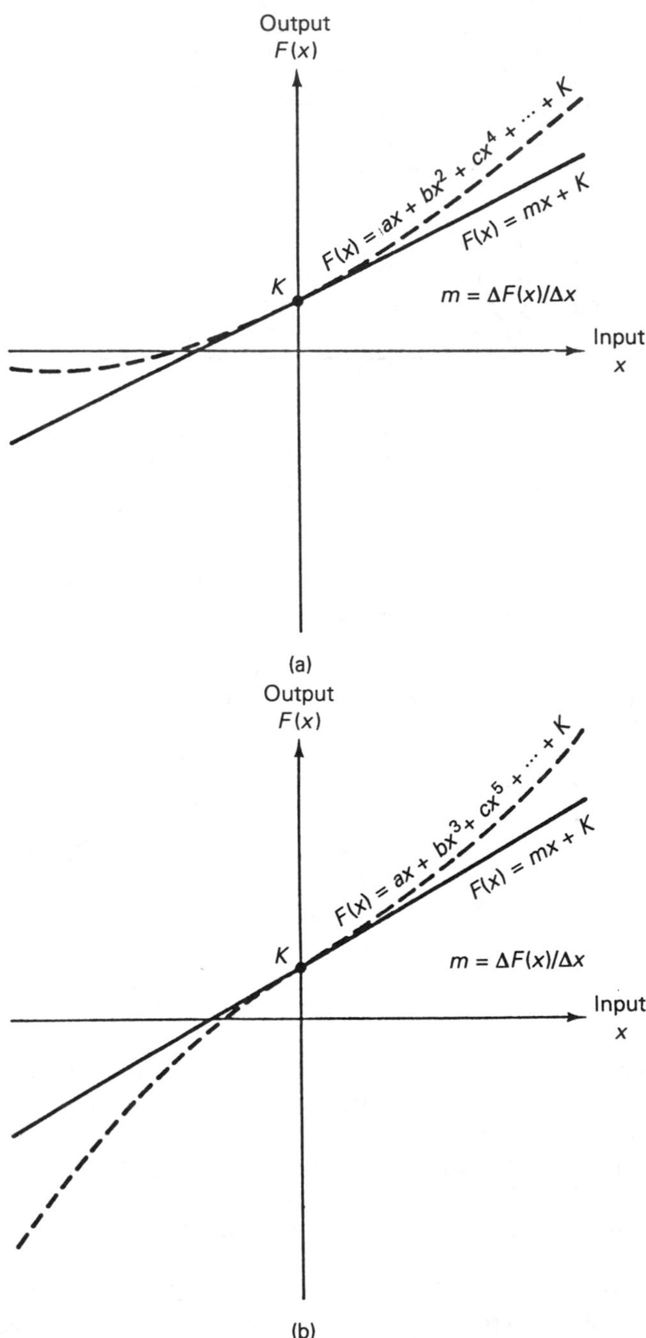

Figure 1–13 Output versus input signal curves showing (a) quadratic error; (b) cubic error.

In the calibration curve of Fig. 1–13(b) the indicated values are *symmetrical* about the ideal $mx + K$ curve. In this case $F(x) = -F(-x)$, and the form of Eq. (1–2) is

$$F(x) = ax + bx^3 + cx^5 + \ldots + K \tag{1–4}$$

In the next section we will examine some of the methods and signal processing criteria that can be adapted to biomedical applications to improve the nature of the data collected from the sensor.

METHODS AND SIGNAL PROCESSING FOR IMPROVED SENSING

The selection of sensors and the circuits that are connected to them are vital for ensuring that the data acquired will accurately represent the physical phenomenon or event being detected.

For proper operation in a dynamic input environment, the sensor selected should have a flat response curve, that is, one that is free of amplitude distortion, phase distortion (which almost invariably causes amplitude distortion), "ringing," or resonances.

The *frequency response* of the sensor and its signal processing system is an important characteristic that plays a role in these problems. Figure 1–14(a) shows a perfectly linear system in which the gain is constant over the entire spectrum of frequencies, that is, in an ideal theoretical system from "dc to daylight" and beyond.* But real systems do not have such characteristics. The specification that determines the ability of the system to handle the frequency spectrum of any given signal or physical parameter is the *bandwidth*. It is usually given in the form of a difference between the maximum and minimum frequencies that must be handled and has units of frequency.

Figure 1–14(b) shows the type of frequency response that might be found in a real system. In this example the gain is flat between two frequencies, and over this region the performance is similar to the ideal case. But beyond these points the gain falls off at a given slope. The breakpoint that defines the flat region is, by convention, taken to be the frequencies (F_L and F_H) at which the gain falls to 70.7 percent of its gain in the flat region. These points are known as the −6 dB points in voltage systems, and the −3 dB points in power systems.

When the frequency response is not entirely flat, phase distortion is likely. Figure 1–14(c) shows the situations where the phase shift of the system is a linear function of frequency (solid line) and also where it is a nonlinear function of frequency (dotted line).

**Note:* Gain is defined as the ratio of the output function to the input function. Because voltage gain A_v is used here for illustration purposes, the gain is V_o/V_{in}.

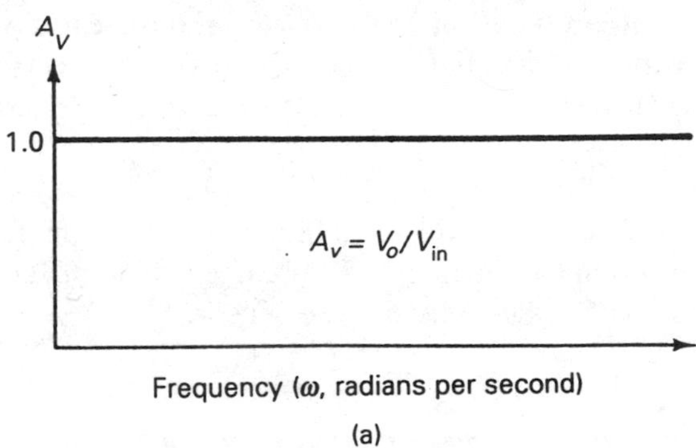

(a)

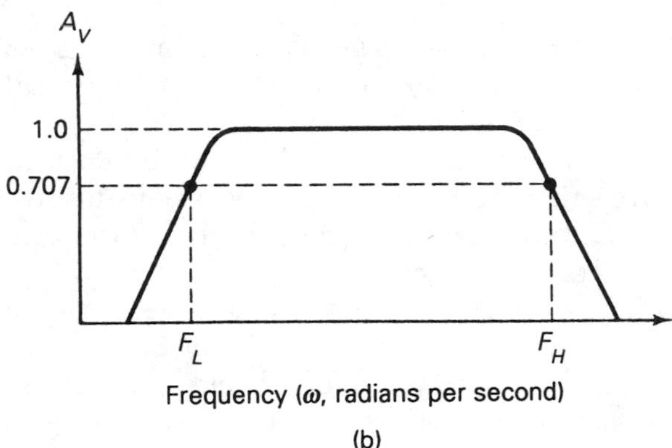

(b)

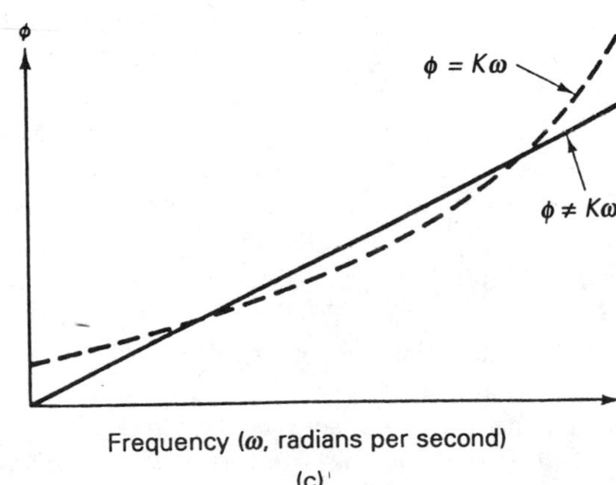

(c)

Figure 1–14 Frequency-response characteristics: (a) wideband; (b) bandpass; (c) typical for a sensor.

We can see the effects of phase distortion in the simplified diagram of Fig. 1–15. Figure 1–15(a) is the applied signal, for example, the output of an ideal sensor in response to step-function changes of the measured input parameter. If the signal processing electronics and the sensor mechanism itself are perfectly ideal, then the only effect of the change will be displacement in time t, as shown in Fig. 1–15(b). There will be no distortion of the shape of the wave. In the presence of phase distortion, however, the wave will be not only time displaced but also distorted. Figures 1–15(c) and 1–15(d) show two forms of distortion that can occur with phase nonlinearity.

A slightly different view of the same phenomenon is shown in Figs. 1–16 and 1–17. Consider a system in which the bandwidth can be varied across several limits, represented by curves *a, b*, and *c* in Fig. 1–16. Curve *c* represents the most restrictive of the three possibilities because it sharply limits both low- and high-frequency response, while curve *a* is the least re-

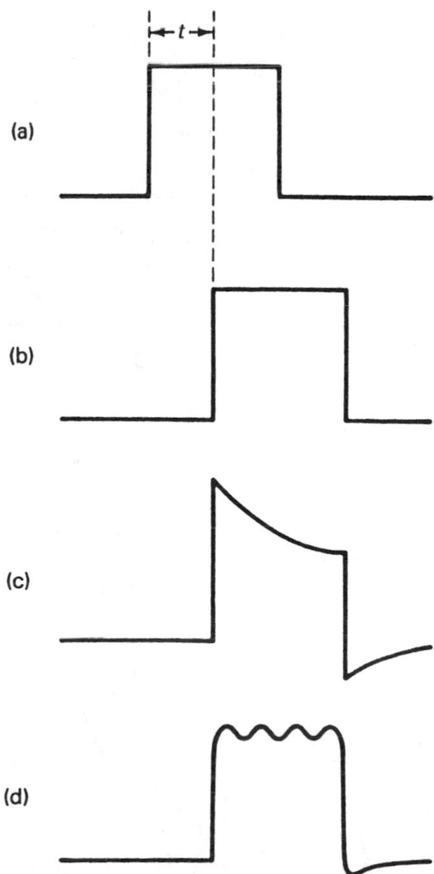

Figure 1–15 (a) Square wave; (b) propagation delay; (c) low-frequency rolloff; (d) ringing.

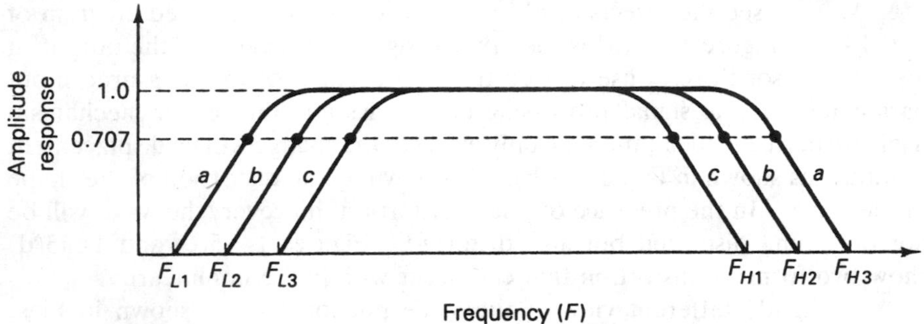

Figure 1–16 Bandpass frequency-response characteristics.

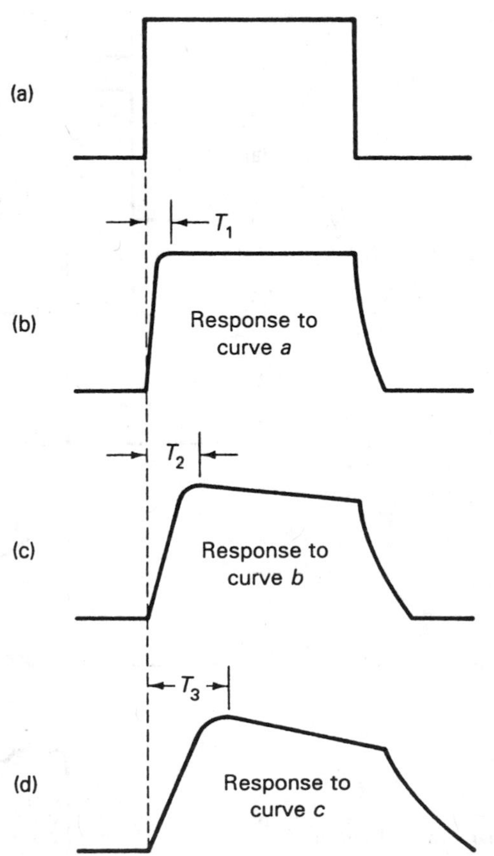

Figure 1–17 Response of curves from Fig. 1–16 to a square wave (a).

strictive. Note in Fig. 1–17 the various responses to the three bandwidths represented in Fig. 1–16.

These curves can be simulated by examining the response to square waves in *RC* filter networks. In fact, one of the problems that one must consider when using electronic filters is the effects of the −6 dB points on the applied waveform.

One might erroneously assume from the preceding discussion that the instrument designer should select amplifiers with as wide a bandwidth as possible. That is not the case, however, because the size of the bandwidth can cause other problems at least as severe as those that are solved. Noise, for example, is proportional to bandwidth. It is possible to eliminate the problems of noise and certain input signal problems such as ringing or resonances by proper selection of the frequency-response cutoff points. Thus, the selection of amplifier bandwidth and phase distortion characteristics is a trade-off between the need to make a high-fidelity recording of the input event and the other problems that can occur in the system.

CONCLUSION

Now that we've discussed some of the basic problems of sensors, we proceed to discussions of sensors and the associated circuits and instruments. The treatment herein is not exhaustive—indeed such would fill a number of very large books—but is intended, instead, to be sufficiently representative to permit a user or designer to gain some comfort with the subject.

RECOMMENDATIONS FOR FURTHER READING

Cobbold, Richard S. C. *Transducers for Biomedical Applications*. New York: John Wiley, 1974.

Geddes L. A., and L. E. Baker. *Principles of Applied Biomedical Instrumentation*. New York: John Wiley, 1968.

Carr, Joseph J., and John M. Brown. *Introduction to Biomedical Equipment Technology*. New York: John Wiley, 1981. Acquired and republished by Prentice Hall, Inc. (1989).

2 Supporting Electronics Circuits: Connection, Grounding, and Shielding

Sensors are used to measure physical parameters by producing a proportional output current or voltage signal that represents that parameter. For example, a thermocouple produces a voltage that is proportional to the temperature applied to its measurement junction. Similarly, a piezoresistive strain gage produces an output voltage that is proportional to strain on a resistance element, which in turn is proportional to an applied displacement, force, or pressure.

Although the number of different sensors and transducers is large and varied, from a circuit point of view there are only a few different basic forms of sensor output circuit configurations. These forms must be properly matched to the input of the circuit that follows the sensor, or trouble can result. In this chapter we examine some of the circuits needed to support sensors.

SENSOR OUTPUT CIRCUIT FORMS

Figure 2–1 shows an array of several different forms of model sensor circuits. In each circuit a current source, source resistance R, and (in some) a voltage source are shown. Figure 2–1(a) shows the standard single-ended grounded sensor.[1] The term *single-ended* means that one side of the sensor circuit is grounded. If neither side is grounded, then the sensor is said to be a single-ended floating sensor [Fig. 2–1(b)]. In the single-ended sensor the output signal is referenced either to ground or to a single common nongrounded point. This form is sometimes subject to serious interference from external fields, especially in the presence of strong audio frequency (AF), radio frequency

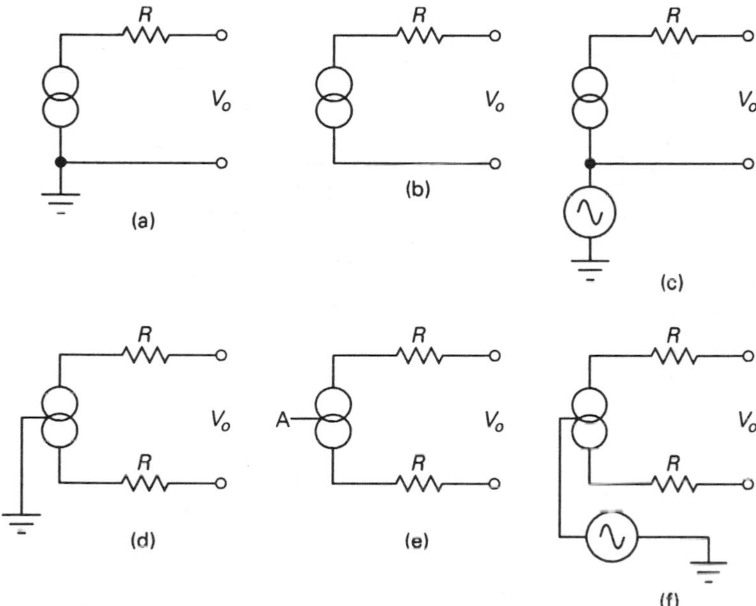

Figure 2–1 Input signal source configurations.

(RF), or 60 Hz power-line fields. A variant on the single-ended floating sensor is the single-ended floating driven-off-ground sensor of Fig. 2–1(c).

If a sensor drives the output through equal resistances, then it is said to be balanced. Figure 2–1(d) shows an example of a balanced grounded sensor. In this form of output circuit the sensor is referenced to ground through two equal resistances (both designated R). The version shown in Fig. 2–1(e) is an example of a balanced floating sensor. That is, the sensor is connected to a nongrounded common point A and outputs through two equal resistances R. The important point to remember about the balanced floating sensor is that it is both balanced and ungrounded. Finally, in Fig. 2–1(f) we see the balanced driven-off-ground sensor.

AMPLIFIER INPUT CIRCUITS

The output circuit of the sensor is usually connected to a signal processing circuit, most frequently an amplifier of some sort (although certain other circuits are also used occasionally). Unfortunately, there are several different types of amplifier input circuits, and not all sensors can be easily interfaced with all of them. Figure 2–2 shows four basic types of input circuits. Figure 2–2(a) depicts the Type I circuit, a single-ended input amplifier. The input circuit is modeled as a resistance to ground. Figure 2–2(b) shows the Type II circuit, which is modeled as a pair of differential inputs that see equal

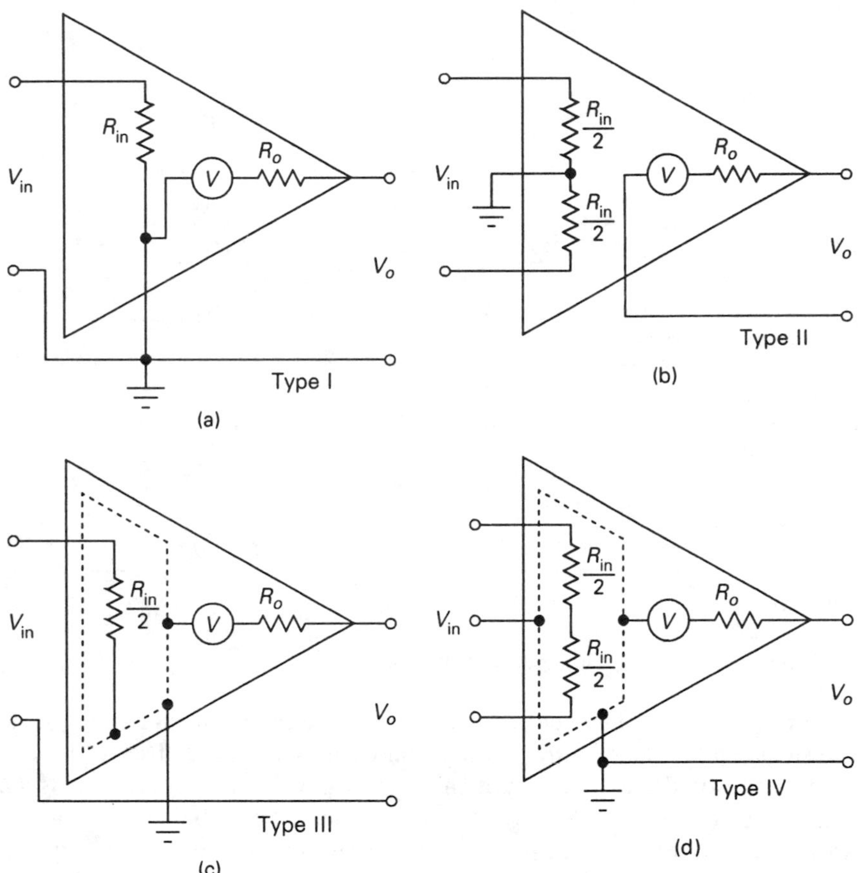

Figure 2–2 (a) Single-ended input equivalent circuit; (b) differential input; (c) single-ended amplifier equivalent circuit; (d) differential amplifier equivalent circuit.

resistances to ground. In both cases the output circuit is a voltage source in series with an output resistance. Figure 2–2(c) shows the Type III input circuit, which is single-ended floating and shielded. The input resembles the regular single-ended input of Fig. 2–2(a), but the input is grounded and protected from interference by a shield. In Fig. 2–2(d) we see the Type IV input circuit. This circuit resembles the Type II, except that the input circuit is protected by a shield, is floating, and is guarded (to be discussed later).

MATCHING SENSORS AND AMPLIFIERS

One cannot simply connect the various forms of sensors to the various types of amplifier input circuits without careful consideration. Figure 2–3 shows a general table relating the sensor and amplifier circuits. A Yes in a block

Input circuit type ⇩	Sensor Signal Source Form					
	A	B	C	D	E	F
I	See text	Yes	No	No	Yes	No
II	See text	Yes	No	Yes	Yes	No
III	Yes	Yes	Yes	No	Yes	No
IV	Yes	Yes	Yes	Yes	Yes	Yes

Figure 2–3 Chart crossing input amplifier types and signal sources.

means the combination (row vs. column) is recommended. A No means that there are problems in that particular combination, so it is not recommended.

There are two combinations that may or may not work, depending on the circumstances, so some degree of caution is required. For example, mixing a Type I input circuit with a Form A sensor output circuit requires consideration of signal levels. This combination should not be used when the output of the sensor is in the microvolt or millivolt range. Also, it is not a good idea to mix two grounds, that is, one each on the amplifier and the sensor. Either one ground should be eliminated or they should be joined together in a *single-point* (also called *star*) ground (to be discussed later). A similar problem occurs when interfacing a Form A sensor and a Type II amplifier input. Some differential amplifiers can be converted into a single-ended amplifier, but one must be certain in each case.

PRACTICAL SENSOR AMPLIFIERS

Sensor amplifiers can easily be constructed from operational amplifier integrated circuits (op-amp ICs) or certain other linear amplifier IC devices. There are three basic forms of amplifiers that are useful for sensor interfacing: single-ended (Fig. 2–4), differential (Fig. 2–5), and isolated (Fig. 2–6).

Single-ended amplifiers. Figure 2–4 shows three variations on the single-ended amplifier. The circuit of Fig. 2–4(a) is an op-amp inverting follower. The voltage gain is $-R_2/R_1$, and the input impedance is the resistance of R_1. As with all op-amp voltage amplifier circuits, the output impedance is low. A noninverting unity-gain follower is shown in Fig. 2–4(b). This circuit has a very high input impedance, a low output impedance (which means that impedance transformation takes place between input and output), and a voltage gain of 1. The power P gain, however, is larger than 1 because V_{in} and V_o are equal, yet $R_{in} >> R_o$. (*Note:* $P = V_2/R$.) Figure 2–4(c)

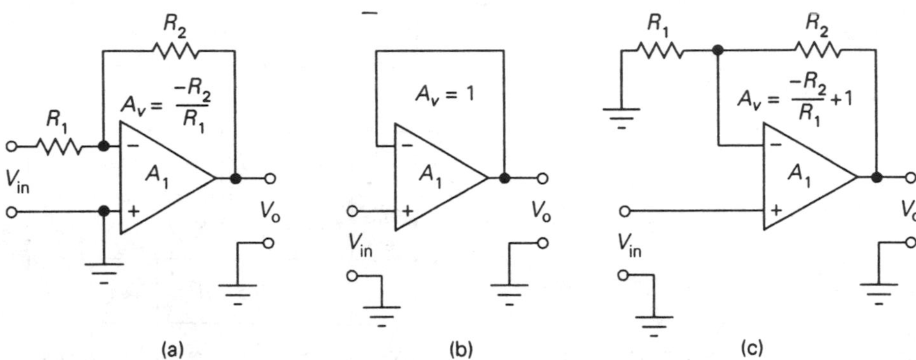

Figure 2–4 (a) Inverting follower cicuit; (b) noninverting unity-gain follower circuit; (c) noninverting follower with gain circuit.

shows the noninverting follower with gain circuit. This circuit has the same attributes as the unity-gain noninverting follower, except that the voltage gain is $[(R_2/R_1) + 1]$.

Differential amplifiers. A *differential amplifier* is one that has two balanced inputs such that one input is inverting $(-)$ and the other input is noninverting $(+)$. The inverting input produces an output signal that is out of phase with the input signal. The noninverting input produces an output signal that is in phase with the input signal.

The main reason for using a differential amplifier in sensor systems is to suppress interference. In many systems the input leads pick up 60 Hz signals from the power-line fields that permeate all electrified buildings. In single-ended amplifiers, the 60 Hz interfering signal is treated as a valid input signal, just like the desired sensor signal. But in differential amplifiers the two leads are affected equally by the field, so they present equal-strength input signals to the $(-)$ and $(+)$ inputs. This type of signal is called a *common mode* signal because it affects both inputs equally. Because these inputs perform opposite each other, the net result is that the two output signal components due to the common mode signal cancel each other. The degree to which common mode signals are suppressed by this mechanism in differential amplifiers is called the *common mode rejection ratio* (CMRR), which is usually expressed in decibels (dB) [*Note:* $\text{dB} = \log(A_{vd}/A_{vcm})$].

There are two basic configurations for differential amplifiers. The dc differential amplifier circuit, using one op-amp device, is shown in Fig. 2–5(a). The input resistors are balanced $(R_1 = R_2)$, as are the feedback resistors $(R_3 = R_4)$. The differential voltage gain of this circuit is R_3/R_1.

If the resistors of the dc differential amplifier are perfectly matched, then the CMRR will be very high—on the order of the inherent CMRR of the op-amp device itself (70 to 120 dB). But any unbalance, even when due

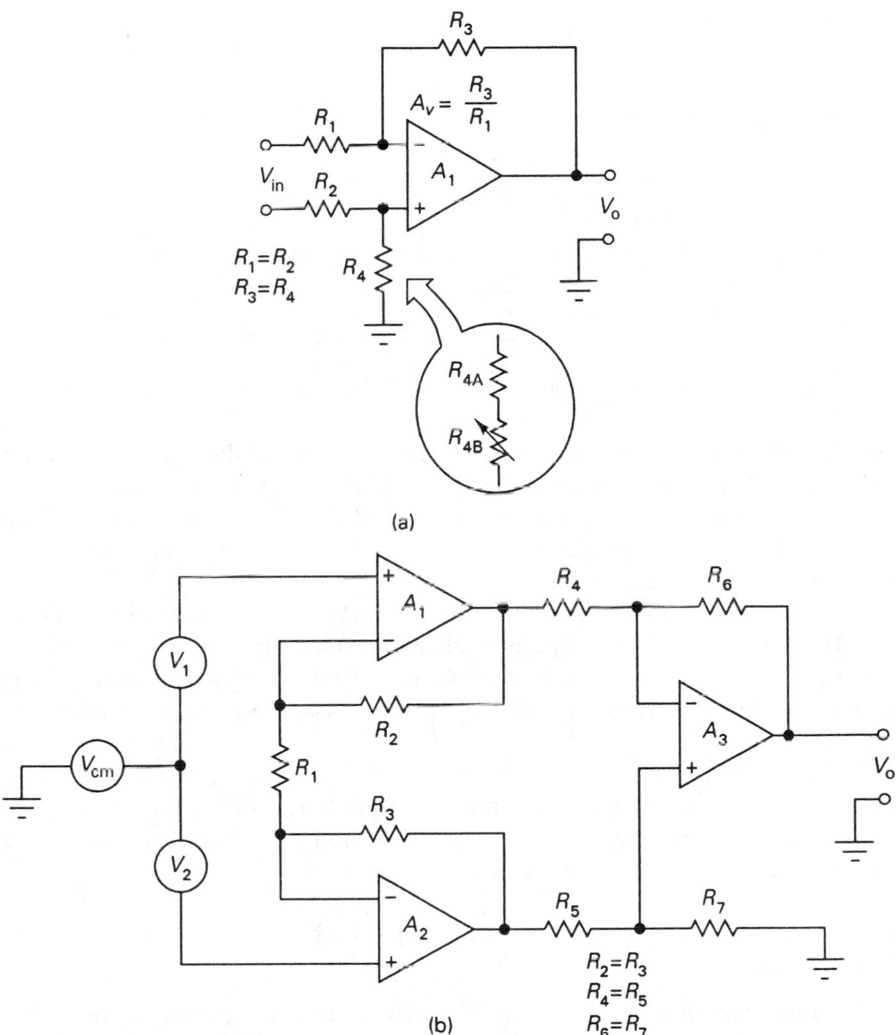

Figure 2–5 (a) Dc differential amplifier with CMRR ADJ control (inset); (b) Instrumentation amplifier circuit.

to resistor tolerances, will upset the balance and force the CMRR lower. This problem can be overcome by replacing R_4 in Fig. 2–5(a) with a series combination of a fixed resistor (R_{4A} in the inset) and a potentiometer (R_{4B}) connected as a rheostat. This network is adjusted by shorting the two inputs together, applying a signal (1 V ac is good in most cases), and adjusting R_{4B} for minimum output signal.

Limitations of the dc differential amplifier of Fig. 2–5(a) include a relatively low maximum gain and a low input impedance (set by R_1 and R_2).

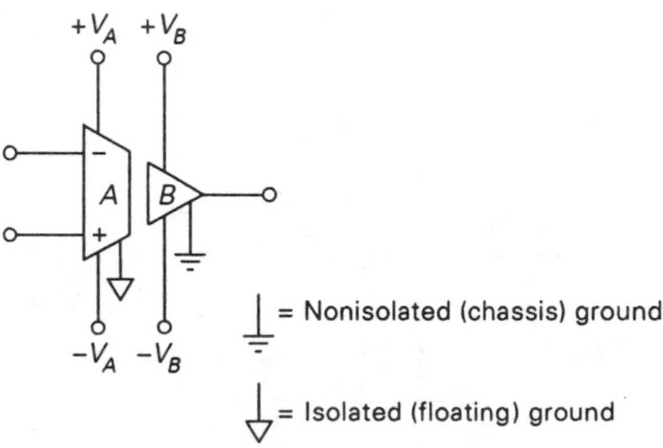

Figure 2–6 Isolation amplifier symbol.

But these problems are overcome in an advanced form of differential amplifier called the *instrumentation amplifier* [Fig. 2–5(b)]. Several semiconductor companies make integrated circuit versions of this circuit in which all three amplifiers are in one package (often called "ICIAs" for integrated circuit instrumentation amplifiers).

The instrumentation amplifier circuit has a very high input impedance, especially when the input amplifiers (A_1 and A_2) are either BiMOS or BiFET types (which use field-effect transistor input stages). The two input amplifiers should be identical types, or (preferably) two sections of a dual or triple op-amp device.

In this amplifier the assumption is that $R_2 = R_3$, $R_4 = R_5$, and $R_6 = R_7$. As in the case of the dc differential amplifier, CMRR adjustment can be provided by making R_7 a series combination of a fixed resistor and a potentiometer. The gain of the circuit is given by

$$A_v = \left(\frac{2R_2}{R_1} + 1\right)\left(\frac{R_6}{R_4}\right) \tag{2–1}$$

It is common to use R_1 as a gain control, but it is important to not allow the value of R_1 to get too low. When $R_1 = 0$, the gain tries to go to infinity, and the amplifier output will saturate with only very tiny input signals. The input signals V_1 and V_2 are referenced to ground, so the differential input signal $V_d = V_2 - V_1$. The common mode signal (if any) is shown as V_{cm}.

Isolation amplifiers. An *isolation amplifier* (shown in Fig. 2–6) is one in which the input circuits (A) are isolated from the output circuits (B) by an extremely high impedance ($>10^{12}$ ohms). These devices are usually variants of differential amplifiers, but the purpose of these devices is to isolate the sensor circuit from the following electronics.

One very common application of these amplifiers is in medical electronics, where patient safety requires such isolation.[2] Other applications are seen in cases where there are high voltages present that could damage, or even destroy, ground-referenced amplifiers. The isolation amplifier provides a local, isolated common point for the high-voltage circuit and a ground-referenced common for the output stages.

GUARD SHIELDING

One of the properties of the differential amplifier, including the instrumentation amplifier, is that its CMRR tends to suppress interfering signals from the environment. When an amplifier is used in a situation where it is connected to an external signal source through wires, those wires are subjected to strong local 60 Hz ac fields from nearby power-line wiring. Fortunately, in the case of the differential amplifier the field affects both lines equally, so the induced interfering signal is canceled out by the common mode rejection property of the amplifier.

Unfortunately, the cancellation of interfering signals is not total. There may be, for example, imbalances in the circuit that tend to deteriorate the CMRR of the amplifier. These imbalances may be either internal or external to the amplifier circuit. Figure 2–7(a) shows a common sensor interface scenario. In this figure we see the differential amplifier connected to shielded leads from the signal source V_{in}. Shielded lead wires offer some protection from local fields, but there is a problem with the standard wisdom regarding shields: It is possible for shielded cables to manufacture a valid differential, but erroneous, signal voltage from a common mode signal!

Figure 2–7(b) shows an equivalent circuit that demonstrates how a shielded cable pair can create a differential signal from a common mode signal. The cable has capacitance between the center conductor and the shield conductor surrounding it. In addition, input connectors and the amplifier equipment internal wiring also exhibit capacitance. These capacitances are lumped together in Fig. 2–7(b) as C_{S1} and C_{S2}. As long as the source resistances and shunt resistances are equal, and the two capacitances are equal, there is no problem with circuit balance. But inequalities in any of these factors (which are commonplace) create an unbalanced circuit in which the common mode signal V_{cm} can charge one capacitance more than the other. As a result, the difference between the capacitance voltages V_{CS1} and V_{CS2} is seen as a valid differential signal.

A low-cost solution to the problem of shield-induced artifact signals is shown in Fig. 2–8(a). In this circuit a sample of the two input signals is fed back to the shield, which in this situation is not grounded. Alternatively, the amplifier output signal is used to drive the shield. This type of shield is called a *guard shield.* Either double shields (one on each input line), as shown, or a common shield for the two inputs can be used.

An improved guard shield example for the instrumentation amplifier is

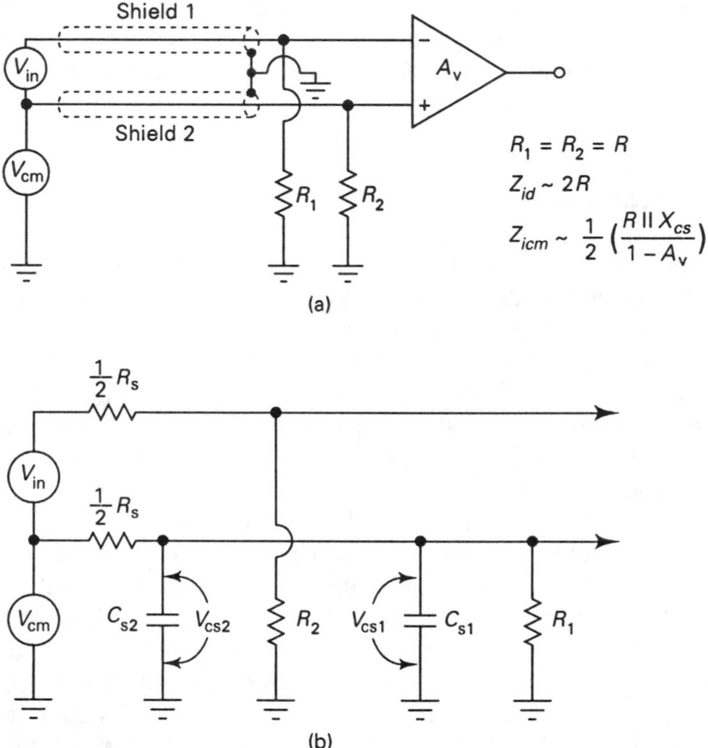

Figure 2–7 (a) Shielded input circuit; (b) equivalent circuit.

shown in Fig. 2–8(b). In this case a single shield covers both input lines, but it is possible to use separate shields. In this circuit a sample of the two input signals is taken from the junction of resistors R_8 and R_9 and fed to the input of a unity-gain buffer/driver "guard amplifier" A_4. The output of A_4 is used to drive the guard shield.

Perhaps the most common approach to guard shielding is the arrangement shown in Fig. 2–8(c). Here we see two shields used; the input cabling is double-shielded insulated wire. The guard amplifier drives the inner shield, which serves as the guard shield for the system. The outer shield is grounded at the input end in the normal manner and serves as an electromagnetic interference suppression shield.

A selectable-gain ICIA (the National Semiconductor LM-363-AD device) is shown in Fig. 2–9; the 16-pin DIP package is shown in Figure 2–9(a), while a typical circuit is shown in Fig. 2–9(b). The type number of this selectable-gain device is LM-363-AD, which distinguishes it from the fixed-gain LM-363-xx devices by the same manufacturer. The gain can be

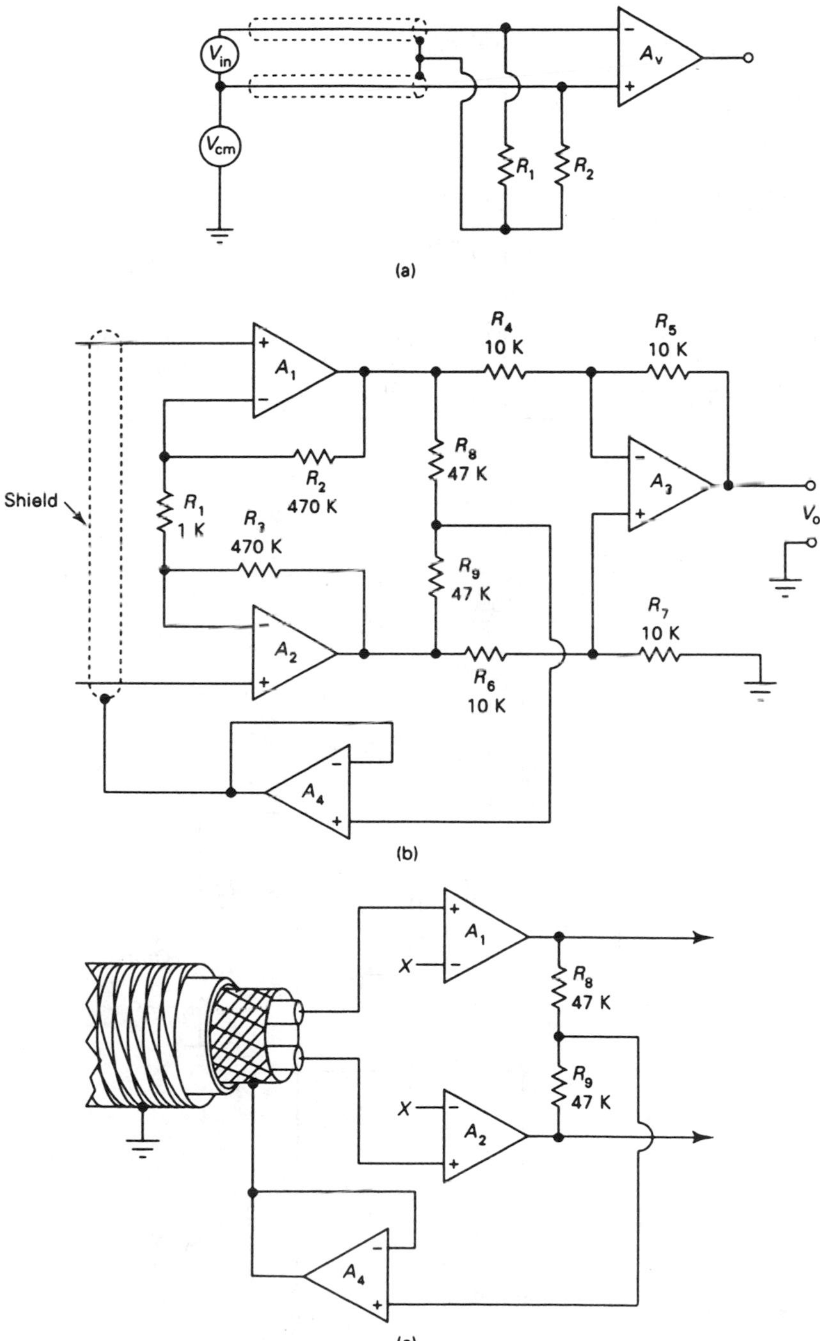

Figure 2–8 (a) Guard shield driver; (b) active driver; (c) multishielded input.

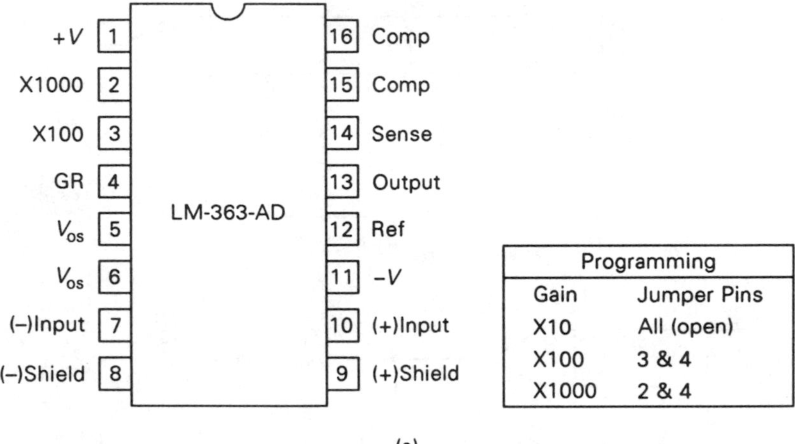

Programming	
Gain	Jumper Pins
X10	All (open)
X100	3 & 4
X1000	2 & 4

(a)

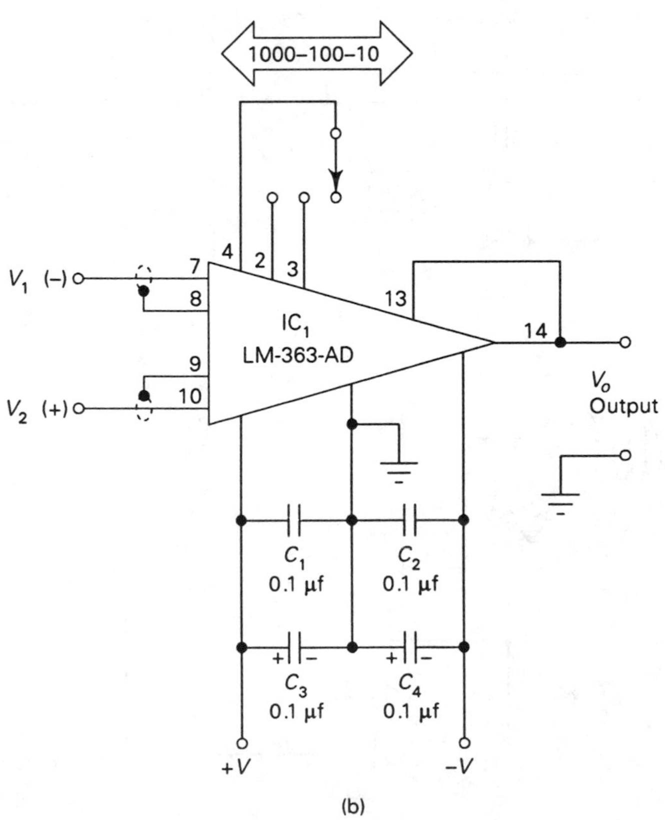

(b)

Figure 2–9 (a) LM-363-AD; (b) LM-363-AD circuit.

×10, ×100 or ×1000 depending on the programming of the gain-setting pins (2, 3, and 4).

Switch S_1 in Fig. 2–9(b) is the GAIN SELECT switch. This switch should be mounted close to the IC device but is quite flexible mechanically. The switch could also be made from a combination of CMOS electronic switches (e.g., 4066).

The dc power supply terminals are treated much the same as the other amplifiers. Again, the 0.1 μF capacitors need to be mounted as close as possible to the body of the LM-363-AD. Pins 8 and 9 are guard shield outputs. These pins are a feature that makes the LM-363-AD more useful for many instrumentation problems than other models. By outputting a signal sample back to the shield of the input lines, we can increase the common mode rejection ratio. This feature is frequently used in bipotential amplifiers and in other applications where a low-level signal must pass through a strong-interference (high-noise) environment.

The LM-363 devices will operate with supply voltages of ±5 V to ±18 V dc, with a CMRR of 130 dB. The 7 $nV/\sqrt{Hz}$ noise figure makes the device useful for low-noise applications (a 0.5 nV model is also available).

GROUNDING AND GROUND LOOP

Impulse noise due to electrical arcs, lightning bolts, electrical motors, and other devices can interfere with the operation of sensors and their associated circuits. Shielding of lines (see previous section) will help somewhat but isn't the entire answer. Filtering (discussed next) is useful, but it is at best a two-edged sword, and it must be done prudently and properly. Filtering on signal lines tends to broaden fast rise time pulses and attenuate high-frequency signals . . . and in some circuits causes as many problems as it solves.

Other nearby electrical devices can induce signals into the instrumentation system, the chief among these sources being the 60 Hz ac power system. It is wise to use only differential amplifier inputs, because of their high CMRR. Signals from the desired sensor source can be connected across the two inputs, and so become a differential signal, while the 60 Hz ac interference tends to affect both inputs equally (so is common mode).

It is sometimes possible, however, to manufacture a differential signal from a common mode signal. We examined this phenomenon earlier in relation to bad shielding practices. In this section we expand on that theme and consider grounds as well as shields.

One source of this problem is called a *ground loop*, as illustrated in Fig. 2–10(a). The differential signal arises from the use of too many grounds. In this example the shielded source, the shielded input lines, the amplifier, and the dc power supply are all grounded to different points on the ground plane. Power supply dc currents (I) flow from the power supply at point A

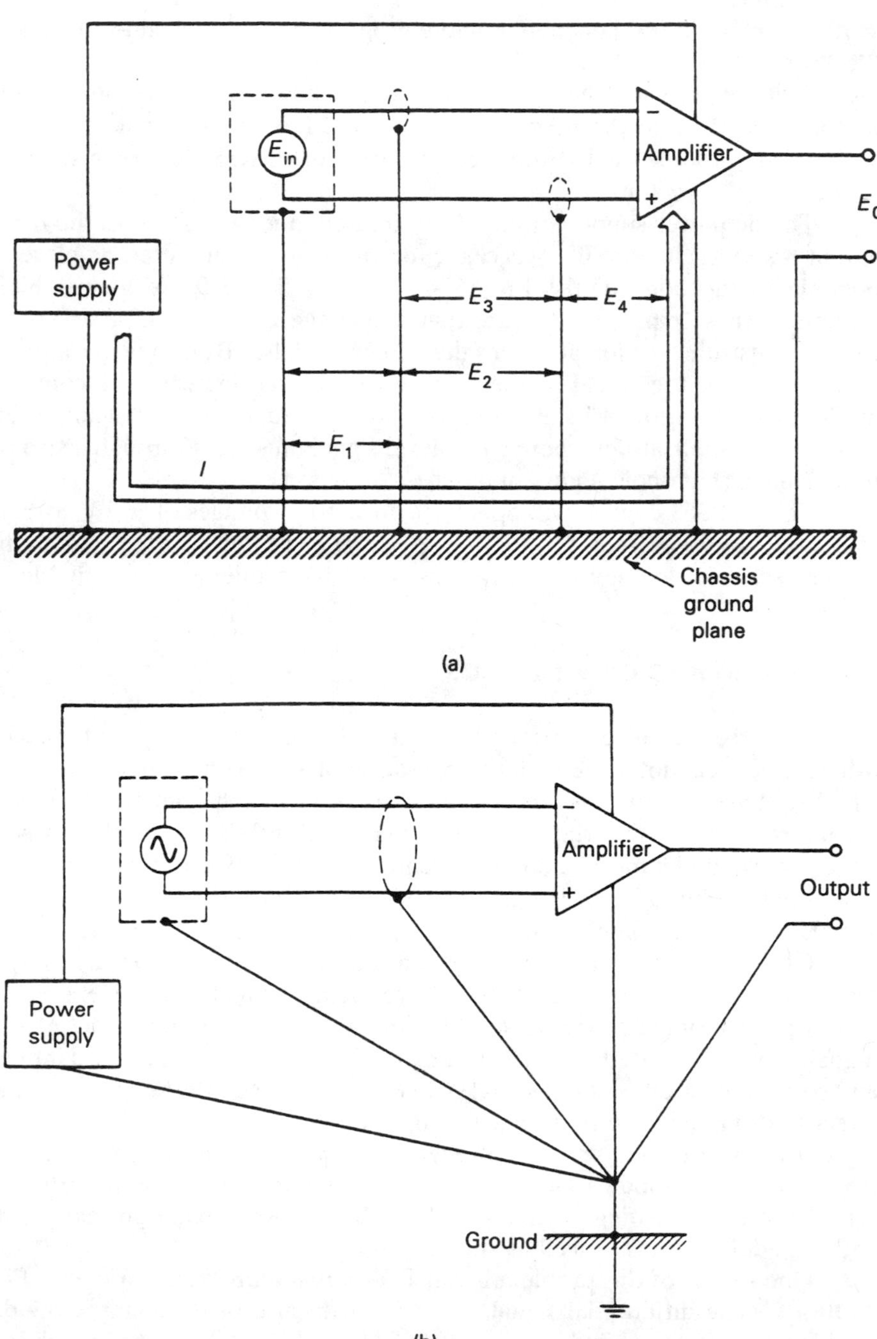

Figure 2–10 (a) Ground loops as the source of error signals; (b) single-point grounding eliminates ground loops.

to the amplifier power common at point E. Since the ground has ohmic resistance, albeit very low resistance, voltage drops E_1 through E_4 are formed. These voltages are seen by the amplifier as valid signals and can become especially troublesome if I is a varying current.

The solution to this problem is to use *single-point grounding*, as shown in Fig. 2–10(b). Some amplifiers used in sensitive instruments, especially those that contain both analog and digital circuits, keep the power and the various signal grounds separate, except at some single, specific common point. In fact, a few models go even further by creating several signal grounds, especially where both analog and digital signals might be present.

In some instances the shield on the input lines must not be grounded at both ends. In those cases it is usually better to ground only the amplifier ends of the cables.

CONCLUSION

Understanding the different types of sensor model circuits and the types of amplifier inputs is important for understanding how sensor interfacing works. Readers who want to learn more about these circuits are referred to any number of texts on operational amplifiers and instrumentation design.

REFERENCES AND NOTES

1. Definitions derived in part from George Klier (Gould, Inc.), "Signal Conditioners: A Brief Outline," *Sensors: The Journal of Machine Perception* 7, no. 1 (Jan. 1990): 44–48. Some of the definitions used in this article originated in the Klier article. For additional information see Daniel H. Sheingold, ed. *Transducer Interfacing Handbook*. Norwood, Mass.: Analog Devices, Inc., 1981.
2. John M. Brown and Joseph J. Carr. *Introduction to Biomedical Equipment Technology*. Formerly published by John Wiley, but now with Prentice Hall, Inc.

3 Resistive, Capacitive, and Inductive Sensors

A large number of different sensors are based on just three basic electrical phenomena: resistance, inductance, and capacitance. In this chapter we examine these phenomena and some of the devices that use them. Although the discussion is centered on the phenomena, and not specific sensors, certain practical examples are provided as illustrations.

RESISTANCE

One of the ways to categorize materials is according to their ability to conduct electrical current. *Conductors* are materials, such as metals, in which the atoms have loosely bound valence electrons, so they have large numbers of "free" electrons to support the flow of current at low levels of electrical potential (voltage). *Insulators* are materials, such as glass or ceramics, that have few free electrons, so they do not easily support the flow of electrical current without the application of extraordinarily high electrical potentials.

Conductors do not conduct an infinite current, because all electrical conductors possess the property of *resistance* R, which is defined as opposition to the flow of electrical current. Resistance is measured in ohms (Ω). The resistance of any specific conductor is directly proportional to its length [Fig. 3–1(a)] and inversely proportional to its cross-sectional area. Resistance is also directly proportional to a property of the conductor material called *resistivity* (ρ). These relationships are stated mathematically in Eq. (3–1) and are depicted in Fig. 3–1(a).

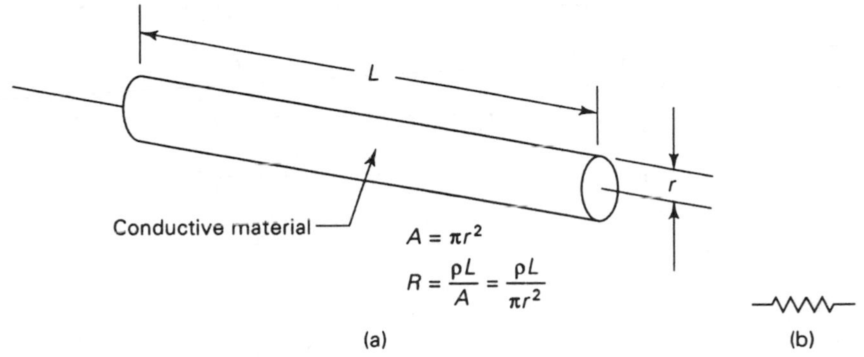

Figure 3–1 (a) Electrical conductor in wire or cylindrical shape; (b) resistance symbol.

$$R = \frac{\rho L}{A} \tag{3–1}$$

where ρ is the resistivity in ohm · centimeters
L is the length in centimeters
A is the cross-sectional area in square centimeters

Although a class of sensors called *strain gages* uses wire resistance elements, most electronic circuits contain discrete *resistors*, which are lumped resistance made of metal film, carbon composition, or other materials. The schematic symbol for a resistor is shown in Fig. 3–1(b).

The current that will flow in a resistance is proportional to the applied voltage E and inversely proportional to the resistance R:

$$I = \frac{E}{R} \tag{3–2}$$

Piezoresistivity

Piezoresistivity denotes the change in resistance that occurs when either the length or the area of a conductor is changed. Figure 3–2(a) shows a cylindrical conductor of initial length L_0 and cross-sectional area A_0. When a compression force is applied, as in Fig. 3–2(b), the length decreases and the cross-sectional area increases. This situation results in a decrease in the electrical resistance. Mathematically,

$$R = (R_0 - \Delta R) \propto \left(\frac{L_0 - \Delta L}{A_0 + \Delta A}\right) \tag{3–3}$$

Similarly, when a tension force is applied [Fig. 3–2(c)] the length in-

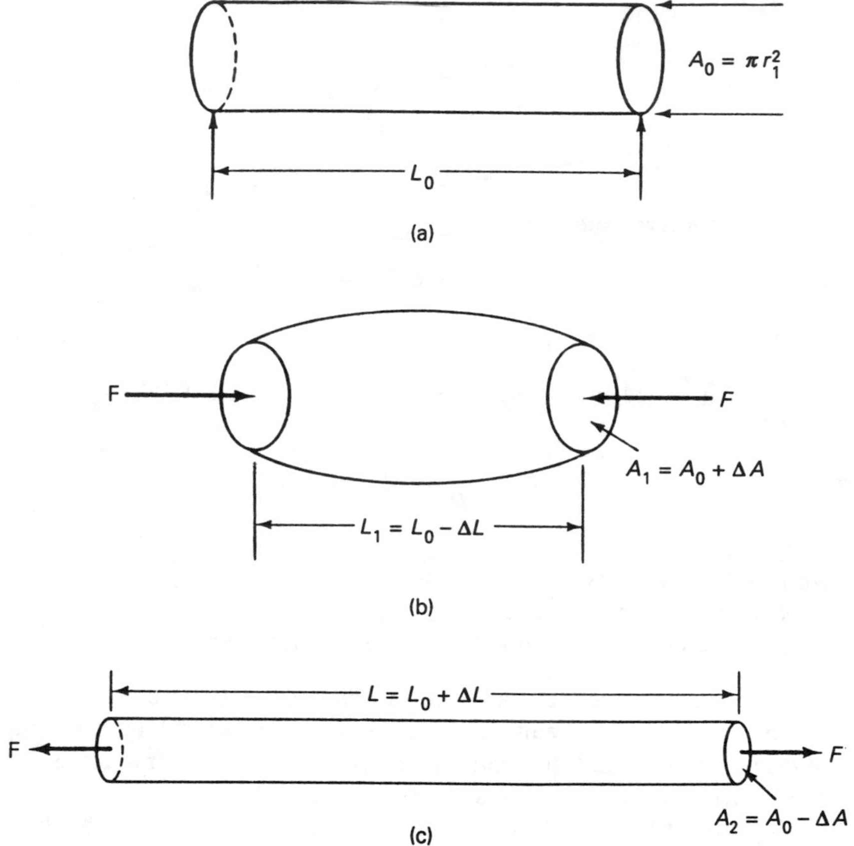

Figure 3–2 Conductor (a) at rest; (b) in compression; (c) in tension.

creases and the cross-sectional area decreases, so the electrical resistance increases:

$$R = (R_0 + \Delta R) \propto \left(\frac{L_0 + \Delta L}{A_0 - \Delta A}\right) \tag{3–4}$$

In both the tension and the compression cases, provided that the physical change is small, the change in electrical resistance is a nearly linear function of the applied force, so it can be used to make measurements of that force.

Strain Gages

A *strain gage* is a piezoresistive element, made of wire, metal foil, or semiconductor material, that is used to measure applied forces. In general, the expression for the change in resistance under force is given by

$$\frac{\Delta R}{R} = \frac{\Delta \rho}{\rho} + \frac{\Delta L}{L} - \frac{\Delta A}{A} \tag{3–5}$$

In the tension case, the length of the cylindrical conductor will decrease according to Poisson's ratio (ν), or

$$\frac{\Delta A}{A} = -2\nu\left(\frac{\Delta L}{L}\right) \tag{3–6}$$

Substituting Eq. (3–6) for $\Delta A/A$ in Eq. (3–5), we obtain

$$\frac{\Delta R}{R} = \frac{\Delta \rho}{\rho} + \frac{\Delta L}{L} - \left[-2\nu\left(\frac{\Delta L}{L}\right)\right] \tag{3–7}$$

or

$$\frac{\Delta R}{R} = \frac{\rho \Delta}{\rho} + (1 + 2\nu)\left(\frac{\Delta L}{L}\right) \tag{3–8}$$

Strain gages can be classified as either *bonded* or *unbonded*. The unbonded strain gage is shown in Fig. 3–3(a). It consists of a wire resistance element stretched taut between two flexible supports. These supports are configured so that they place a tension or compression force on the taut wire when external forces are applied. In the particular example shown, the supports are mounted on a thin metal diaphragm that flexes when a force is applied. Force F_1 will cause the flexible supports to spread apart, exerting increased tension on the wire and thereby increasing its resistance. Alternatively, when force F_2 is applied, the ends of the supports tend to move closer together, effectively exerting a compression force on the wire element and thereby reducing its resistance. The wire's resting condition is tautness, which implies a tension force. Thus, F_1 increases the normal tension, while F_2 decreases it.

The bonded form of strain gage is shown in Fig. 3–3(b). In this type of device a wire, foil, or semiconductor element is cemented to a thin metal diaphragm. When the diaphragm is flexed, the element deforms to produce a resistance change.

The linearity of both types of strain gages can be quite good, provided that the elastic limits of the diaphragm and element are not exceeded and the change of length is only a small percentage of the resting length.

In the past it was commonly thought that bonded strain gages were more rugged but less linear than unbonded models. Although this may have been the situation at one time, recent experience has shown that modern manufacturing techniques can produce rugged, linear, reliable units of both types of construction.

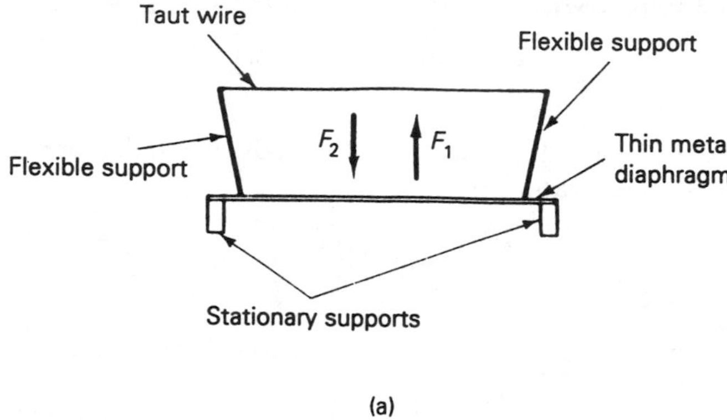

(a)

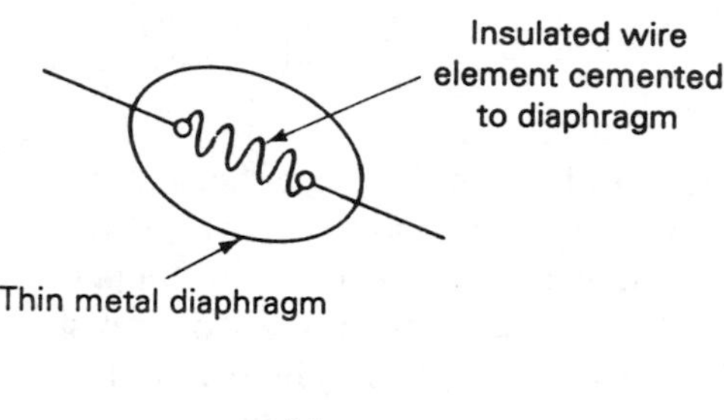

(b)

Figure 3–3 (a) Unbonded strain gage; (b) bonded strain gage.

Strain-Gage Sensitivity

The *sensitivity* of the strain gage is expressed in terms of unit change of electrical resistance per unit change of length and is most often given in the form of the *gage factor S* for the element:

$$S = \left(\frac{\Delta R/R}{\Delta L/L}\right) \tag{3–9}$$

The factor $\Delta L/L$ is sometimes represented by ε, so Eq. (3–9) is also seen in the form

$$S = \frac{\Delta R/R}{\varepsilon} \tag{3–10}$$

STRAIN-GAGE CIRCUITRY

To be useful the resistive strain gage (or other form of resistive sensor) must be connected into a circuit that will convert its resistance changes into a current or voltage output. Most applications use voltage output circuits.

Figure 3–4 shows several popular forms of circuits. The circuit in Fig. 3–4(a) is both the simplest and least useful (although not useless). It is sometimes called the *half-bridge* or *voltage-divider* circuit. The strain-gage element of resistance R is placed in series with a fixed resistor R_1 across a stable dc voltage E. The output voltage E_o is found from the simple voltage-divider equation:

$$E_o = \frac{ER}{R + R_1} \tag{3-11}$$

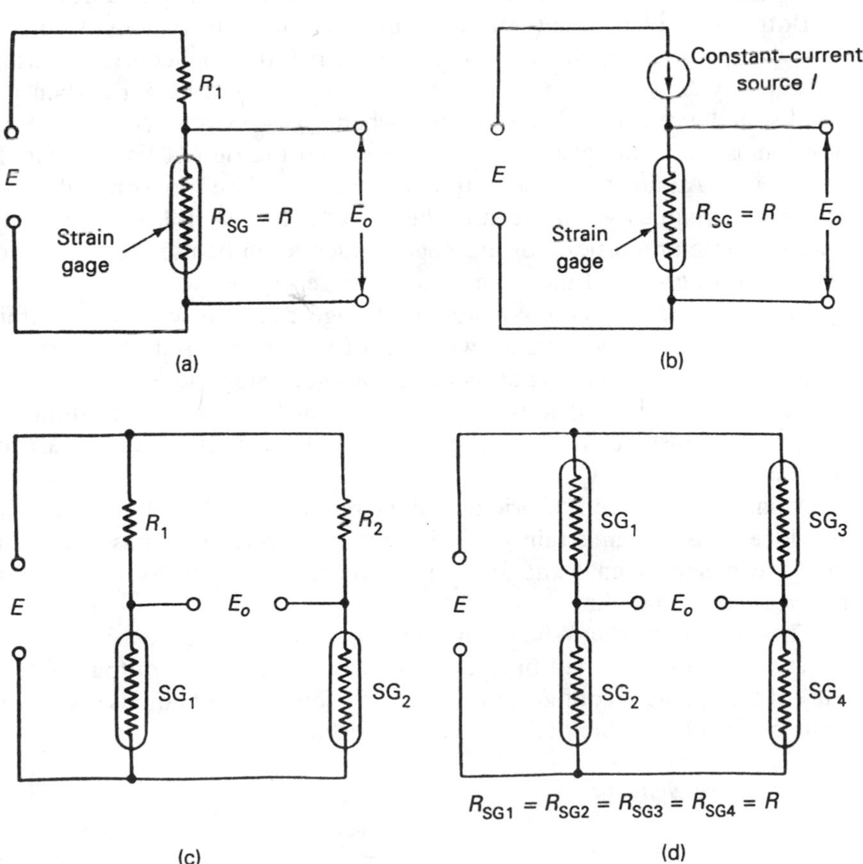

Figure 3–4 Circuits for using strain gages: (a) half-bridge; (b) current-sourced half-bridge; (c) two-element Wheatstone bridge; (d) four-element Wheatstone bridge.

Equation (3–11) describes the output voltage E_o when the sensor is at rest (i.e., nothing is stimulating the resistive element). When the element is stressed, however, its resistance changes by a small amount, ΔR. The output voltage in that case is

$$E_o = \frac{E(R + \Delta R)}{(R \pm \Delta R) + R_1} \qquad (3\text{–}12)$$

Another form of half-bridge circuit is shown in Fig. 3–4(b), but in this case the strain gage is connected in series with a constant-current source (CCS), which maintains current I at a constant level regardless of changes in the strain-gage resistance. In this case $V_o = I(R \pm \Delta R)$.

One major problem with all resistive sensors is that their resistance changes with changes in temperature (a fact that makes thermal resistors possible). In some sensors, the temperature is stabilized by insulation or other means. In others, temperature-compensation circuitry is used.

Both half-bridge circuits suffer from a serious defect: Output voltage V_o is always present regardless of the value of the force applied to the sensor. Ideally in any sensor system, the output voltage should be zero when the applied stimulus is zero. For example, when a gas-pressure sensor is open to the atmosphere, the gage pressure is zero, so the output voltage should also be zero. Additionally, the output voltage should be proportional to the value of the stimulus when the stimulus is not zero. A *Wheatstone bridge* circuit has these properties. Strain-gage elements can be used for one, two, three, or all four arms of the Wheatstone bridge.

Figure 3–4(c) shows a Wheatstone bridge circuit in which two strain gages (SG_1 and SG_2) are used in two arms of the bridge, and fixed resistors R_1 and R_2 form the remaining arms of the bridge. SG_1 and SG_2 are usually configured so that their actions oppose each other; that is, under stimulus, SG_1 will have resistance $R + \Delta R$, and SG_2 will have resistance $R - \Delta R$, or vice versa.

A linear form of sensor bridge is the circuit of Fig. 3–4(d), in which all four bridge arms contain strain-gage elements. In most such sensors all four strain-gage elements have the same resistance, which usually has a value between 50 and 1000 ohms.

The output from a Wheatstone bridge is the difference between the voltages across the two half-bridges. We can calculate the output voltage for any of the standard configurations from the following equations (assuming all four bridges have nominally the same resistance R):

One active element:

$$E_o = \frac{E\Delta R}{4R} \qquad (3\text{–}13)$$

(Accurate to ± 5 percent if $h < 0.1$)

Two active elements:

$$E_o = \frac{E\Delta R}{2R} \tag{3–14}$$

Four active elements:

$$E_o = \frac{E\Delta R}{R} \tag{3–15}$$

where E_o is the output potential in volts
E is the excitation potential in volts
R is the resistance of all bridge arms
ΔR is the change in resistance in response to the applied stimulus

SENSOR SENSITIVITY

The sensitivity factor Φ of a Wheatstone bridge sensor circuit relates the output voltage V_o to the applied stimulus value Q and the excitation voltage.* In most cases the sensor maker will specify a number of microvolts (or millivolts) *output potential per volt of excitation potential per unit of applied stimulus*:

$$\Phi = V_o/V/Q_o \tag{3–16}$$

or, written another way,

$$\Phi = \frac{V_o}{VQ_o} \tag{3–17}$$

where Φ is the sensor sensitivity in microvolts per volt per unit of stimulus
V_o is the output potential in volts
V is the excitation potential in volts
Q is one unit of applied stimulus

If we know the sensitivity factor, then we can calculate the output potential as follows:

$$V_o = \Phi VQ \tag{3–18}$$

Equation (3–18) is the one that is most often used in circuit design.

3–1

A certain fluid-pressure sensor is often used for measuring human and animal blood pressures through an indwelling catheter. It has a sensitivity Φ of 5 μV/V/torr, or 5 μV output potential is generated per volt of excitation potential per torr of pressure.

*Both V and E are used to represent voltage. This usage may be a little inconsistent, but it represents actual practice.

Find the output potential when the excitation potential is +7.5 V dc and the pressure is 400 torr (the usual high-end limit for such sensors):

$$V_o = \Phi VQ$$

$$V_o = \left(\frac{5\ \mu\text{V}}{\text{V}\cdot\text{torr}}\right) \times (7.5\ \text{V}) \times (400\ \text{torr})$$

$$V_o = (5 \times 7.5 \times 400)\mu\text{V} = 15{,}000\ \mu\text{V}$$

(which is 15 mV, or 0.015 V).

BALANCING AND CALIBRATING A BRIDGE SENSOR

Few, if any, Wheatstone bridge sensors meet the ideal condition in which all four bridge arms have exactly equal resistances at rest. In fact, the bridge resistance specified by the manufacturer is only a nominal value, and the actual value may vary quite a bit from the specified value. There will inevitably be an offset voltage (i.e., V_o is not zero when Q is zero). Figure 3–5 shows two circuits that will balance the bridge when the stimulus is zero.

In Fig. 3–5(a) the balancing potentiometer is placed between the excitation potential and one of the excitation nodes. The resistance balance of the potentiometer is varied between the two legs of the bridge, nullifying any differences between them. The potentiometer is usually a precision type with 5 to 15 turns to cover the entire range.

The purpose of the potentiometer in Fig. 3–5(b) is to inject a balancing current I into the bridge circuit at one of its nodes. With the stimulus at zero, R_1 is adjusted for zero output voltage.

Another application for this type of circuit is for injecting an intentional offset potential. For example, on an electronic scale, one that uses a strain gage to measure weight, such a circuit is used to adjust for the *tare* of the scale, which is the sum of the weight of the platform and all other weights acting on the sensor when nobody is standing on the scale. This is also sometimes called *empty-weight compensation*.

Calibration of a bridge sensor can be accomplished either the hard way or the easy (and less accurate) way. The hard way is to set the sensor up in a system, apply the stimulus, measure the stimulus, and compare the result with the sensor output. For example, in testing a pressure sensor, a manometer (pressure-measuring device containing a column of mercury) is connected, and the pressure is measured directly. The result is compared with the sensor output. All sensors should be tested in this manner when initially placed in service and then periodically thereafter.

The easy (and less accurate) way is to connect a calibrating resistor in parallel with one leg of the bridge [R_3 in Fig. 3–5(b)] to create an offset that is equal to some standard stimulus. A CAL switch (S_1) will insert the resistor in the circuit whenever a quick calibration or test is needed.

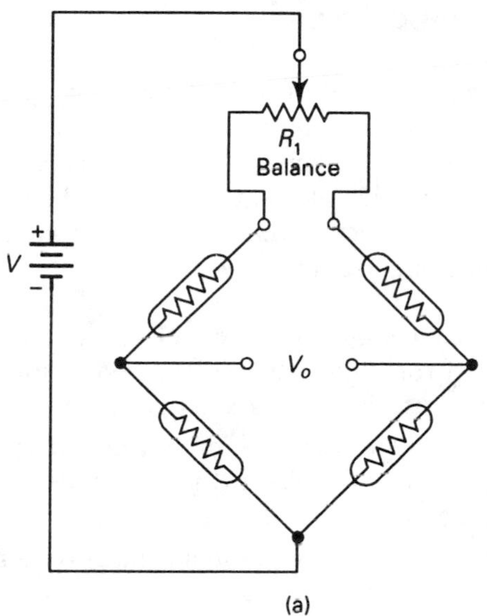

(a)

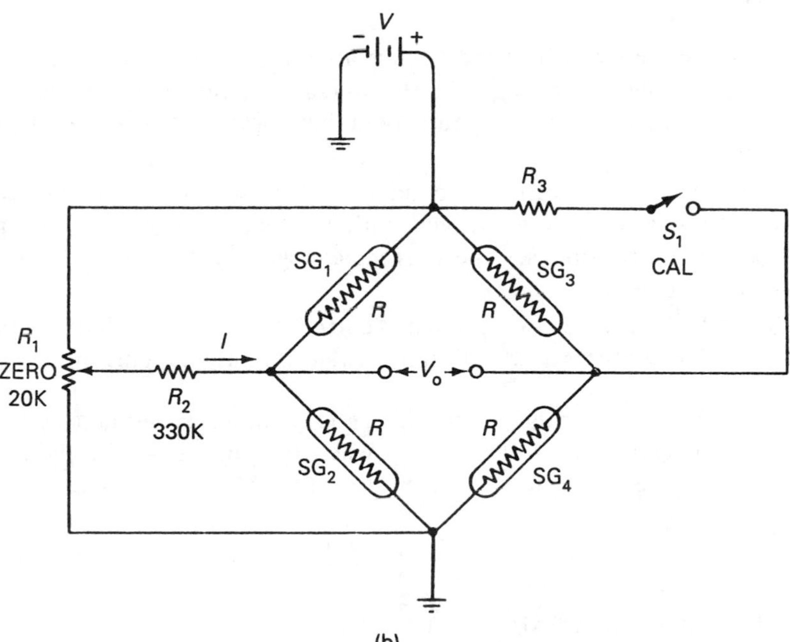

(b)

Figure 3–5 (a) Wheatstone bridge with balance control added; (b) circuit with ZERO and CAL controls added.

INDUCTORS AND INDUCTANCE

Inductance L is a property of electrical circuits that opposes *changes* in the flow of current (note the word *changes*; it is important). Inductance is therefore somewhat analogous to the concept of inertia in mechanics. An inductor stores energy in the magnetic field that is created whenever current flows in a conductor (an important fact to remember). To understand the concept of inductance we must understand three physical facts:

1. When an electrical conductor moves relative to a magnetic field, a current is generated (or "induced") in the conductor. An *electromotive force* (EMF or "voltage") appears across the ends of the conductor.
2. When a conductor is in a magnetic field that is changing, a current is induced in the conductor. As in the first case, an EMF is generated across the conductor.
3. When an electrical current moves in a conductor, a magnetic field is set up around the conductor.

According to *Lenz's law*, the EMF created in a circuit is in a direction that opposes the effect that produced it. This phenomenon has the following consequences:

1. A current induced by either the relative motion of a conductor and a magnetic field or changes in the magnetic field always flows in the direction that sets up a magnetic field that opposes the original magnetic field.
2. When a current flowing in a conductor changes, the magnetic field that it generates changes in a direction that induces a further current in the conductor that opposes the current change that caused the magnetic field to change.
3. The EMF generated by a change in current has a polarity that is opposite the polarity of the potential that created the original current.

The unit of inductance is the *henry* (H), which is defined as the inductance that creates an EMF of 1 V when the current in the inductor is changing at a rate of 1 ampere (A) per second. Mathematically,

$$V = L\left(\frac{\Delta I}{\Delta t}\right) \tag{3–19}$$

where V is the induced EMF in volts
L is the inductance in henrys
I is the current in amperes
t is the time in seconds

The henry is the appropriate unit for inductors such as the smoothing filter chokes used in dc power supplies but is far too large for sensor circuits. In those circuits the subunits *millihenrys* (mH) and *microhenrys* (μH) are used. These are related to the henry by: 1 henry = 1,000 millihenrys = 1,000,000 microhenrys. Thus, 1 mH = 10^{-3} H and 1 μH = 10^{-6} H.

The phenomenon described in point 2 of the preceding list is called *self-inductance*. When the current in a circuit changes, the magnetic field generated by that current change also changes. This changing magnetic field induces a counter current in the direction that opposes the original current change. This induced current also produces an EMF (discussed in point 3), which is called the *counter electromotive force* (CEMF). As with other forms of inductance, self-inductance is measured in henrys and its subunits.

Self-inductance can be increased by forming the conductor into a multiturn coil (Fig. 3–6) in such a manner that the magnetic fields in adjacent turns reinforce one another. This requirement means that the turns of the coil must be insulated from one another. A coil wound in this manner is called an *inductor*, or simply *coil*, in radio circuits. The inductor pictured in Fig. 3–6 is properly called a *solenoid-wound coil* if the length l is greater than the diameter d. The inductance of the coil is actually self-inductance, but in common usage the *self-* is usually dropped in favor of simple *inductance*.

Several factors affect the inductance of a coil. Perhaps the most obvious are the length, the diameter, and the number of turns N in the coil. Also affecting the inductance is the nature of the *core* material and its cross-sectional area. In the example of Fig. 3–6 the core is simply air, and the cross-sectional area is directly related to the diameter, but in many sensor circuits the core is made of powdered iron or ferrite materials.

For an air-core solenoid-wound coil in which the length is greater than $0.4d$, the inductance can be approximated by

$$L = \frac{d^2N^2}{18d + 40l} \tag{3–20}$$

where L is in microhenries.

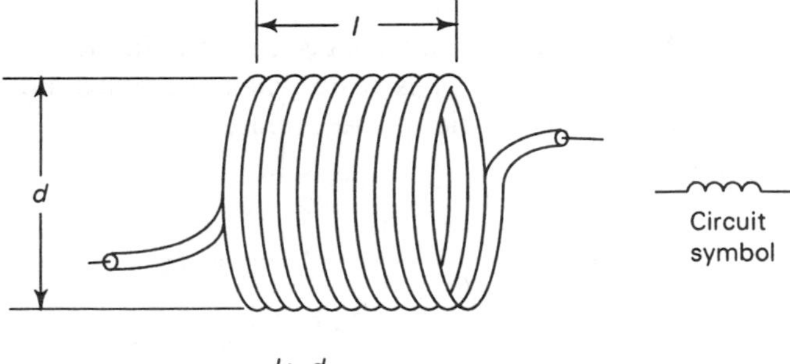

Figure 3–6 Solenoid-wound inductor.

The core material has a certain *magnetic permeability* μ, which is the ratio of the number of lines of flux produced by the coil with the core inserted to the number of lines of flux with an air core (i.e., core removed). The inductance of the coil is multiplied by the permeability of the core.

Combining inductors. When inductors are connected together in a circuit, the calculation of their combined inductance is similar to the calculation of the resistance of several resistors connected in parallel or series. For inductors in which their respective magnetic fields *do not interact*:

(a) Series-connected inductors:

$$L_{total} = L_1 + L_2 + L_3 + \ldots + L_n \tag{3–21}$$

(b) Parallel-connected inductors:

$$L_{total} = \frac{1}{\left(\frac{1}{L_1} + \frac{1}{L_2} + \frac{1}{L_3} + \ldots + \frac{1}{L_n}\right)} \tag{3–22}$$

Or, in the special case of two inductors in parallel:

$$L_{total} = \frac{L_1 L_2}{L_1 + L_2} \tag{3–23}$$

If the magnetic fields of the inductors in the circuit *interact*, then the total inductance becomes somewhat more complicated to express mathematically. For the simple case of two inductors in series, the expression becomes

(a) Series inductors:

$$L_{total} = L_1 + L_2 \pm 2M \tag{3–24}$$

where M is the *mutual inductance* caused by the interaction of the two magnetic fields. (*Note:* $+M$ is used when the fields aid each other, and $-M$ is used when the fields are opposing.)

(b) Parallel inductors:

$$L_{total} = \frac{1}{\left(\frac{1}{L_1 \pm M}\right) + \left(\frac{1}{L_2 \pm M}\right)} \tag{3–25}$$

In general, for coils in close proximity:

1. Maximum interaction between the coils occurs when the coils' axes are parallel to each other.
2. Minimum interaction between the coils occurs when the coils' axes are at right angles to each other.

For the case where the coil axes are along the same line, the interaction depends on the distance between the coils.

Adjustable coils. One practical problem with the standard fixed coil is that the inductance cannot easily be adjusted to account for the tolerances in the circuit.

Air-core coils are difficult to adjust. They can be lengthened or shortened; the number of turns can be changed; or a tap or series of taps can be established on the coil to permit an external switch to select the number of turns that are allowed to be effective. None of these methods is very elegant, and although all have been used in one application or another, they are basically useless for sensor applications.

The solution to the adjustable inductor problem that was developed relatively early in the history of mass-produced radios, and still used today, was to insert a powdered iron or ferrite core (or *slug*) inside the coil form (Fig. 3–7). The permeability of the core increases or decreases the inductance according to how much of the core is inside the coil. If the core is made with either a hexagonal hole or screwdriver slot, then the inductance of the coil can readily be adjusted by moving the core in or out of the coil. These coils are called *slug-tuned inductors*.

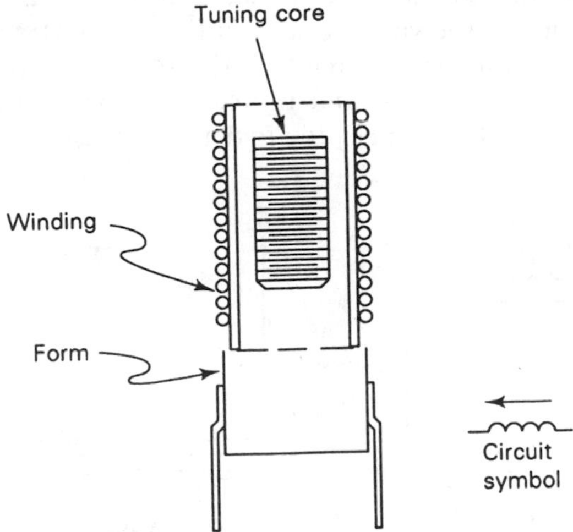

Figure 3–7 Slug-tuned coil.

Inductors in AC Circuits

Impedance (Z) is the total opposition to the flow of alternating current (ac) in a circuit, and as such it is analogous to resistance in dc circuits. The impedance is made up of a resistance component R and a component called *reactance* X. Like resistance, reactance is measured in ohms. If the reactance is produced by an inductor, then it is called *inductive reactance* X_L, and if by a capacitor, it is called *capacitive reactance* X_c. Inductive reactance is a function of the inductance and the frequency of the ac source:

$$X_L = 2\pi f L \tag{3-26}$$

where X_L is the inductive reactance in ohms
f is the ac frequency in hertz
L is the inductance in henrys

In a purely resistive ac circuit [Fig. 3–8(a)] the current I and voltage V are said to be *in-phase* with each other; that is, they rise and fall at exactly the same times in the ac cycle. In vector notation [Fig. 3–8(b)], the current and voltage vectors are along the same axis, which is an indication of the 0° phase difference between the two.

In an ac circuit that contains only an inductor [Fig. 3–9(a)] and is excited by a sine-wave ac source, the change in current is opposed by the inductance. As a result, the current I in an inductive circuit lags behind the voltage V by 90°. This is shown vectorially in Fig. 3–9(b), and as a pair of sine waves in Fig. 3–9(c).

The ac circuit that contains a resistance and an inductance [Fig. 3–10(a)] exhibits a phase shift θ, shown vectorially in Fig. 3–10(b), that is other than the 90° seen in purely inductive circuits. The phase shift is proportional to the voltage across the inductor and the current flowing through it. The impedance of this circuit is found by the Pythagorean rule, which is also called

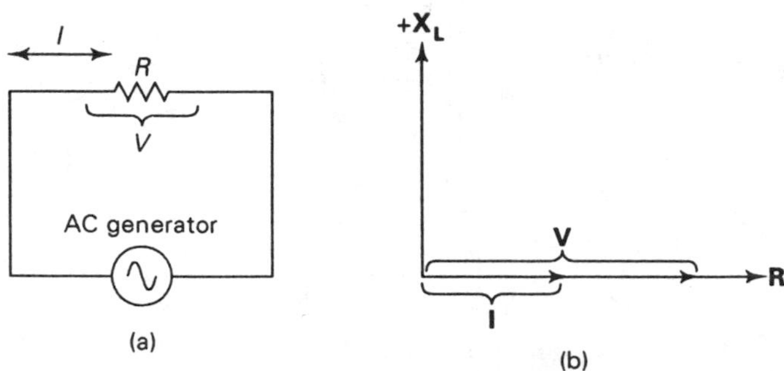

Figure 3–8 (a) Resistive circuit; (b) phase relationships (shown vectorially).

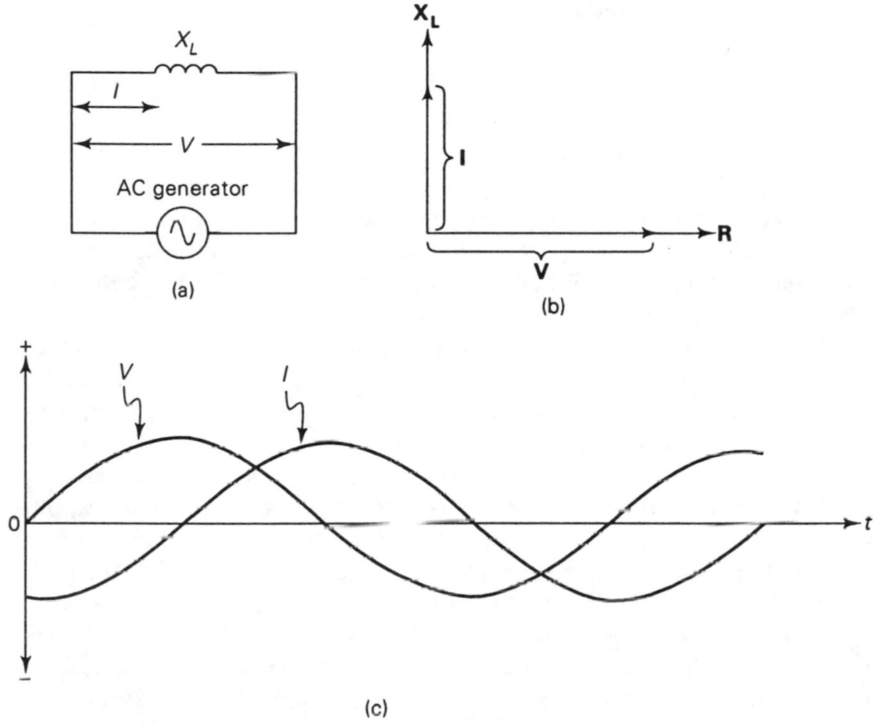

Figure 3–9 (a) Inductive circuit; (b) phase relationships; (c) current lags voltage by 90° (shown vectorially).

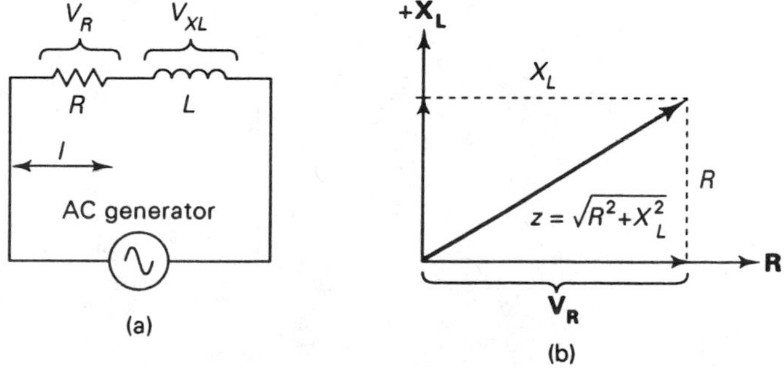

Figure 3–10 (a) Resistor-inductor (*RL*) circuit; (b) phase relationships.

the *root of the sum of the squares method*:

$$|Z| = \sqrt{R^2 + (X_L)^2} \tag{3–27}$$

Inductive Sensors

In the following two examples the sensor consists of a pair of inductors wound around a movable, permeable core. The core helps determine the inductance of each coil. When the core is evenly spaced between the two coils, then the respective inductances are equal.

Figure 3–11 shows an inductive Wheatstone bridge sensor. The arms of the bridge comprise fixed resistors R_1 and R_2 plus inductors L_1 and L_2. The inductive reactances X_L of L_1 and L_2 are a function of the applied ac excitation frequency and the inductance of L_1 and L_2. When $L_1 = L_2$—for example, when the sensor is at rest—then the bridge is at null, and V_g is zero. When the coil core is moved, as when a pressure or force is applied, the relative inductances of L_1 and L_2 change, so the reactances are no longer equal, and the bridge is unbalanced by an amount proportional to the mutual coupling caused by the applied stimulus.

A *linear variable differential transformer* (LVDT) sensor is shown in Fig. 3–12. In this case there are three coils because ac excitation is applied to the system via L_1. Coils L_2 and L_3 are equal, and when the core is placed equally between the two, their inductances are also equal. But when the core moves, the inductances become unequal. The operation of the LVDT depends on the fact that the two output coils are connected in series-opposing

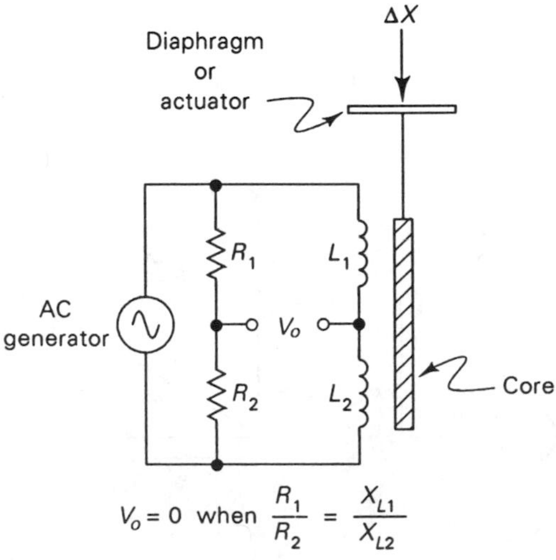

Figure 3–11 Inductive-bridge Δx displacement sensor.

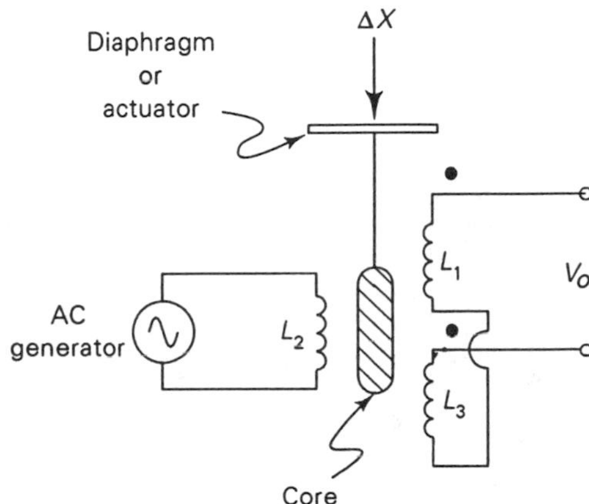

Figure 3–12 Linear voltage differential transformer (LVDT) sensor.

fashion, so that the total output voltage is the algebraic difference. When $V_{L2} = V_{L3}$, the total voltage output is zero. Only when a stimulus is applied, when these voltages are not equal, will there be an output voltage.

Inductance sensors can be used as displacement sensors, but they have a very small range that is limited to the range of travel of the internal core. This distance can be increased, however, by using a lever arm, reducing gears (angular displacement cases), or a mechanical linkage.

CAPACITORS AND CAPACITANCE

Capacitors are the other component used in radio tuning circuits and in many other applications. Like the inductor, the capacitor is an energy storage device. Whereas the inductor stores electrical energy in a magnetic field, the capacitor stores energy in an *electrical* (or *electrostatic*) field; electrical charge Q is stored in the capacitor.

A basic capacitor consists of a pair of facing metallic plates separated by an insulating material (e.g., mica, ceramic, mylar, or dry air) called a *dielectric*. This arrangement is shown schematically in Fig. 3–13(a). The fixed capacitor shown in Fig. 3–13(b) consists of a pair of square metal plates separated by a dielectric. Layers of glass and foil are sandwiched together to form a high-voltage capacitor.

The *capacitance* C of the capacitor is a measure of its ability to store electrical charge. The principal unit of capacitance is the *farad* (F). One farad is the capacitance that will store one *coulomb* (C) of electrical charge (6.28×10^{18} electrons) at an electrical potential of 1 V.

$$C = \frac{Q}{V} \tag{3–28}$$

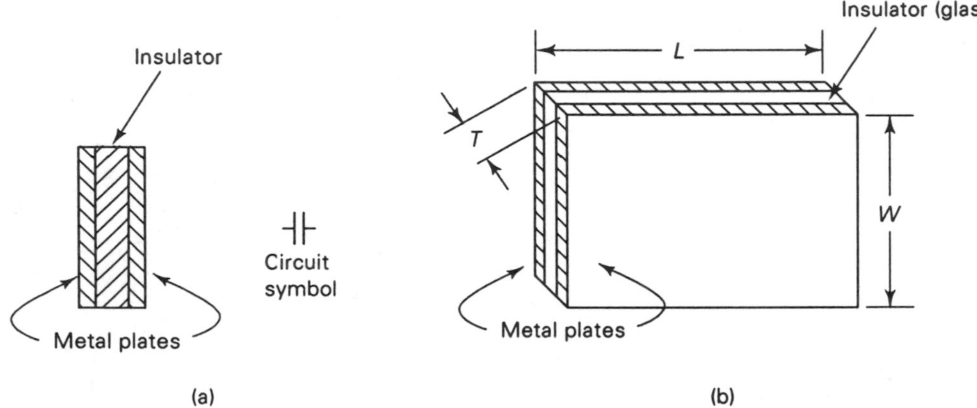

Figure 3–13 (a) Basic capacitor and symbol; (b) parallel-plate capacitor and its dimensions.

where C is in farads
Q is in coulombs
V is in volts

The farad is far too large for practical electronics work, so the subunits *microfarads* (μF or mF) and *picofarads* (pF) are used. A 1-ft^2 capacitor made of ⅛-in. thick glass and foil has a capacitance of about 2000 pF.

Capacitance is directly proportional to the area of the plates [in terms of Fig. 3–13(b) $L \times W$], inversely proportional to the thickness T of the dielectric (or the spacing between the plates, if you prefer), and directly proportional to the *dielectric constant* K of the dielectric.

Dielectric constant is a property of the insulator material used for the dielectric. The dielectric constant is a measure of the material's ability to support electrical flux and is thus analogous to the permeability of a magnetic material. The standard of reference for dielectric constant is a perfect vacuum, which is assigned the value of $K = 1.00000$. The values of K for some common materials are given:

Vacuum	1.0000
Dry air	1.0006
Paraffin (wax) paper	3.5
Glass	5–10
Mica	3–6
Rubber	2.5–35
Dry wood	2.5–8
Pure (distilled) water	81

The capacitance of any given capacitor is found from the relationship

$$C = \frac{0.0885KA(N - 1)}{T} \tag{3-29}$$

where C is the capacitance in picofarads
K is the dielectric constant
A is the area of one of the plates ($L \times W$), assuming that the two plates are identical
N is the number of identical plates
T is the thickness of the dielectric

Breakdown Voltage

A capacitor operates by supporting an electrical field between two metal plates. If this potential (voltage) gets too large, however, free electrons in the dielectric material (there are a few, but not many, in any insulator) may flow. If a stream of electrons gets started, then the dielectric may break down and allow a current to pass between the plates. The capacitor is then said to be *shorted*. The maximum breakdown voltage of the capacitor must not be exceeded. However, for practical purposes there is a smaller voltage called the *dc working voltage* (WVDC) that defines the maximum safe voltage that can be applied to the capacitor. Typical values found in common electronic circuits range from 8 WVDC to 1000 WVDC.

Circuit Symbols for Capacitors

The circuit symbols used to designate fixed-value capacitors are shown in Fig. 3–14(a), and for variable capacitors in Fig. 3–14(b). Both types of symbol are common. In certain types of capacitors, the curved plate shown on the left in Fig. 3–14(a) is usually the outer plate, that is, the one closest to the outside package of the capacitor. This end of the capacitor is often indicated with a color band next to the lead attached to that plate.

The symbols for the variable capacitor shown in Fig. 3–14(b) are the symbols for the fixed-value capacitor with the addition of an arrow drawn through the plates. Small variable capacitors used as trimmers and padders are often denoted by the symbol in Fig. 3–14(c). The variable set of plates is designated by the curved arrow.

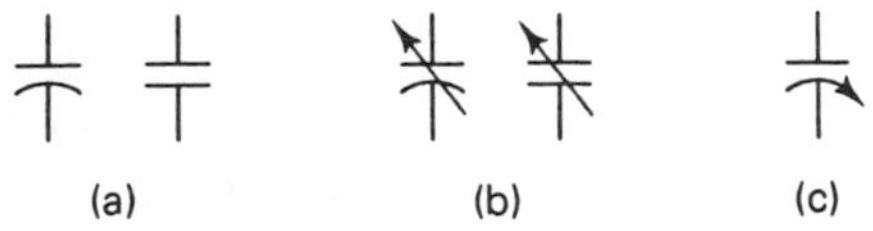

(a) (b) (c)

Figure 3–14 Capacitor circuit symbols: (a) fixed; (b) variable; (c) padder.

Fixed Capacitors

Several types of fixed capacitors are found in typical electronic circuits. They are classified by dielectric type: *paper*, *mylar*, *ceramic*, *mica*, *polyester*, and others.

The old-fashioned paper capacitor consists of a strip of paraffin wax paper sandwiched between strips of metal foil. The strip sandwich is rolled into a tight cylinder that is then packaged in a hard plastic, bakelite, or paper-and-wax case. If the case cracks or the wax end plugs become loose, the capacitor must be replaced even if it tests good, because it won't last long. Paper capacitors come in values from about 300 pF to about 4 μF. The working voltages range from 100 WVDC to 600 WVDC.

In modern capacitors of this type a mylar dielectric is used in place of the wax paper. A mylar unit with exactly the same capacitance rating and a WVDC rating that is equal to or greater than the original WVDC rating is equivalent to a paper capacitor.

Ceramic capacitors come in values from a few picofarads up to 0.5 μF. The working voltages range from 400 WVDC to more than 30,000 WVDC. The common disk ceramic capacitors are usually rated at either 600 WVDC or 1000 WVDC. Tubular ceramic capacitors typically have much smaller values than disk or flat capacitors. They are used extensively in VHF and UHF circuits for blocking, decoupling, bypassing, coupling, and tuning.

The feedthrough type of ceramic capacitor is used to pass dc and low-frequency ac lines through a shielded panel. These capacitors are often used to filter or decouple lines that run between circuits that are separated by a shield to reduce electromagnetic interference (EMI) reduction.

Ceramic capacitors are often rated as to *temperature coefficient*. This specification gives the change of capacitance per change of temperature in degrees Celsius. A *P* prefix indicates a positive temperature coefficient, an *N* indicates a negative temperature coefficient, and the letters *NPO* indicate a zero temperature coefficient (NPO stands for "negative positive zero"). (*Note:* Do not deviate from these ratings when servicing a piece of electronic equipment. Use a capacitor with exactly the same temperature coefficient as the original manufacturer used.) Nonzero temperature coefficients are often used in oscillator circuits to temperature-compensate the oscillator's frequency drift.

The mica capacitor consists of either a sheet of mica between metal plates or a sheet of mica that is silvered with a metal on both sides. The range of values for mica capacitors tends to be 50 pF to 0.02 μF at voltages in the range of 400 WVDC to 1000 WVDC.

Other Capacitors

Today the equipment designer has a number of different dielectric capacitors available that were not commonly available (or available at all) a few years ago. Polycarbonate, polyester, and polyethylene capacitors are used in a wide variety of applications where the previously discussed capacitors once ruled supreme. In digital circuits we find tiny 100 WVDC capacitors carrying ratings of 0.01 μF to 0.1 μF. These are used for decoupling the noise on the +5 V dc power supply line. In some circuits the leakage resistance across the capacitor becomes extremely important, so a polyethylene capacitor might be the best choice.

Capacitors in AC Circuits

When an electrical potential is applied across a capacitor, current will flow as charge is stored in the capacitor. As the charge in the capacitor increases, the voltage across the capacitor plates rises until it equals the applied potential. At this point if the voltage change ceases, the capacitor is fully charged, and no further current will flow, but in ac circuits the voltage changes continuously.

Figure 3–15 shows an analogy for the capacitor in an ac circuit. The actual circuit is shown in Fig. 3–15(a). It consists of an ac source connected in parallel across the capacitor (C). The mechanical analogy is shown in Fig. 3–15(b). The "capacitor" C consists of a two-chamber cylinder in which the upper and lower chambers are separated by a flexible membrane or diaphragm. The "wires" are pipes to the "ac source" (which is a pump). As the pump moves up and down, pressure is applied to first one side of the diaphragm then the other, alternately forcing fluid to flow into and out of the two chambers of the "capacitor."

The mechanical analogy is not perfect, but it works for our purposes. Now let's apply these ideas to the electrical case. In Fig. 3–16 we see a

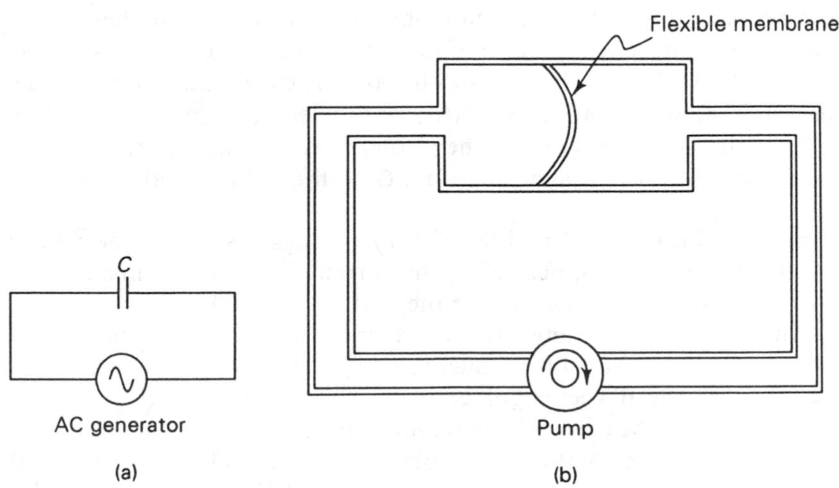

Figure 3–15 (a) Capacitor in ac circuit; (b) plumbing analogy.

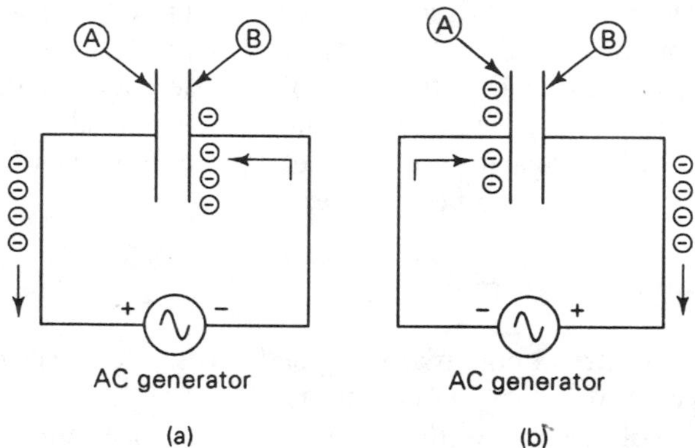

Figure 3–16 Capacitors on alternating cycles of an ac wave: (a) positive; (b) negative.

capacitor connected across an ac (sine-wave) source. In Fig. 3–16(a) the ac source is positive, so negatively charged electrons are attracted from plate *A* to the ac source, and electrons from the negative terminal of the source are repelled toward plate *B* of the capacitor. On the alternate half-cycle [Fig. 3–16(b)], the polarity is reversed, so electrons from the new negative pole of the source are repelled toward plate *A* of the capacitor, and electrons from plate *B* are attracted toward the source. Thus, current will flow in and out of the capacitor on alternating half-cycles of the ac source.

Voltage and Current in Capacitor Circuits

Consider the circuit in Fig. 3–17(a): an ac source *V* connected in parallel with the capacitor *C*. It is the nature of a capacitor to oppose these changes in the applied voltage (the inverse of the action of an inductor). As a result, the voltage *V* lags behind the current *I* by 90°. These relationships are shown in terms of sine waves in Fig. 3–17(b), and in vector form in Fig. 3–17(c).

The following mnemonic will help you remember the difference between the action of inductors *L* and capacitors *C* on the voltage and current.

ELI the ICE man. "ELI the ICE man" suggests that in the inductive *L* circuit, the voltage *E* comes before the current *I*—ELI, and in a capacitive *C* circuit the current *I* comes before the voltage *E*—ICE.

An ac circuit containing a resistance and a capacitance is shown in Fig. 3–18(a). As in the case of the inductive circuit, there is no phase shift across the resistor, so the **R** vector points in the "east" direction [Fig. 3–18(b)]. The voltage across the capacitor, however, is phase-shifted −90°, so its vector points "south." The total resultant phase shift θ is found using the Pythagorean rule to calculate the angle between V_r and V_t.

The impedance of the *RC* circuit is found in exactly the same manner

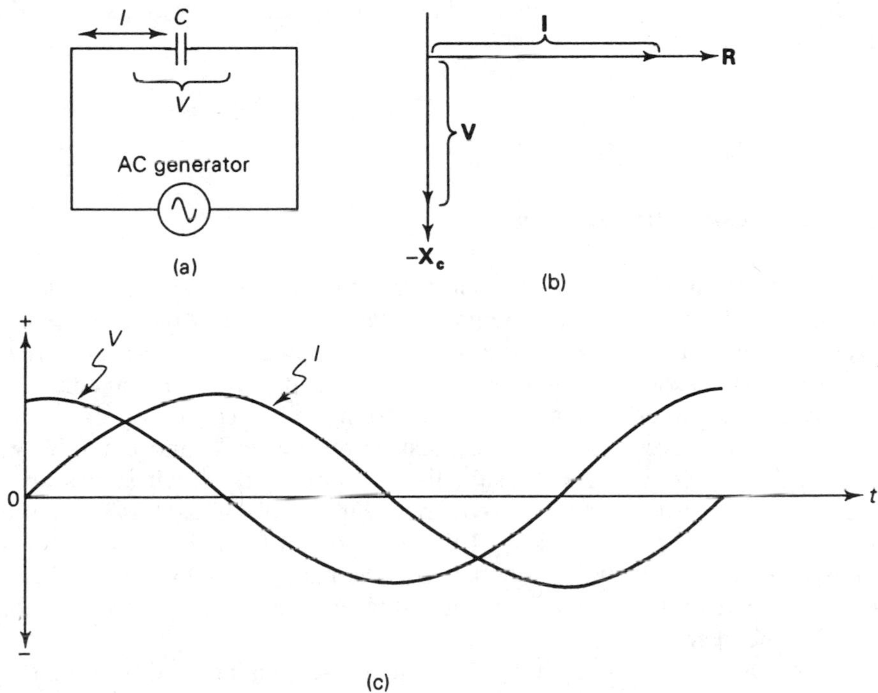

Figure 3–17 (a) Capacitor circuit; (b) vector phase relationship; (c) phase shown as sine waves.

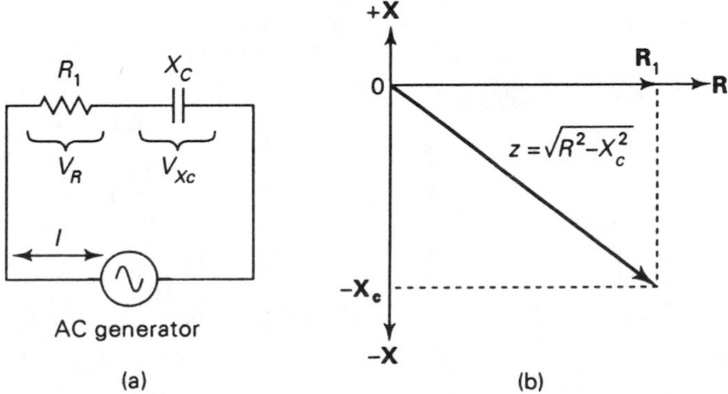

Figure 3–18 (a) Resistor-capacitor circuit; (b) phase relationship.

as the impedance of an *RL* circuit, that is, using the root of the sum of the squares:

$$|Z| = \sqrt{R^2 + X_c^2} \tag{3-30}$$

LC RESONANT CIRCUITS

The combination of an inductor *L* and a capacitor *C* in the same circuit forms an *LC resonant circuit*, also sometimes called a *tank circuit* or *resonant tank circuit*. There are two basic forms of *LC* resonant tank circuit: *series* [Fig. 3–19(a)] and *parallel* [Fig. 3–19(b)]. These circuits have much in common, but much makes them fundamentally different from each other.

Resonance occurs when the capacitive reactance X_c and inductive reactance X_L are equal. As a result, the resonant tank circuit shows up as purely resistive at the resonant frequency [Fig. 3–19(c)], and as a complex impedance at other frequencies. The *LC* resonant tank circuit operates by an oscillatory exchange of energy between the magnetic field of the inductor and the electrostatic field of the capacitor, with a current between them carrying the charge.

Because the two reactances are frequency-dependent and because they are inverses of each other, the resonance occurs at only one frequency, f_r.

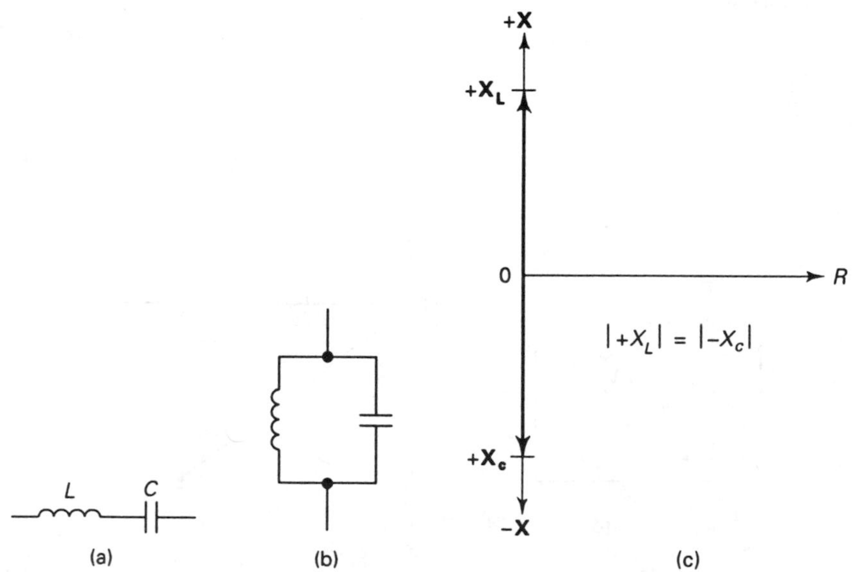

Figure 3–19 Inductor-capacitor circuits: (a) series; (b) parallel; (c) conventional way to show respective phase relationships.

We can calculate the standard resonance frequency by setting the two reactances equal to each other and solving for f:

$$f_r = \frac{1}{2\pi\sqrt{LC}} \tag{3-31}$$

Series-resonant circuits. The series-resonant circuit [Fig. 3–20(a)], like other series circuits, is arranged so that the terminal current I from the source V flows in both components equally. The vector diagrams of Figs. 3–20(a) through 3–20(c) show the situation under three different conditions:

Figure 3–20(a). The inductive reactance is larger than the capacitive reactance, so the excitation frequency is greater than f_r. Note that the voltage drop across the inductor is greater than that across the capacitor, so the total circuit looks as if it contains a small inductive reactance.

Figure 3–20(b). The situation is reversed: The excitation frequency is less than the resonant frequency, so the circuit looks slightly capacitive to the outside world.

Figure 3–20(c). The excitation frequency is at the resonant frequency, so $X_c = X_L$, and the voltage drops across the two components are equal but of opposite phase.

In a circuit that contains a resistance, an inductive reactance, and a capacitive reactance, there are three vectors ($\mathbf{X_L}$, $\mathbf{X_c}$, and $\mathbf{R}$) to consider (Fig. 3–21), plus a resultant vector. As in Fig. 3–18, the "north" direction represents $\mathbf{X_L}$, the "south" direction represents $\mathbf{X_c}$, and the "east" direction represents $\mathbf{R}$. Using the parallelogram method, we first construct a resultant for the $\mathbf{R}$ and $\mathbf{X_c}$, which is shown as vector $\mathbf{A}$. Next, we construct the same kind of vector ($\mathbf{B}$) for $\mathbf{R}$ and $\mathbf{X_L}$. The resultant $\mathbf{C}$ is made using the parallel-

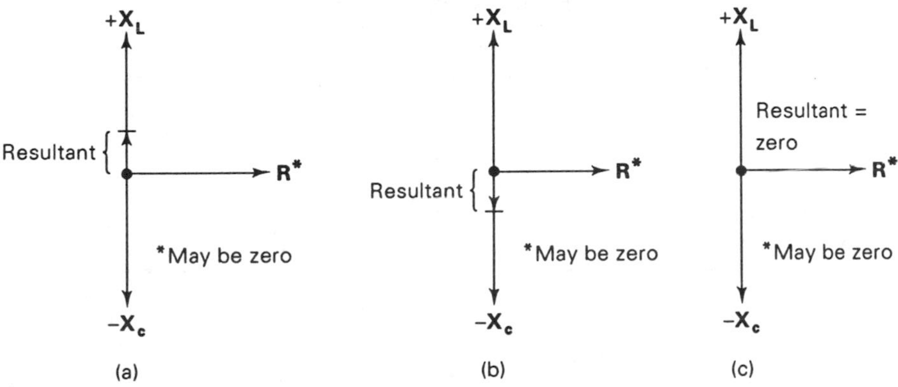

Figure 3–20 Three different cases: (a) $X_c > X_L$; (b) $X_c < X_L$; (c) $X_c = X_L$.

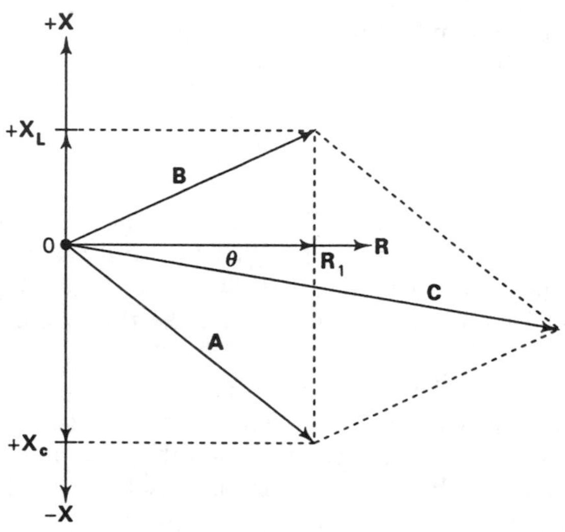

Figure 3–21 Vector relationships in *RLC* circuit.

ogram method on **A** and **B**. Vector **C** represents the impedance of the circuit: The magnitude is represented by the length, and the phase angle by the angle between **C** and **R**.

Figure 3–22(a) shows a series-resonant *LC* tank circuit, and Fig. 3–22(b) shows the current and impedance as a function of frequency.

A *series-resonant circuit has a low impedance at its resonant frequency and a high impedance at all other frequencies.*

As a result, the line current *I* from the source is maximum at the resonant frequency, and the voltage across the source is minimum.

Parallel-resonant circuits. The parallel-resonant tank circuit [Fig. 3–23(a)] is the inverse of the series-resonant circuit. The line current *I* from the source splits and flows in the inductor and the capacitor separately.

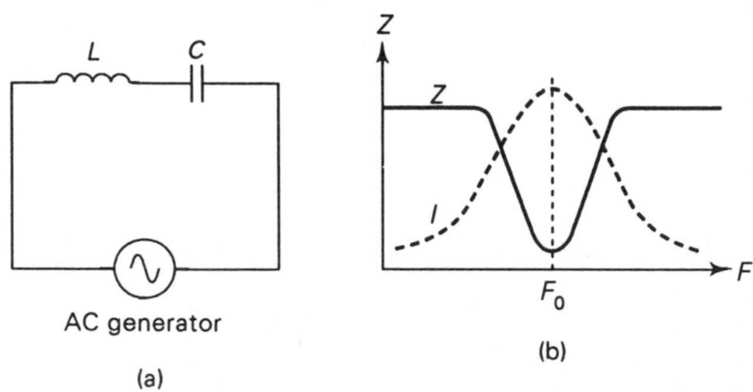

Figure 3–22 (a) *LC* series circuit; (b) impedance as a function of frequency.

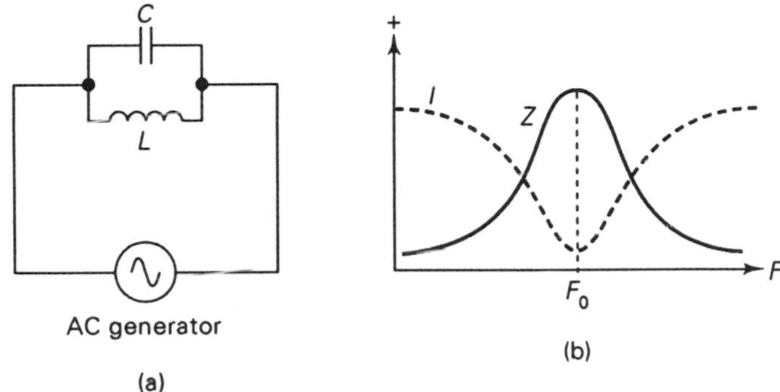

Figure 3–23 (a) parallel resonant circuit; (b) I and Z as a function of F.

The *parallel-resonant circuit has its highest impedance at the resonant frequency and a low impedance at all other frequencies.* Thus, the line current from the source is minimum at the resonant frequency [Fig. 3–23(b)], and the voltage across the LC tank circuit is maximum. This fact is important in radio tuning circuits, as we shall see later.

VARIABLE CAPACITORS

Like all capacitors, variable capacitors consist of two parallel sets of metal plates separated by a dielectric of air, mica, ceramic, or a vacuum [Fig. 3–24(a)]. The difference between variable and fixed capacitors is that in variable capacitors the plates are constructed in such a way that the capacitance can be changed. There are two principal ways to vary the capacitance: by varying either the spacing between the plates or the cross-sectional area of the plates that face each other. Figure 3–24(b) shows the construction of a typical variable capacitor used for the main tuning control in radio receivers. The capacitor consists of two sets of parallel plates. The *stator* plates are fixed in their position and are attached to the frame of the capacitor. The *rotor* plates are attached to the shaft that is used to adjust the capacitance.

Another form of variable capacitor found in radio receivers is the *compression capacitor* shown in Fig. 3–24(c). It consists of metal plates separated by sheets of mica dielectric. To increase the capacitance, the manufacturer may increase the area of the plates and mica or the number of layers (alternating mica and metal) in the assembly. The entire capacitor is mounted on a ceramic or other form of holder. If mounting screws or holes are provided, they will be part of the holder assembly.

Still another form of variable capacitor is the *piston capacitor* shown in Fig. 3–24(d). This type of capacitor consists of an inner cylinder of metal coaxial with, and inside, an outer cylinder of metal. An air, vacuum, or (as

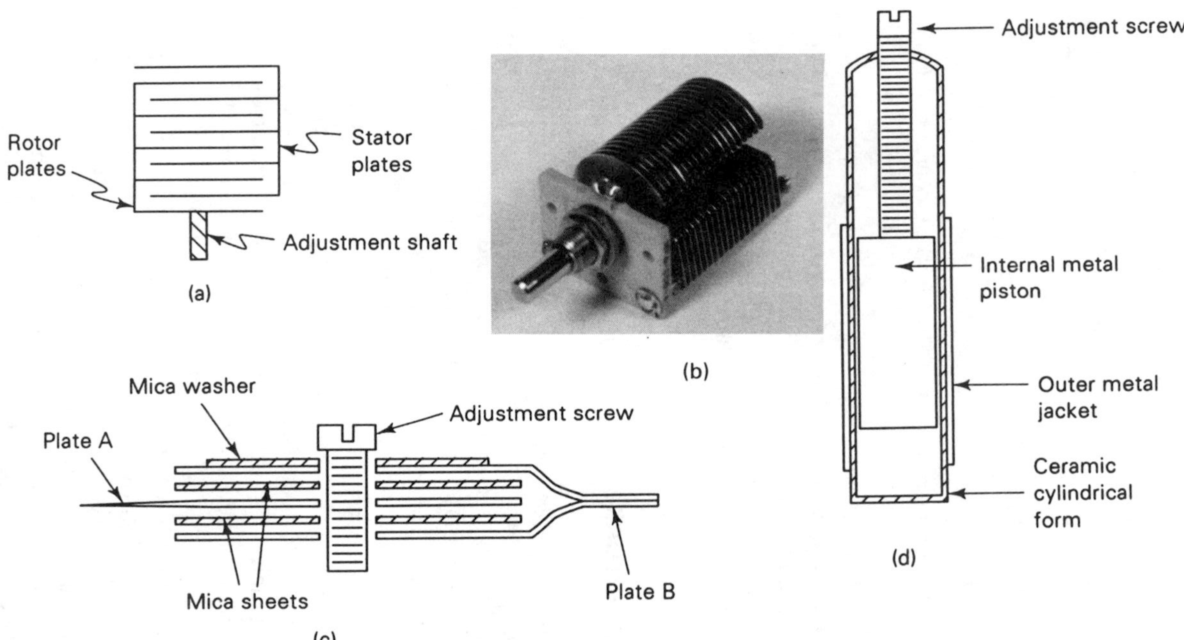

Figure 3–24 (a) Diagram of a variable, air-dielectric capacitor; (b) actual capacitor; (c) diagram of a mica compression trimmer capacitor; (d) piston trimmer capacitor.

shown) ceramic dielectric separates the two cylinders. The capacitance is increased by inserting the inner cylinder farther into the outer cylinder.

The small compression or piston-style variable capacitors are sometimes combined with air variable capacitors. The smaller capacitor used in conjunction with the larger air variable is called a *trimmer capacitor*. These capacitors are often mounted directly on the air variable frame [Fig. 3–25(a)] or very close by in the circuit. In many radios the "trimmer" is actually part of the air variable capacitor.

There are actually two uses for small variable capacitors in conjunction with the main tuning capacitor in radios. First, there is the true trimmer, that is, a small-valued variable capacitor in *parallel* with the main capacitor [Fig. 3–25(b)]. This capacitor is used to trim the exact value of the main capacitor. The other form of small capacitor is the *padder* capacitor, which is connected in *series* with the main capacitor. The error in terminology referred to in the preceding paragraph is calling both series and parallel capacitors "trimmers," when only the parallel-connected capacitor is properly so called.

Air variable main tuning capacitors. The capacitance of an air variable capacitor at any given setting is a function of how much of the rotor plate set is shaded by the stator plates. In Fig. 3–26(a), the rotor plates are completely outside the stator plate area. Because the shading is zero, the capacitance is minimum. In Fig. 3–26(b), however, the rotor plate set has been slightly meshed with the stator plate, so some of its area is shaded by the stator. The capacitance in this position is at an intermediate value. Finally, in Fig. 3–26(c) the rotor is completely meshed with the stator, so the cross-sectional area of the rotor that is shaded by the stator is maximum. Therefore, the capacitance is also maximum. The following two rules are useful:

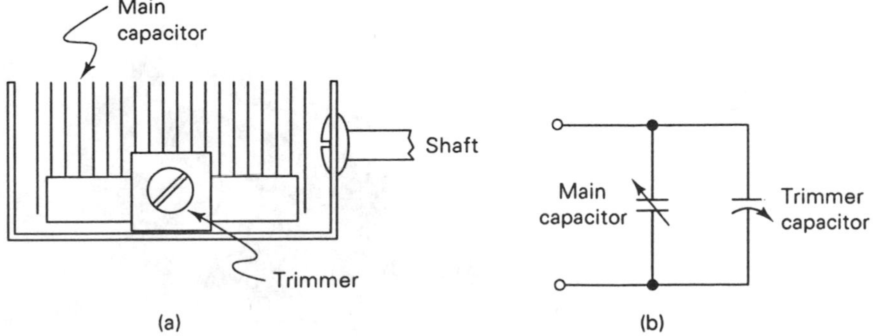

Figure 3–25 (a) Trimmer attached to an air variable capacitor; (b) circuit.

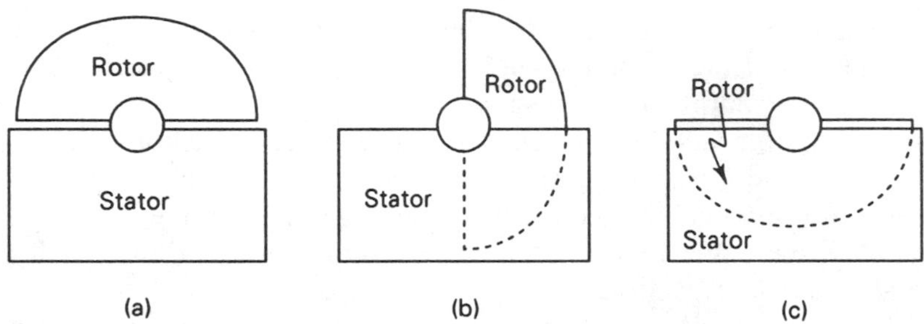

Figure 3–26 Variable capacitor at (a) minimum capacitance; (b) midrange; (c) maximum capacitance.

1. *Minimum capacitance is found when the rotor plates are completely unmeshed with the stator plates.*
2. *Maximum capacitance is found when the rotor plates are completely meshed with the stator plates.*

Figure 3–27 shows a typical single-section variable capacitor. The stator plates are attached to the frame of the capacitor, which in most circuits is grounded. Front and rear plates have bearing surfaces to ease the rotor's action. These capacitors were often used in early (1920s) multi-tuning-knob radio receivers. These ganged variable capacitors are basically two or (in the case of Fig. 3–28) three variable capacitors mechanically ganged on the same rotor shaft.

In Fig. 3–28 all three sections of the variable capacitor have the same capacitance, so they are identical to each other. If this capacitor is used in a superheterodyne radio, the section used for the local oscillator (LO) tuning must be padded with a series capacitance to reduce the overall capacitance.

Figure 3–27 Large air variable.

Figure 3–28 Three-section air variable capacitor.

Straight-Line Capacitance versus Straight-Line Frequency Capacitors

The variable capacitor shown on the right in Fig. 3–29 has the rotor shaft in the geometric center of the rotor plate half-circle. The capacitance of this type of variable capacitor varies directly with the rotor shaft angle. (Thus, it can be used as an angular-position sensor.) As a result, this type of capacitor is called a *straight-line capacitance* model. Unfortunately, the

Figure 3–29 Straight-line frequency (left) and straight-line capacitance (right) air variables.

frequency of a tuned circuit based on inductors and capacitors is not a linear (straight-line) function of capacitance. If a straight-line capacitance unit is used for the tuner, then the frequency units on the calibration dial will be cramped at one end and spread out at the other (you've probably seen this phenomenon on some older radios that have an analog dial). Some air variable capacitors have an offset rotor shaft (unit on left in Fig. 3–29) that compensates for the nonlinearity of the tuning circuit. The shape of the plates and the location of the rotor shaft are designed to produce a linear relationship between the shaft angle and the resonant frequency of the tuned circuit in which the capacitor is used.

SPECIAL VARIABLE CAPACITORS

In the preceding section we discussed the standard forms of variable capacitors. These capacitors are largely used for tuning radio receivers, oscillators, signal generators, and other variable-frequency *LC* oscillators. They are also used for determining angular position and other applications in the sensor world. In this section we examine some special forms of variable capacitors.

Split-stator capacitors. The split-stator capacitor is one in which two variable capacitors are mounted on the same shaft. The symbol for the split-stator capacitor is shown in Fig. 3–30. The split-stator capacitor normally uses a pair of identical capacitors turned by the same shaft. The rotor is common to both capacitors. Thus, the capacitor will tune two tuned circuits at the same time.

Differential capacitors. Although some differential capacitors are often mistaken for split-stator capacitors, they are actually quite different. The split-stator capacitor is tuned in tandem, that is, both capacitor sections have the same value at any given shaft setting, whereas the differential capacitor (Fig. 3–31) is arranged so that one capacitor section increases in capacitance while the other section decreases in exactly the same proportion. Figure 3–31(a) gives the circuit symbol for a differential capacitor, while Fig.

Figure 3–30 Symbol for split-stator capacitor.

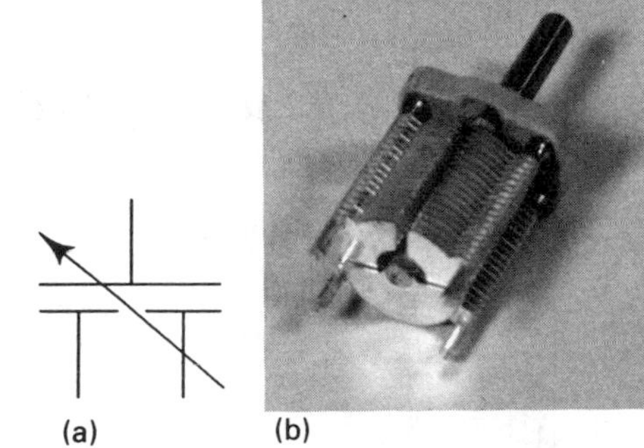

Figure 3–31 Differential capacitor: (a) symbol; (b) example.

3–31(b) shows a typical example. Note that the rotor plate is set to shade stator A and stator B equally. If the shaft is moved clockwise, it will shade more of stator B and less of stator A, so C_a will decrease, and C_b will increase by exactly the same amount. (Note: The total capacitance C_t is constant no matter what position that rotor shaft takes, only the proportion between C_a and C_b changes.)

4 Temperature Sensors

Temperature is one of the "big four" of sensor technology. It is one of the most often measured physical parameters and is important to nearly every engineering, scientific, or medical application of sensor technology. In this chapter we discuss the different forms of temperature sensors and the thermometry (measurement of temperature) circuits that support them.

TEMPERATURE SCALES

Among the different scales used in the measurement of temperature are the familiar *Fahrenheit* and *Celsius* (also called *centigrade*) and the less familiar *Kelvin* and *Rankine* scales. The Celsius and Fahrenheit scales are defined so that 0°C is equivalent to 32°F. The intervals on both scales were originally defined by the *ice point* (the freezing point of water at standard temperature and pressure) and the *steam point* (the boiling point of water at standard temperature and pressure). Temperatures on the two scales can be interconverted by the equation

$$°F - 32 = 1.8°C \tag{4–1}$$

The Kelvin scale uses the same size degree as the Celsius scale, but it is defined by *triple point* of water (the temperature at which the three phases of water are in equilibrium at constant pressure). This temperature was assigned the value 273.16 K. A second fixed point on the Kelvin scale, 0 K or *absolute Zero* (the point where molecular activity ceases), is implied but not defined. Thus, 0 K ≃ −273.16°C. Celsius and Kelvin temperatures can be interconverted by the equations

$$K = °C + 273.16 \tag{4–2}$$

$$°C = K - 273.16 \tag{4–3}$$

On the Rankine scale a degree is the same size as a Fahrenheit degree, but the zero point is at absolute zero on the Fahrenheit scale. Thus, 0°R ≃ −459.7°F.

Several different sensors are commonly used to measure temperature: thermal resistors (RTDs and thermistors), thermocouples, and PN semiconductor junctions. Although applications for three different forms of sensors overlap, key parameters and other factors often favor one over another.

THERMAL RESISTORS

Thermal resistors are electrically conductive elements that are designed to change electrical resistance in a predictable manner with changes in applied temperature. The two basic classes of thermal resistors are resistance temperature devices (RTDs) and thermistors.

The amount of resistance change is designated by the *temperature coefficient* α of the material, which is measured in ohms of resistance change per ohm of resistance per degree Celsius. A *positive temperature coefficient* (PTC) device exhibits increased resistance with increases in temperature. Alternatively, a *negative temperature coefficient* (NTC) device shows decreased resistance with increases in temperature. Typical curves for NTC and PTC thermistors and a platinum RTD are shown in Fig. 4–1. The usual circuit symbols for thermal resistors are shown in Fig. 4–2. The indirectly heated variety uses an internal heating element, whereas the directly heated form absorbs heat from the environment.

Resistance Change with Temperature

Most thermal resistors have a nonlinear characteristic curve when it is plotted over a wide temperature range. But when operation is limited to narrow temperature ranges, the linearity is considerably better (again see Fig. 4–1). When using such devices, it is necessary to ensure that the temperature will remain within the permissible linear range. Methods for linearizing the devices will be discussed in a later section.

Thermal resistors are among the oldest temperature sensors known. The temperature sensitivity of electrical resistance in silver sulfide was noted by physicist Michael Faraday in 1833, and that of iron wire was equally well known at early dates. There are several different types of thermal resistors,

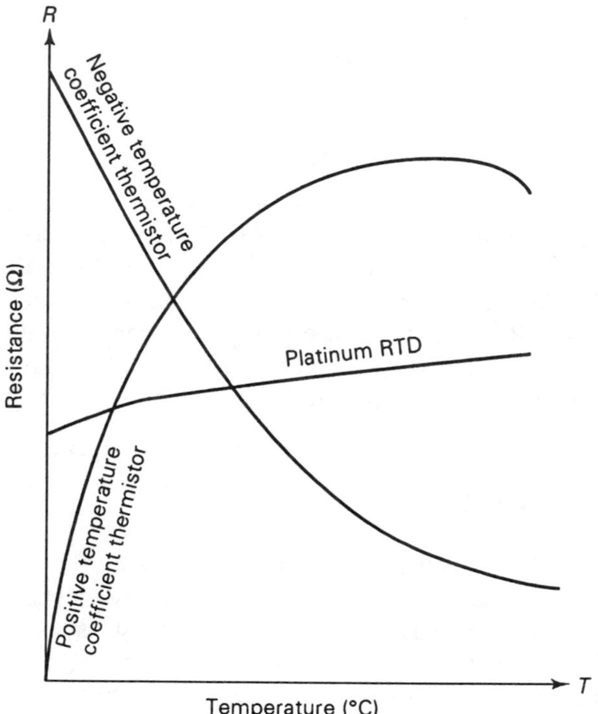

Figure 4–1 Thermistor characteristic curves.

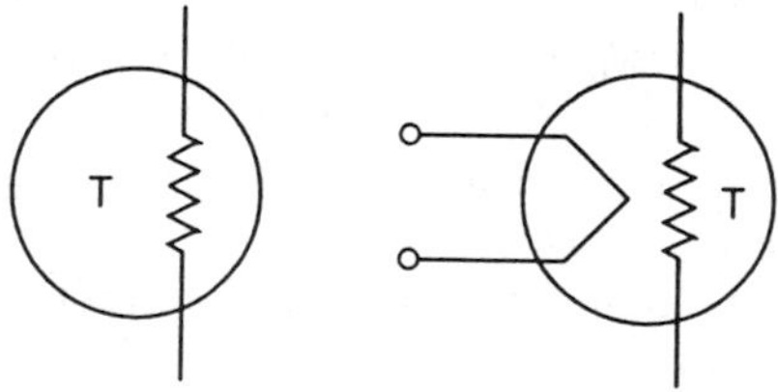

Figure 4–2 Circuit symbols for thermistors.

but the simplest is the *resistance temperature device* (*RTD*). The RTD may be either a wire element or a thin metallic film.

Simple RTD elements are based on the tendency of materials to change physical dimensions with changes in temperature. Metals, for example, tend to expand when heated, so their electrical resistance increases. The resistance of metal wires is directly proportional to the length of the sample. Most metals have a positive temperature coefficient ($\alpha > 0$). Copper, for example, has a temperature coefficient of $+0.004$.

Not all materials have positive temperature coefficients, however. Some materials, like carbon and some ceramics, have a negative temperature coefficient (for carbon $\alpha = -0.0005$). Other materials, including certain metal alloys or oxides, have temperature coefficients that are very low. For example, the temperature coefficient for manganin and constantan is approximately +0.00002, and for nichrome it is +0.00017. Table 4–1 gives the temperature coefficients of some common materials.

The change in resistance caused by changes in temperature is a function of α and the value of temperature change. For a wire element the new resistance is found from the equation

$$R_{T2} = R_{T1}[1 + \alpha(T_2 - T_1)] \tag{4–4}$$

where R_{T1} is the resistance at the initial temperature
R_{T2} is the resistance at the final temperature
α is the temperature coefficient
T_1 is the initial temperature
T_2 is the final temperature

[Note: In some sources the term R_{T1} is replaced by R_0 when temperature T_1 is 0°C. In that case, T_1 is relabeled T_0].

Wire elements are sometimes used as thermal resistors. Platinum-wire RTD elements, for example, are frequently used as sensors at temperatures down to 20 K, and rhodium is used below 20 K. In platinum RTDs the resistance change with temperature ($\Delta R/T$) is about 0.377%/°C, with linearity of ±0.2% over the range 0°C to 100°C and accuracies of ±0.001°C.

The platinum RTD is used as an international standard in thermometry. One International Platinum Temperature Scale (IPTS) defines a standard platinum RTD over the range from the boiling point of oxygen, −183.962°C, to the melting point of antimony, +630.74°C. In 1968 a new standard (IPTS-68) was specified in which the reference end points are the triple point of hydrogen (13.81 K) and the freezing point of antimony (903.89 K). The following general expression applies:

$$R_T = R_0[1 + a_1T + a_2T^2 + a_3T^3 + \ldots + a_nT^n] \tag{4–5}$$

TABLE 4–1

Material	Resistivity (ρ) ($\mu\Omega \cdot$ cm)	Temperature Coefficient (α) ($\Omega/\Omega/$°C)
Carbon (C)	3496	−0.0005
Iron (Fe)	10	+0.005
Nichrome	101	+0.00017
Platinum	10	+0.00377
Silver (Ag)	1.628	+0.0038
Aluminum (Al)	2.828	+0.0036
Annealed copper (Cu)	1.724	+0.0039
Gold (Au)	2.4	+0.004
Nickel (Ni)	6.84	+0.0067
$10^{16}/cm^3$ Si (N-type)	1.43	+0.007
10^{16}/cm Si (P-type)	0.62	+0.007

For the limited temperature range 0°C to 100°C, a simplified equation is often used:

$$R_T = R_0(1 + a_1 T) \tag{4–6}$$

and from 0°C to 630°C:

$$R_T = R_0 + \alpha R_0 T + \beta R_0 T^2 \tag{4–7}$$

where R_T is the thermistor resistance at temperature T
R_0 is the resistance of the thermistor at the ice point (0°C)
T is the temperature being measured
α is a constant (3.98×10^{-3} Ω/°C for platinum)
β is a constant (-0.586×10^{-6} Ω/°C for platinum)

Another expression that is frequently used in thermal resistor calculations over the old IPTS range (−183°C to +630°C) is the *Callendar–Van Dusen equation*:

$$R_T = R_0\alpha[T - \sigma(0.01T - 1)(0.01T) - \beta(0.01T - 1)(0.01T)^3] \tag{4–8}$$

where R_T is the resistance at temperature T
R_0 is the resistance at the ice point (0°C)
α is the resistance at 100°C
β is the resistance at −183.96°C
σ is the resistance at 444.7°C

(Typically, $\alpha = 3.9 \times 10^{-3}$, $\beta = 0$ if $T > 0$ and 0.11 if $T < 0$, and $\sigma = 1.49$ when $R_0 = 100$.)

Thermistors

Another type of thermal resistor (the *thermistor*) is made of evaporated films, carbon or carbon compositions, or ceramiclike semiconductors formed of oxides of copper, cobalt, manganese, magnesium, nickel, titanium, or uranium. Unlike the basic RTD device, thermistors can be molded or compressed into a variety of clever shapes to fit a wide range of applications (Fig. 4–3). These devices have a resistance change characteristic of 4% to 6%/°C and generally a negative temperature coefficient. A special class of thermistors, called *posistors*, which are made of barium titanate or strontium titanate ceramics, have a positive temperature coefficient. Positive temperature coefficients are also found in silicon thermistors in which the Si semiconductor is doped to a density of about $10^{16}/\text{cm}^3$.

At least one major supplier of thermistor products uses the *Steinhart-Hart equation* to describe thermistor behavior to within 0.01°C accuracy:

$$\frac{1}{T} = a_0 + a_1 \ln R + a_2 \ln R^3 \tag{4–9}$$

where T is temperature in kelvins

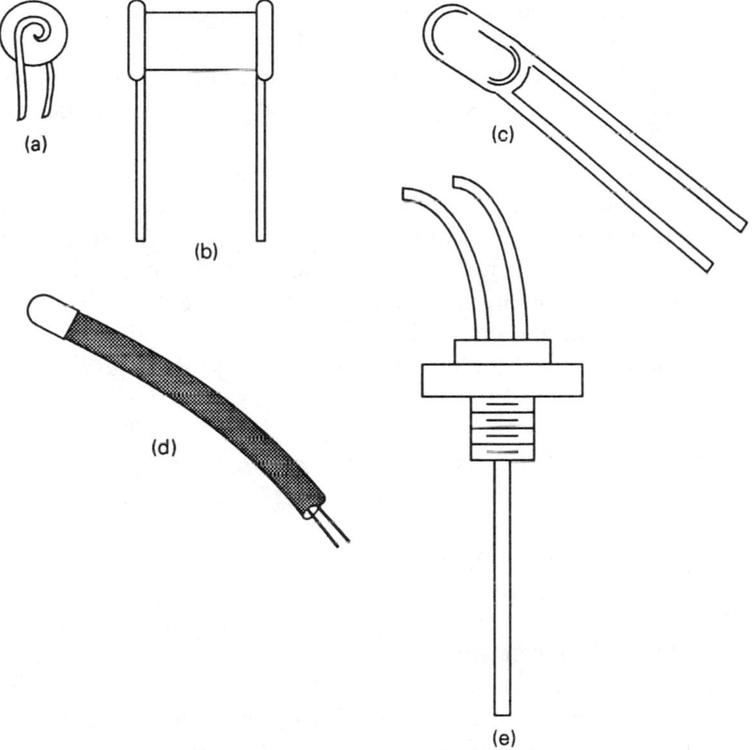

Figure 4–3 Body styles of various thermistors.

or, when the temperature range is limited:

$$T = \frac{a_1}{\ln R - a_0} - a_2 \qquad (4\text{–}10)$$

The constants a_0, a_1, and a_2 are found empirically by solving three simultaneous equations with known resistance and temperature values.

Thermistor Parameters

To successfully use thermistors of either type it is necessary to first understand some of the basic properties of the thermistor, which are expressed in the form of certain standard parameters.

Cold (zero-power) resistance. This parameter is the resistance of the thermistor at a standard reference temperature [usually either room temperature (25°C) or the ice point of water (0°C)] under conditions of no self-heating power dissipation. This parameter is the cold resistance that is listed in the specifications sheet as the *nominal resistance*. For example, a device listed as a "1000 ohm thermistor" has a resistance of 1000 ohm at the standard reference temperature (25°C unless otherwise specified). The conditions under which the thermistor is operated for measurement of the cold resistance include a requirement that the current through the device be sufficiently low to prevent self-heating.

Hot resistance. The hot resistance of the thermistor is measured when the device is operated at a temperature higher than the cold resistance temperature. The higher temperature is due to ambient temperature, the current flow through the thermistor, the applied heater current (indirectly heated types only), or a combination of all these factors. Equation (4–4) can be modified as follows to find the hot resistance of wire elements:

$$R_T = R_0[1 + \alpha(T - T_0)] \tag{4–11}$$

For other forms of thermistor the expression is

$$R_T = R_0 e^{\beta(1/T - 1/T_0)} \tag{4–12}$$

where T_0 is the reference temperature (25°C)
T is the new temperature
R_0 is the thermistor resistance at the reference temperature
R_T is the resistance at temperature T
α is the coefficient of resistance
β is a materials factor called the *characteristic temperature*, in kelvins (usually between 1500 K and 7000 K, typically 4000 K)

Resistance versus temperature. This parameter is an expression of the characteristic curve shown in Fig. 4–1. The exact shape of the curve is a function of the thermistor in question but is of the form shown in Fig. 4–1 and is quite nonlinear in certain parts of its range.

Resistance ratio R_T/R_0. The resistance ratio is essentially a simplified expression of the R-vs.-T curve. It gives the ratio of the thermistor resistance at a specified temperature (50°C, 100°C, or 125°C) to the cold (25°C) resistance.

Voltage versus current. Directly heated thermistors have an unusual V-vs.-I curve (Fig. 4–4) that includes both ohmic and negative resistance regions. At a constant ambient temperature, an increase in current through the thermistor will cause a linear increase in voltage drop across the thermistor. Because this behavior is in accordance with Ohm's law, $V = IR$, that portion of the curve is called the *ohmic* or *positive-resistance region*. At a certain point, however, internal self-heating becomes dominant and begins to alter the resistance of the thermistor. At this point the voltage drop begins to *decrease with increasing current flow*. In other words, in this region the thermistor is a *negative-resistance device*. Notice also in Fig. 4–4 that the curve for a thermistor immersed in water has a longer linear region than the curve for the same thermistor in air. This is due to the superior heat-transfer properties of water compared with air.

Current versus time. A typical thermistor *I*-vs.-*t* curve is shown in Fig. 4–5. The thermistor current ideally snaps to the level V/R_T when a step-

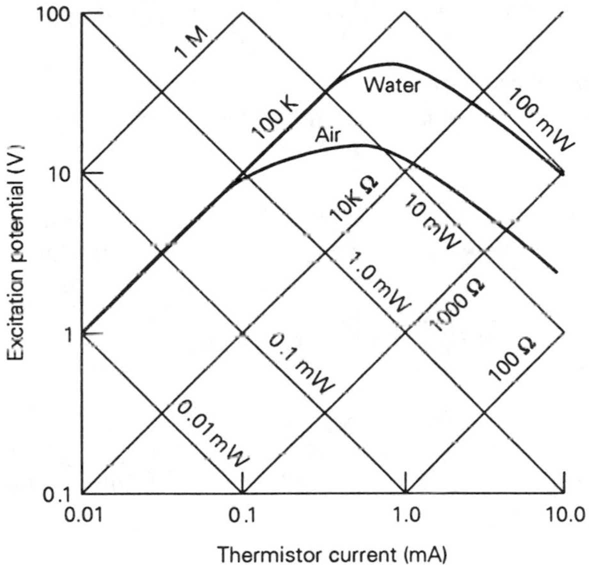

Figure 4–4 Self-heating curves for water- and air-cooled thermistors.

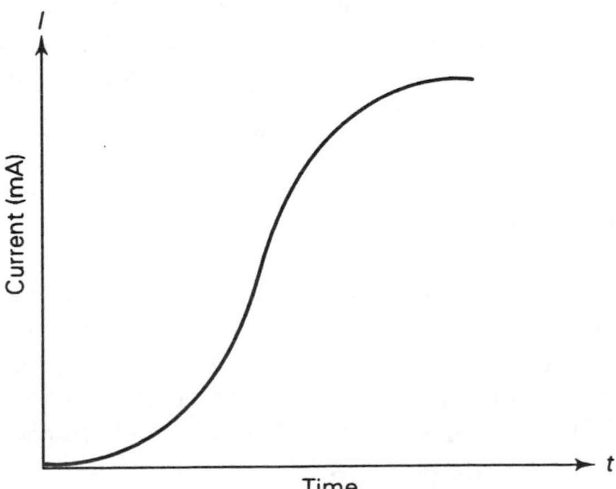

Figure 4–5 Current-vs.-time characteristics.

function voltage is applied (or the applied level is changed). However, because there is always a small amount of self-heating involved in any thermistor, this response is not linear. There is a lag between the time a change is made in the applied voltage and the time the current in the thermistor reaches the level mandated by that new voltage for the thermistor resistance.

Maximum power P_{max}. This parameter is the maximum allowable constant power level that the thermistor can handle without destruction, permanent alteration of its characteristics, or degradation of its performance.

Dissipation constant δ. This factor is the ratio of the change in power dissipation for small changes in the body temperature of the thermistor ($\delta = dP_d/dT_B$).

Sensitivity γ. This parameter is the ratio of resistance change to temperature change dR/dT, expressed as a percent change per degree of temperature. Because the R-vs.-T curve (Fig. 4–1) is nonlinear over most of its range, sensitivity values are valid over only a limited range. Typical values for gamma run from 0.5%/°C to 4%/°C.

Temperature range. The thermistor's characteristics are specified over only a limited temperature range, T_{min} to T_{max}. The value of T_{min} is typically −200°C, while T_{max} is typically +650°C (although there are devices with a narrower range).

Thermal time constant τ. The body temperature of a thermistor does not change instantaneously in response to a step-function change in ambient temperature. If T_i is the initial temperature, and T_f is the final temperature, then the thermal time constant is the time required for the body temperature of the thermistor to change by 63.2 percent of the range between these two temperatures. The value 63.2 is derived from $(1 - e^{-t})$ when $t = 1$ s.

Linearizing Thermistors

The R-vs.-T curve seen earlier in Fig. 4–1 is nonlinear over much of its range. For some measurements it is therefore necessary either to restrict the use of the device to a limited range of temperatures or to linearize the curve. There are several ways to linearize the curve. Some of them involve electronic circuits, so we shall discuss them in detail after we have discussed the circuits involved. There are, however, two methods that involve only simple resistors or other thermistors. Figure 4–6(a) shows several linearization networks used by a thermistor manufacturer. Although the network

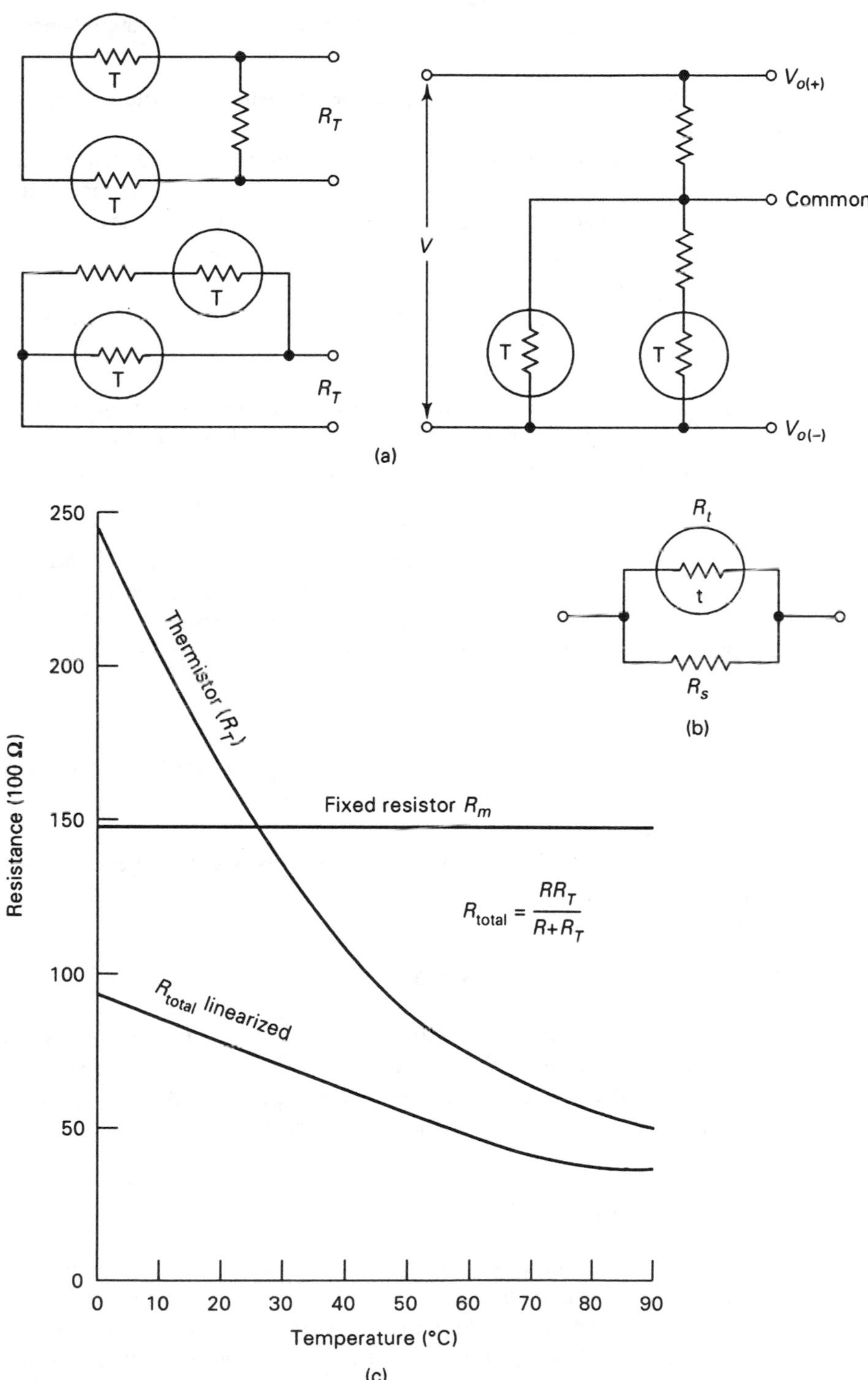

Figure 4–6 (a) Linearization circuits for thermistors; (b) simple linearization circuit; (c) curves for linearized circuit (b).

appears to function like a single two-terminal thermistor, it actually consists of a network of resistors and thermistors.

A relatively easy method for linearizing a thermistor is shown in Fig. 4–6(b). This method involves shunting a low-temperature-coefficient resistor R_s across the thermistor R_t. The total value of the network is the parallel resistance of the two elements:

$$R_{\text{total}} = \frac{R_t R_s}{R_t + R_s} \tag{4–13}$$

The value of R_s is the mean value R_m of R_t over the temperature range of interest. Suppose, for example, that it is necessary to linearize a thermistor over the human physiological temperature range (i.e., 30°C to 45°C). The value of R_m in this case, hence the value of R_s, is the thermistor resistance at a temperature of (30 + (45 − 30)/2)°C, or 37.5°C. Figure 4–6(c) shows the relationship between the unlinearized thermistor resistance R_t and the parallel fixed resistor R_m. Note that R_{total} is linear over a considerably larger portion of its temperature range than the unlinearized version.

Modifying Eq. (4–4) gives us the expression for the value of the total resistance R_t':

$$R_t' = \frac{R_m}{2}\left[1 + \frac{\alpha}{2}(T - T_m)\right] \tag{4–14}$$

There are other methods for linearizing thermistors, but these are not discussed here.

Thermal Resistor Thermography

Thermography is the art of measuring temperature. Thermal resistors (RTDs or thermistors) utilize both *passive* and *active* circuits. The principal difference between the two is that in an active circuit an operational amplifier is intimately interconnected with the thermal resistor.

Passive Thermography Circuits

A thermistor is intended to pass current when it operates in circuits. When the current I flows in a resistance R, a voltage drop V is created: $V = IR$. When the resistance is a thermistor, the voltage is proportional to the applied temperature within the limits of the linearity and accuracy of the device. Passive circuits can be classified as *voltage-excited* or *current-excited* circuits.

Voltage-Excited Circuits

A voltage-excited thermography circuit uses a voltage-regulated dc power supply to provide the excitation to the sensor. Perhaps the simplest voltage-excited thermal resistor circuit is the *half-bridge* or *voltage-divider* circuit of Fig. 4–7. The output voltage V_o is a fraction of the excitation voltage V. The ratio of the fixed resistor to the thermistor resistance determines the output voltage:

$$V_o = V \frac{R_t}{R + R_t} \tag{4–15}$$

or, when the thermistor resistance changes (ΔR) in response to a temperature change (ΔT):

$$V_o = V \frac{R_t \pm \Delta R}{R + (R_t \pm \Delta R)} \tag{4–16}$$

The half-bridge circuit is easy to construct and low in cost, but it suffers from a serious drawback. The output voltage of the half-bridge is never zero unless the thermistor resistance is zero. In other words, an offset voltage exists at all temperatures. The ideal would be to make the output voltage zero when the temperature is zero. A solution to this problem is to use a pair of half-bridge circuits in parallel and then take the difference between their respective output voltages as the output voltage of the network. This type of circuit is called a *Wheatstone bridge* [Fig. 4–8(a)].

The Wheatstone bridge consists of four resistance arms forming two voltage dividers. We can combine the two half-bridge equations of these voltage dividers to find the two half-bridge output voltages V_A and V_B

$$V_A = \frac{VR_t}{R_1 + R_t} \tag{4–17}$$

$$V_B = \frac{VR_3}{R_2 + R_3} \tag{4–18}$$

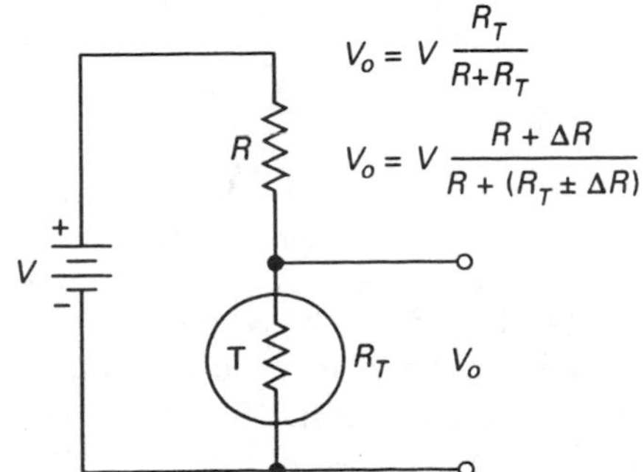

Figure 4–7 Half-bridge circuit.

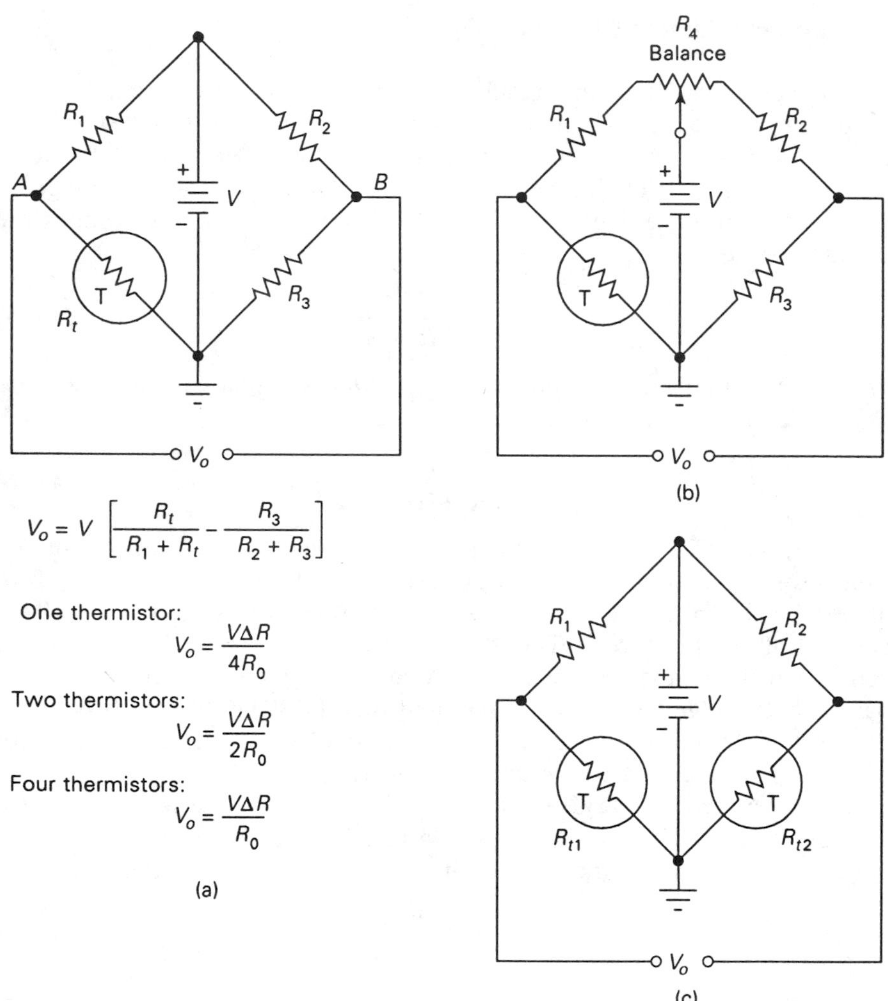

Figure 4–8 (a) Wheatstone bridge with thermistor element; (b) Wheatstone bridge with balancing potentiometer; (c) two-thermistor differential temperature bridge.

to form the expression

$$V_o = V_A - V_B \tag{4–19}$$

$$V_o = V\left(\frac{R_t}{R_1 + R_t} - \frac{R_3}{R_2 + R_3}\right) \tag{4–20}$$

Equation (4–19) shows that the output voltage V_o is zero when V_A and V_B are equal. We can therefore adjust the output voltage to zero by balancing the bridge for the 0°C condition. Balance can be achieved when $R_1 = R_2 = R_3 = R_0$, where R_0 is the ice-point (0°C) resistance of the thermistor. Note that another way to achieve the null condition is to make the ratio of the resistances in the two half-bridges equal. This condition exists in Fig. 4–8(a) when

$$\frac{R_1}{R_t} = \frac{R_2}{R_3} \tag{4–21}$$

Equation (4–21) describes most practical bridge designs.

A version of this expression that is sometimes more helpful is found by solving for R_t:

$$R_t = \frac{R_1[R_2/(R_2 + R_2) + V_o/V]}{1 - (V_o/V) - R_3(R_2 + R_3)} \tag{4–22}$$

or, in the simplified case where all bridge arms are equal ($R_1 = R_2 = R_3 = R$):

$$R_t = \frac{R(V + 2V_o)}{V - 2V_o} \tag{4–23}$$

If the null condition is not met because of errors in the values of bridge resistors, then there will be an offset in the value of V_o. The bridge can be forced to null by using a BALANCE control such as potentiometer R_4 in Fig. 4–8(b). The potentiometer is adjusted for $V_o = 0$ when the thermistor is equilibrated at 0°.

There are two basic uses for the Wheatstone bridge in thermography. (1) The bridge can be nulled ($V_o = 0$) at some reference temperature (e.g., the ice point), and the output voltage can then be measured to find the temperature at which R_t changes. (2) One of the bridge elements (e.g., R_1 or R_3) can be made into a potentiometer with a precision calibrated dial on its shaft. The dial is calibrated in degrees of temperature. When the potentiometer is adjusted to balance the bridge ($V_o = 0$), the resistance reading of the potentiometer dial will indicate the temperature.

Differential temperature measurements attempt to measure the difference between two temperatures and present an output indication that is proportional to that difference. Figure 4–8(c) shows a Wheatstone bridge differential thermometer circuit. The differential aspect is achieved by replacing R_3 of Fig. 4–8(b) with another thermistor, R_{T2}.

Thermal Resistor Errors

Thermal resistors used in thermography are sensitive to certain forms of error. The self-heating error was discussed earlier. It is controlled by keeping the current in the thermistor below the critical value at which self-heating occurs. Self-heating is also controlled by using a higher current but presenting it in the form of low-duty-cycle pulses rather than as dc level. At the output of the thermography circuit the output voltage can be sampled synchronously with the pulses, or the pulses can be integrated to form a dc level that is proportional to the average of the pulse-train amplitudes.

Another source of error is *thermoelectric potentials* in the circuit. Any time two different metallic surfaces are brought into contact, as when wires and connection terminals are joined together, a small EMF is created. We shall see later that this EMF is the basis for *thermocouple* temperature sensors, but in a thermal resistor circuit the potential must be controlled. Perhaps the best way to handle this EMF, where it is a factor, is to measure it and then subtract it out of the final result. Some electronic circuits do this automatically.

Another potential error exists because of the resistance of the leads that make up the thermal resistor assembly. Unfortunately, many thermal resistors, especially RTDs, have very low resistance values, and these values are close to the typically very low resistances R_L found in the leads that connect the thermal resistor to its circuits. Equation (4–23) must be modified to the form

$$R_t = \frac{R(V + 2V_o)}{V - 2V_o} + \xi \qquad (4\text{–}24)$$

where ξ is the error term

$$\xi = \frac{4R_L V_o}{V - 2V_o} \qquad (4\text{–}25)$$

Note in this equation that the errors tend to drop when the bridge is near the null condition ($V_o = 0$) and also when the lead resistance R_L is very much smaller than the thermal resistor value at the temperatures being measured ($R_L << R_t$).

The lead resistance error can be reduced by separating the lead carrying the excitation current from the output voltage lead. A three-wire bridge circuit such as that shown in Fig. 4–9 will accomplish this job. All three wires must be heavy with respect to the amount of current that they carry in order to reduce the voltage drops associated with R_L.

Current-Excited Circuits

Current-excited thermography circuits use a *constant-current source* (CCS) to provide dc power to the thermistor circuit. These circuits (e.g., Fig. 4–10) are basically current regulators. They keep the output current at the set level despite variations in the output load resistance. Figure 4–10(a) is a simple junction field-effect transistor (JFET) CCS circuit. The JFET is operated above its saturation knee, so the output current is constant. The circuit of Fig. 4–10(b) is an op-amp version that is voltage-controlled by a zener diode. The inverting input of the op-amp senses the current level and

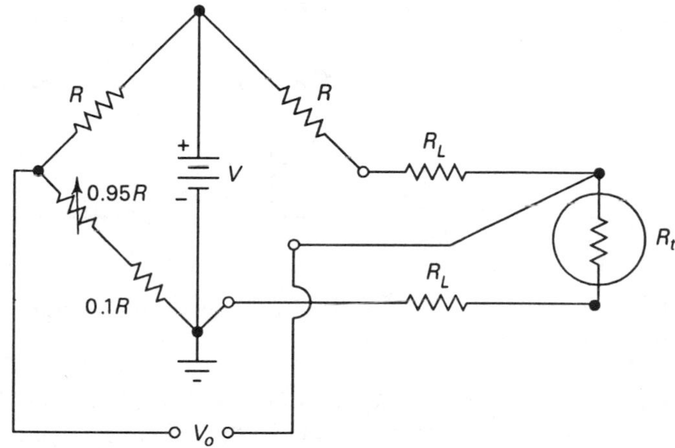

Figure 4–9 Remote thermistor bridge.

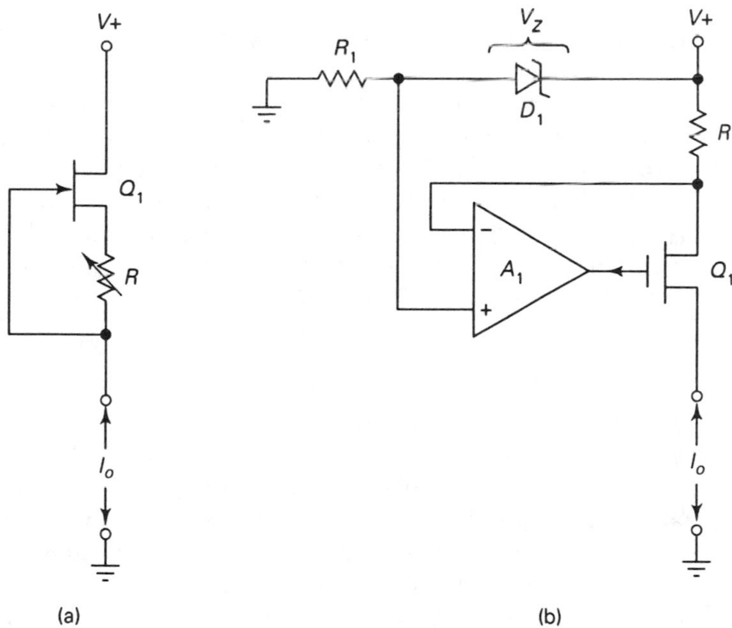

Figure 4–10 (a) JFET constant-current source; (b) active constant-current source.

adjusts the channel resistance of the MOSFET transistor Q_1 to keep it at the set point.

The basic current-excitation circuit is shown in Fig. 4–11(a). The current source I_1 is connected across the thermistor. When the current flows in the thermistor, output voltage V_o is created that is proportional to the resistance: $V_o = I_1R_t$. This circuit will work for low values of I_1, and short,

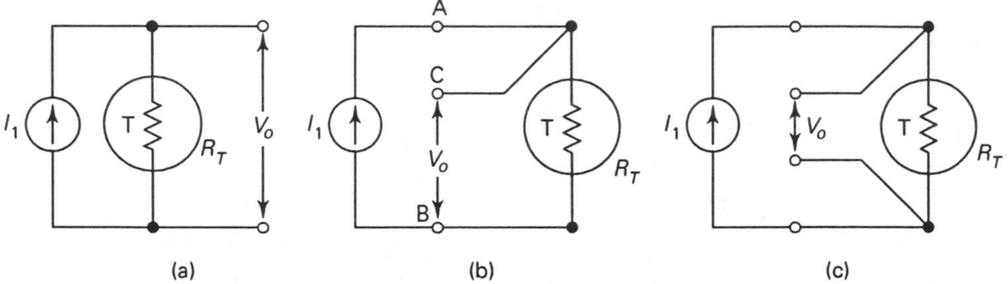

Figure 4–11 (a) thermistor excited by a constant-current source; (b) better circuit model with leads; (c) both leads accounted for in four-terminal circuit.

low-resistance runs of thermal resistor connection leads, but it can suffer from the same lead resistance problem as the voltage-excited circuits. Figures 4–11(b) and 4–11(c) show the equivalent three- and four-wire circuits for reducing these problems in current-excited circuits.

Active Thermal Resistor Thermography Circuits

Active thermal resistor thermography circuits differ from passive circuits by having an amplifier or other electronic device process the output signal. Some of these circuits may be simple dc differential amplifiers such as the one in Fig. 4–12. In this case the input resistances R_1 and R_2 of the amplifier must be very large relative to the Thevenin equivalent resistance of the Wheatstone bridge. On bridges where all elements are equal, this value is the resistance R of any one arm. The voltage gain of the circuit is given by

$$A_v = \frac{R_3}{R_1} \tag{4–26}$$

Unfortunately, this situation is often difficult to achieve in thermistor circuits, although with low-resistance RTD sensors it is somewhat easier. It is possible, even desirable, to replace the dc differential amplifier with an instrumentation amplifier (IA) circuit. The IA circuit has a very high input impedance, so it does not suffer from the problems that would exist in a dc differential amplifier with an input resistance that is too low.

A circuit based on a simple op-amp inverting follower circuit is shown in Fig. 4–13. In any inverting follower the voltage gain of the circuit is the ratio of the feedback resistor to the input resistor: $A_v = -R_f/R_{in}$. For the circuit of Fig. 4–13, therefore, we can state that

$$V_o = -V_{ref}\frac{-R_f}{R_{in}} \tag{4–27}$$

$$V_o = -V_{ref}\frac{-R_t}{R_1 + R_2} \tag{4–28}$$

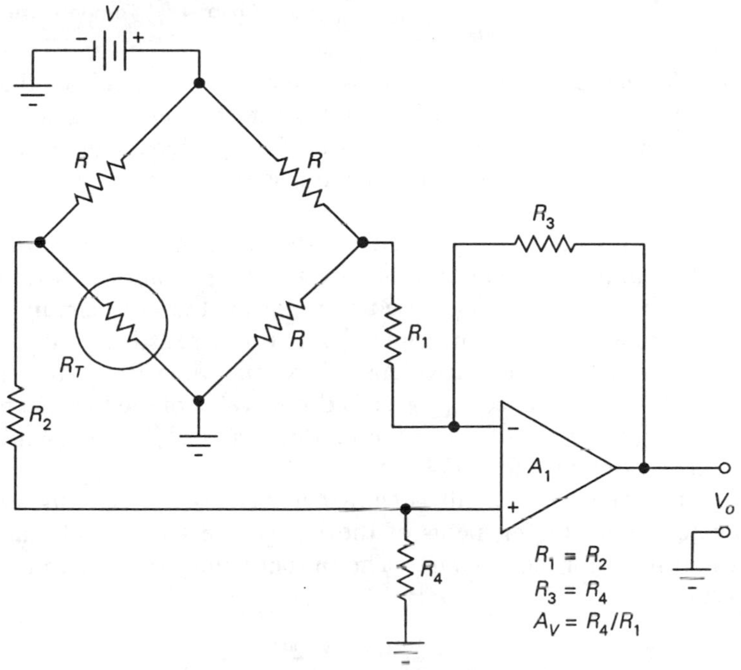

Figure 4–12 Differential amplifier used to amplify thermistor bridge circuit output voltage.

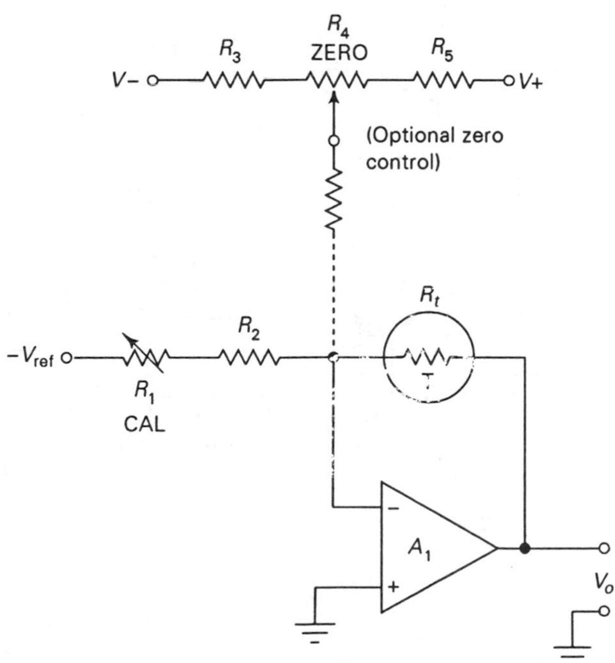

Figure 4–13 Thermistor used with inverting follower amplifier.

Like the half-bridge circuit discussed earlier, the circuit of Fig. 4–13 suffers from the problem of an offset voltage in the output signal any time that the thermal resistor value is nonzero. That problem can be overcome by using a zero offset control potentiometer as shown in Fig. 4–13 (labeled "optional").

A different form of bridge circuit is shown in the active thermometer of Fig. 4–14. In this case a unity-gain, noninverting op-amp is used to buffer the bridge. The gain can be increased by replacing the short circuit between the output terminal and the inverting input with a resistive voltage-divider feedback network. The output voltage V_o is zero when $R_1 = R_2$, $R_3 = R_4$, and the calibration resistance R_{set} is set to the R_0 value of the thermal resistor. This bridge can be used in both of the modes discussed for any Wheatstone bridge: null seeking or null initialized.

Yet another bridge circuit is shown in Fig. 4–15. In this amplified thermometer circuit the elements of the bridge are an integral part of the gain-setting circuit for the op-amp. The output voltage V_o is found from the relationship

$$V_o = \frac{V\Delta R[(R + 2R_f)/R]}{4R_x} \tag{4–29}$$

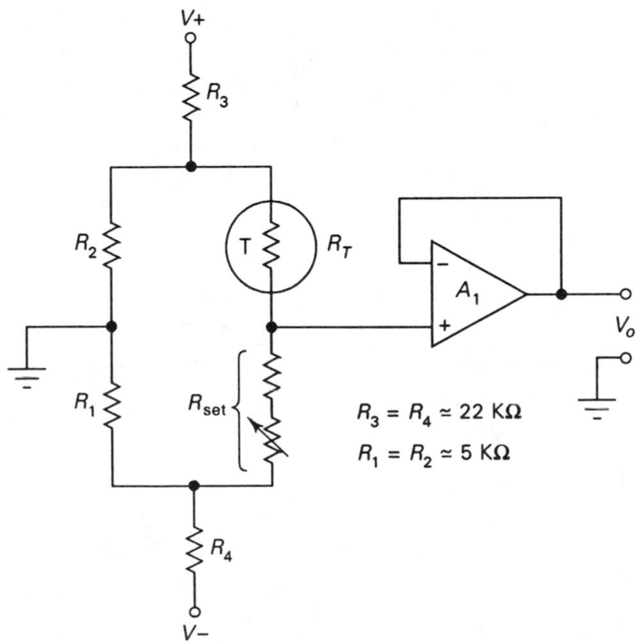

Figure 4–14 Thermistor used with noninverting follower amplifier.

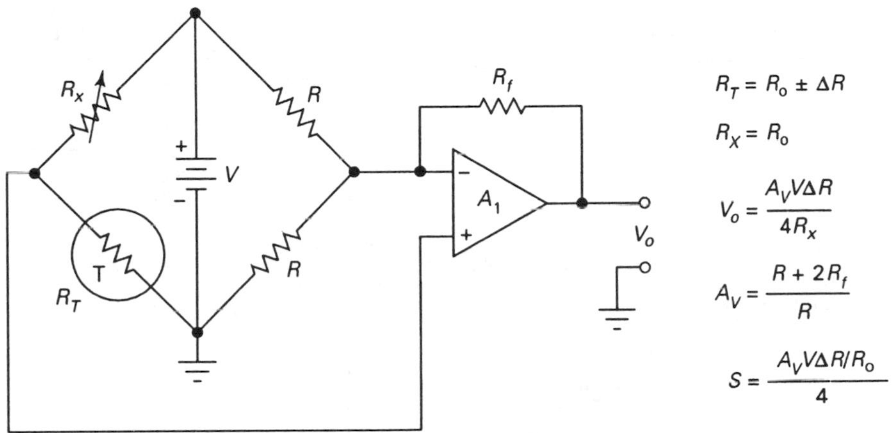

Figure 4–15 Differential amplifier and thermistor bridge.

This circuit has the advantage that lead resistances are automatically "servoed out" by the feedback action of the op-amp.

The thermistor is easy to use and is reasonably well behaved within the temperature range for which it is rated. But when a wider temperature range is needed, especially when the temperature measurement is in a very hot environment, then the sensor of choice may well be the thermocouple.

THERMOCOUPLES

Thermoelectricity is electricity generated by heat. Thermoelectricity was discovered by Thomas J. Seebeck in 1821 during experiments on electromagnetism in circuits containing bismuth and either copper or antimony.[1] Seebeck's discovery was that a junction made of two different metals will produce an EMF called the *Seebeck potential* V_s when it is heated [Fig. 4–16(a)]. Within a decade A. C. Becquerel discovered that thermoelectricity could be used to measure temperature. Today we call the Seebeck-Becquerel junction a *thermocouple*. This type of transducer consists of two dissimilar metals or other materials (some ceramics and semiconductors are used) that are fused together at one end. When several thermocouples are connected in series, the combination is called a *thermopile* [Fig. 4–16(b)]. Several thermocouples can be connected in parallel to find the arithmetic mean of several temperatures [Fig. 4–16(c)]. When a thermocouple system is connected to a resistive load, a current will flow that is proportional to the thermoelectric potential and inversely proportional to the resistance. By convention the current in the circuit is positive if it flows from hot to cold.

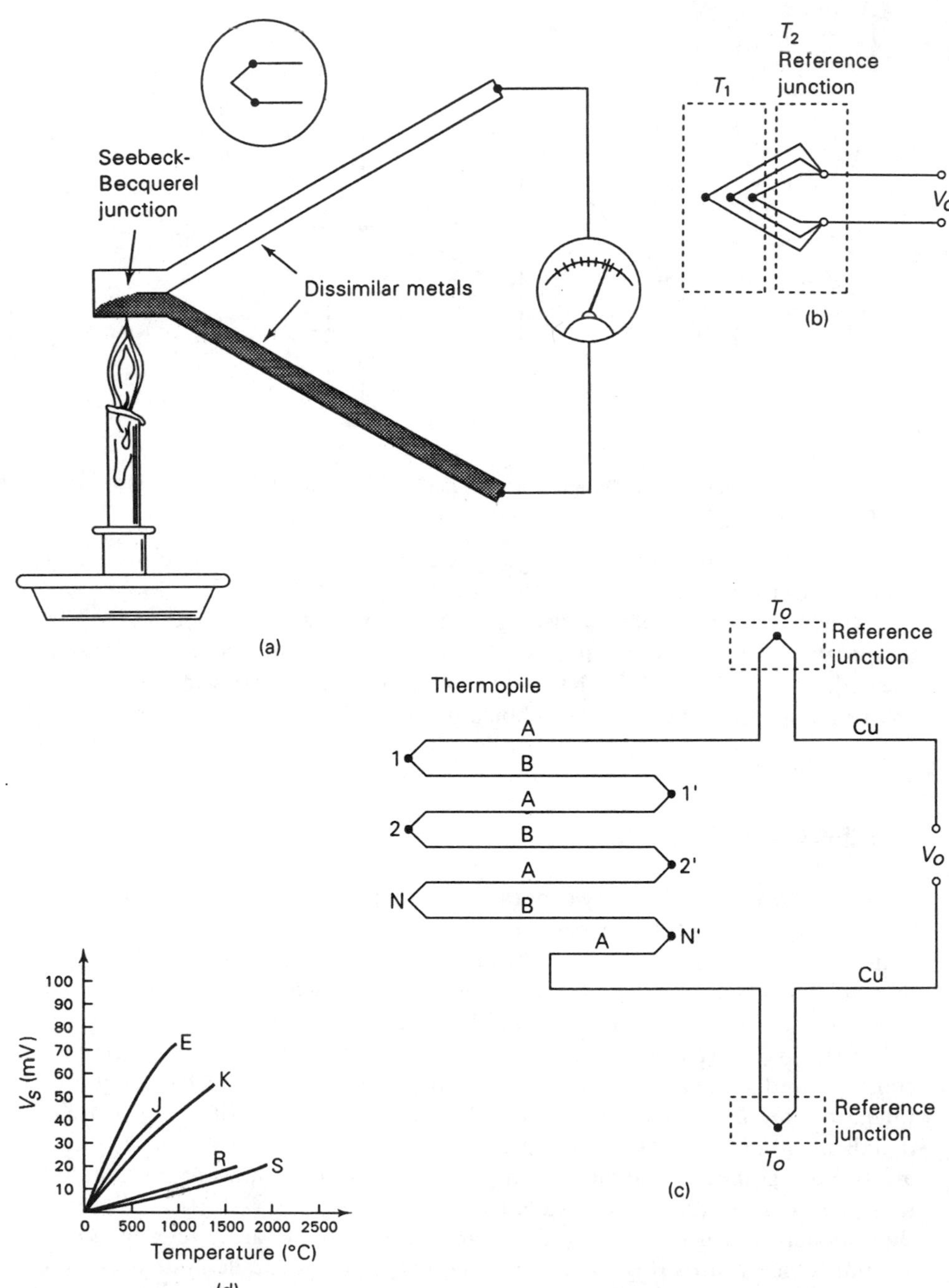

Figure 4–16 (a) Thermocouple and heat source; (b) average temperature thermocouple circuit; (c) thermopile using multiple thermocouples for greater output voltage; (d) characteristic curves of several different types of thermocouples.

Several standard thermocouples are defined, and their curves are shown in Fig. 4–16(d).

The Seebeck potential is related to the temperature and to the difference in the *work functions* (difficulty in stripping electrons from their associated metallic atoms) of the two metals. The potential is approximately linear with changes of temperature over small ranges, although over very large ranges of temperature (for any given pair of materials) nonlinearity increases markedly. The Seebeck potential varies from about 6 μV/°C to about 90 μV/°C and is found by integrating the expression

$$dV_s = \alpha_{a,b}\, dT \qquad (4\text{–}30)$$

where V_s is the Seebeck potential
$\alpha_{a,b}$ is the Seebeck coefficient of the system
T is the temperature

The Seebeck coefficient is determined empirically by measuring the coefficients for both materials (A and B) against a third material, called the reference material, at a standard reference temperature and then calculating their algebraic sum. We can further define the Seebeck coefficient in the form[2]

$$\alpha_{a,b} = \frac{dV_s}{dt} \qquad (4\text{–}31)$$

Another expression for the Seebeck potential is

$$V = \alpha(T_1 - T_2) + \gamma(T_1^2 - T_2^2) \qquad (4\text{–}32)$$

where α and γ are constants

The *sensitivity* or *thermoelectric power* of the thermocouple is expressed by

$$S = \frac{dV}{dT} = \alpha + 2\gamma T \qquad (4\text{–}33)$$

A subsequent study by Jean Charles Althanase Peltier (1834–1835) revealed an interesting phenomenon. When Peltier inserted a current into the Seebeck circuit, he found that one junction absorbed heat while the other gave up heat. This effect is used to operate certain modern solid-state refrigerators in recreational vehicles or campers and in certain areas where the noise and bulk of a compressor-operated refrigerator are unacceptable. Several companies make styrofoam electrical coolers that are exactly the right size for a six-pack of beer or soda. In practical Peltier devices the thermocouple consists of two regions of semiconductor differentially doped with dissimilar impurities.

Thermocouples are typically used in pairs [Fig. 4–17(a)] or even threes. One junction is used as the measurement thermocouple, while the other is

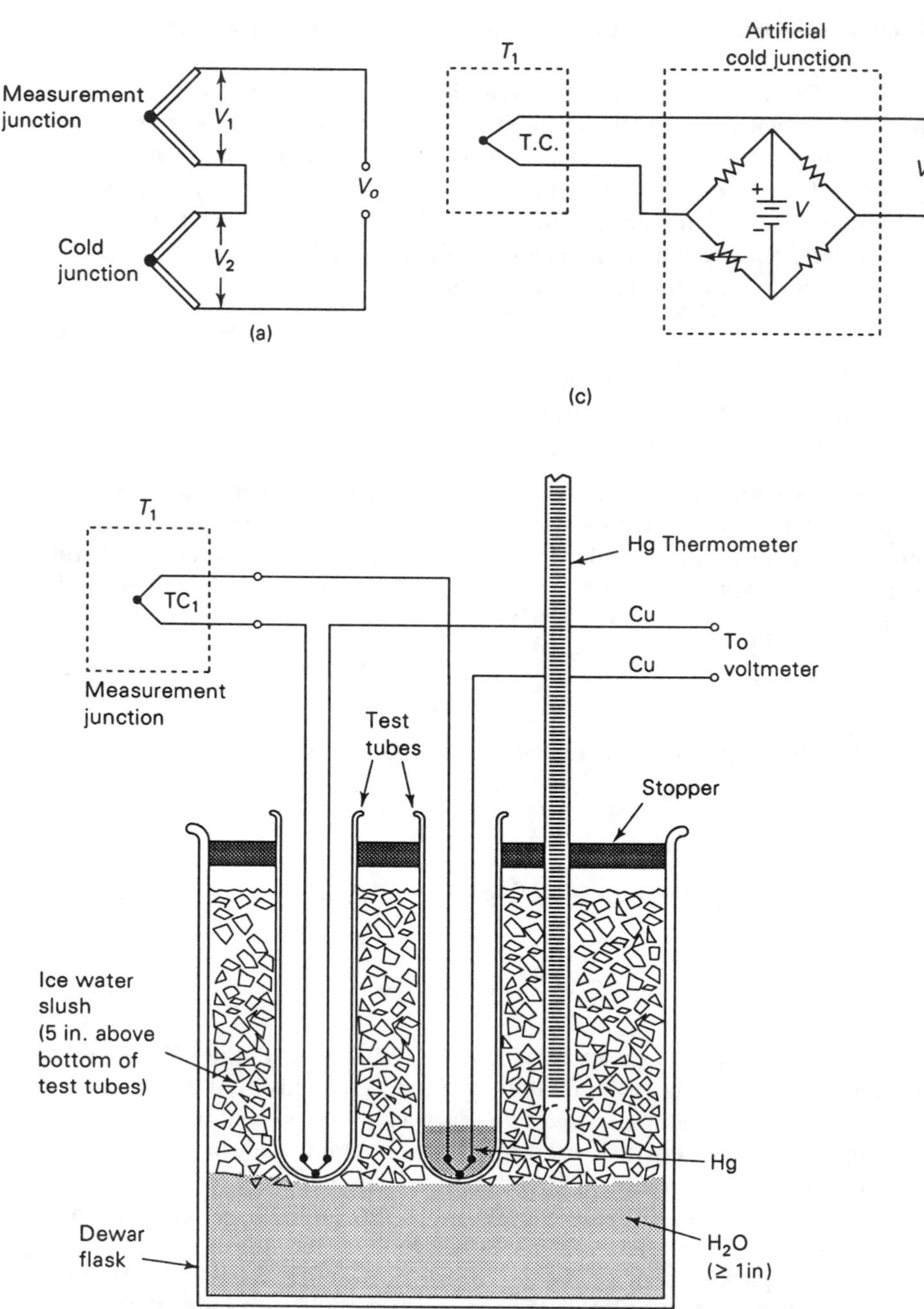

Figure 4–17 (a) Thermocouple using reference cold junction for compensation; (b) ice bath for cold-junction compensation; (c) resistor cold-junction compensation.

the *cold junction*. The cold junction may be the ice point (0°C) or the triple point (0.01°C) of water. The former gives an accuracy of 0.05 ±0.001°C, whereas the latter is capable of 0.01 ±0.0005°C. In some low-accuracy systems, room temperature may be used.

The difference in Seebeck potentials produced by the two junctions is used as a measure of the temperature difference between them. In laboratory settings where the ice point is used, the cold junction might be an ice-water bath [Fig. 4–17(b)]. A Dewar flask is filled with water and ice chips and then stoppered. A measurement thermistor and a reference junction are immersed in the bath to produce the cold junction potential. Although it can be successfully argued that the ice-water bath is the best cold junction for high-accuracy work, the use of an artificial cold junction [Fig. 4–17(c)] is popular. In the artificial cold junction, a potential equal to the expected ice-point potential is applied to the circuit in lieu of the second thermocouple.

Thermoelectric Laws

Over the nearly two centuries since Thomas Seebeck discovered thermoelectricity a series of "laws" have been developed regarding the phenomenon.

Law of homogeneous materials. This law states that a thermoelectric current will not flow in a circuit made of the same material. In other words, unless the materials of the conductors are *dissimilar*, that is, have different work functions, then no current is generated by heating the junction.

Law of intermediate materials. This law pertains to circuits containing a third material (Fig. 4–18). In this circuit a pair of thermocouples (TC_1 and TC_2) are formed of materials *A* and *B*, at two different temperatures (T_1 and T_2). A third material *C* is introduced into the circuit. According to this law, if the third material is completely within a homogeneous temperature environment T_3, then it will not produce a change in the net EMF. In other words, in terms of Fig. 4–18, there will be two *AC* junctions entirely within a constant-temperature region. That means their respective thermoelectric potentials null each other.

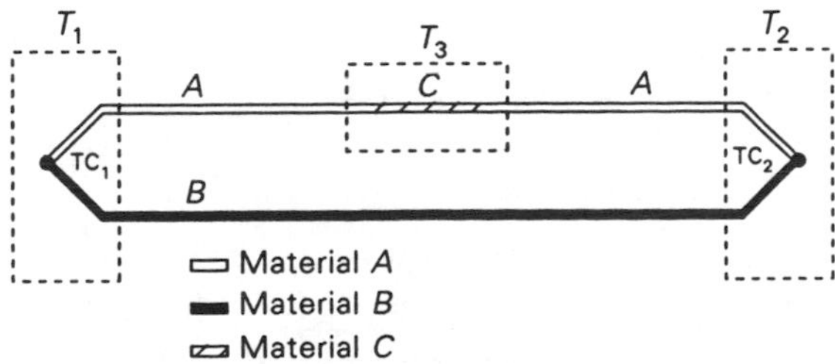

Figure 4–18 Thermocouples at different temperatures.

Figure 4–19 shows another consequence of this law. The properties of a thermocouple of materials A and C can be deduced from observation of the respective behaviors of these materials against a third, reference, material B. The reference material selected is usually platinum or palladium. If potentials V_1 and V_2 are the thermoelectric potentials of AB and BC, respectively, then $V_1 - V_2$ is the total system potential of the two thermocouples together.

Law of Intermediate Temperature Summation

Consider a system of thermocouples (TC_1 and TC_2) made of materials A and B. In Fig. 4–20(a) the two thermocouple junctions are at temperatures T_1 and T_2 and produce a thermoelectric potential V_1. In Fig. 4–20(b) the junctions are at temperatures T_2 and T_3 and produce a different potential, V_2. Now consider Fig. 4–20(c). The junctions are now at T_1 and T_3, and the new potential is the algebraic sum of the two other cases: $V_s = V_3 = V_1 + V_2$.

There are two main uses for this law. First, it allows us to connect extension wires to the system if the wires have the same thermoelectric values as the thermocouple wires. Second, it allows us to use the thermocouple system at temperatures other than its reference or calibration temperature, provided that a correction factor is added into the system.

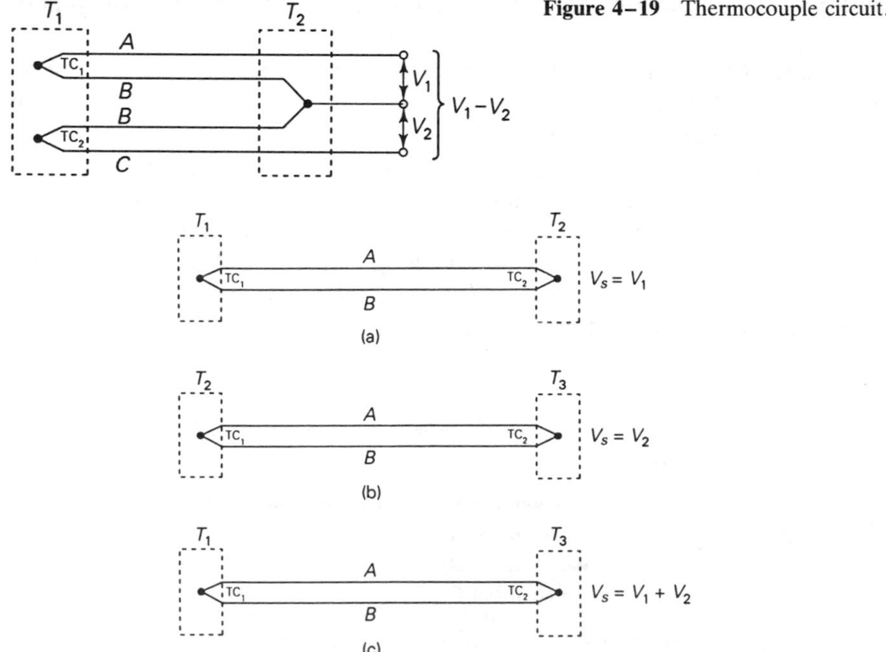

Figure 4–19 Thermocouple circuit.

Figure 4–20 Thermocouple pair under three different circumstances: (a) $V_s = V_1$; (b) $V_s = V_2$; (c) $V_s = V_1 + V_2$.

Thermocouple Output Voltage

The differential voltage between the two thermocouple junctions is proportional to the temperature difference and is used as the output voltage. This potential is found from the equation

$$V = a + bT + cT^2 + dT^3 + eT^4 + fT^5 \qquad (4\text{–}34)$$

where E is the output potential in volts
T is the temperature of the measurement junction
a, b, c, d, e, and f are constants that are a function of the materials used in the thermocouple

In many practical cases only the first three terms are used (making the equation a quadratic), whereas in others the first four terms are used.

Linearizing a Thermocouple

Equation (4–34) demonstrates a strong nonlinear dependence of thermocouple output voltage on temperature. As explained, in some cases an approximation of the output voltage is made using just the quadratic version of the equation (cubic and higher terms are deleted or approximated with an additional constant). This practice was especially reasonable when better linearization methods were not easily available. Analog circuits to solve the quadratic equation are, after all, somewhat less complicated than circuits for the cubic equation. But today computer linearization is possible.

As with other systems, there are several ways to linearize a thermocouple. For example, a *diode breakpoint generator* circuit can be used to piecewise linearize a circuit. This circuit uses a series of diodes, each biased to a different point, in the gain-setting network of an amplifier. When a diode is turned on by an input voltage that overcomes the bias, the gain of the amplifier is altered. The gain change linearizes the input circuit. But that system is both cumbersome and subject to thermal drift in the breakpoint generator diode circuit (this same phenomenon forms the basis for our next category of sensors—*semiconductors*).

It is also possible to use a computer or computerlike circuit for linearization. The two computer methods involve (1) a look-up table to correct the value of output voltage for any given temperature and (2) an algorithm that will solve Eq. (4–35) for T at any given output voltage. In both cases the computer can be programmed with information on the specific type of thermocouple being used so that it can select either the correct look-up table or the correct values of the coefficients of Eq. (4–35). An example of a table of coefficients is shown in Table 4–2.

$$T = a_0 + a_1V + a_2V^2 + a_3V^3 + \ldots + a_nV^n \qquad (4\text{–}35)$$

TABLE 4–2

Type	a_0	a_1	a_3	a_4	a_5
J	−0,04886825	19873.145	−218614.535	−264917530	2018441300

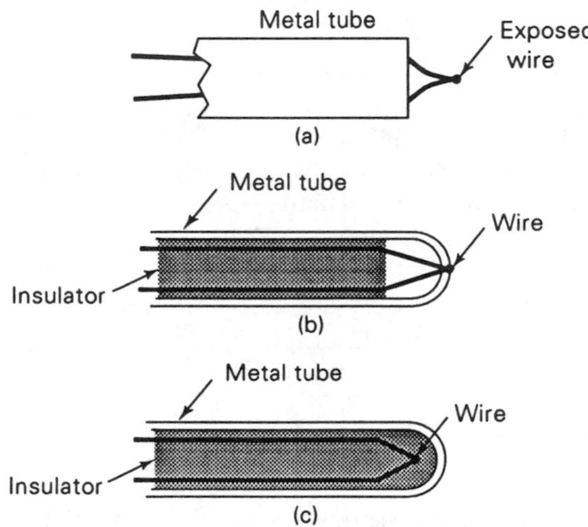

Figure 4–21 Thermocouple packaging; (a) exposed-tip; (b) covered-tip; (c) embedded.

Construction of Practical Thermocouple Sensors

The basic thermocouple is made by fusing together two or more wires with different properties to form a junction. In some cases this simple junction is used. In other cases, however, the thermocouple must be packaged. Figure 4–21 shows three basic forms of thermocouple package. In Fig. 4–21(a) is the *exposed-tip thermocouple.* The wires of the thermocouple are passed through (and insulated by a manganese oxide sheath from) a metallic tube (which provides mechanical protection and electrical shielding), but are exposed to the environment at the tip. In the *grounded thermocouple* of Fig. 4–21(b) the wires are insulated from the metal tube except at the end. In Fig. 4–21(c) we see the *insulated* or *floating thermocouple.* The construction is similar to that in Fig. 4–21(b), except that the thermocouple wires are not attached to the metal shield. In this example the insulating material fills the entire tube housing.

Thermocouple Interface Circuits

The traditional method for measuring the thermoelectric potential output from a thermocouple system is with a precision potentiometer. These null-seeking instruments are designed to match the unknown potential V_x

with a known reference potential V_{ref} by looking for the condition $V_{diff} = V_x - V_{ref} = 0$. Although this method is now more practical in electronic instruments than it once was because of computerization [and devices like digital-to-analog (A/D) converters], it is generally only practical in certain laboratory settings. Figure 4–22 shows three circuits for using thermocouples in practical electronic instruments.

The circuit of Fig. 4–22(a) is based on the inverting follower op-amp configuration. Because the thermocouple has a very low ohmic resistance, the input resistor of the amplifier does not have to be very high. The gain of this circuit is $-R_2/R_1$, and values of R_1 can be >100 ohms. Gains up to 500 or so are reasonable with ordinary op-amp devices and require only minimal precautions to prevent oscillation of the amplifier. Other circuit techniques must be used when the required gain is higher. In Fig. 4–22(b) the noninverting op-amp configuration is used. In this circuit the thermocouple is connected to the noninverting (+) input of the op-amp. The gain of this circuit is $[(R_2/R_1) + 1]$. In both circuits the gain is set by the ratio of the two feedback resistors. The gains are approximately equal at high gains, but at low gains the additional "1" in the noninverting case becomes more dominant.

Figure 4–22(c) shows a method of cold junction compensation recommended in the Analog Devices, Inc. literature. A reference potential of +2.5 Vdc is provided by the AD-580, while the actual cold junction compensation is provided by linking a second thermocouple intimately with an AD-590 temperature sensor.

Analog Devices, Inc. also makes a thermocouple integrated circuit called the AD-595 [Fig. 4–22(d)]. This device contains the necessary gain and cold junction compensation. The AD-595 also provides a binary-level alarm function for overtemperature conditions.

Whichever amplifier scheme is used, the needed voltage gain must be considered. Thermocouple output voltages are on the order of 6 to 90 μV/°C, so are typically very small. For example, at 300°C (a temperature not uncommon in simple gas furnaces), the output of a 40 μV/°C thermocouple would be 12,000 μV, or 12 mV. If an attempt is made to display this voltage on a standard 0–1999 mV meter, only a very small amount of the range is used. A similar problem exists if an A/D converter is used to convert the thermocouple output to a form that can be used in a computer. The standard 8-bit A/D converter divides its input range into 256 discrete levels. If the range is 0 to 10 V, the minimum discernible step, called the *least significant bit* (or 1-LSB) level, is 10 V/256, or 0.039 V (39 mV). In other words, the 1-LSB step value is larger than the expected value of the thermocouple output. As a result, measurement resolution is lost. The amplifier gain should be scaled to fill up the available range when the input parameter is at a maximum point. For example, if the 40 μV/°C thermocouple is used in a system that has a maximum possible temperature of 450°C, the expected maximum voltage

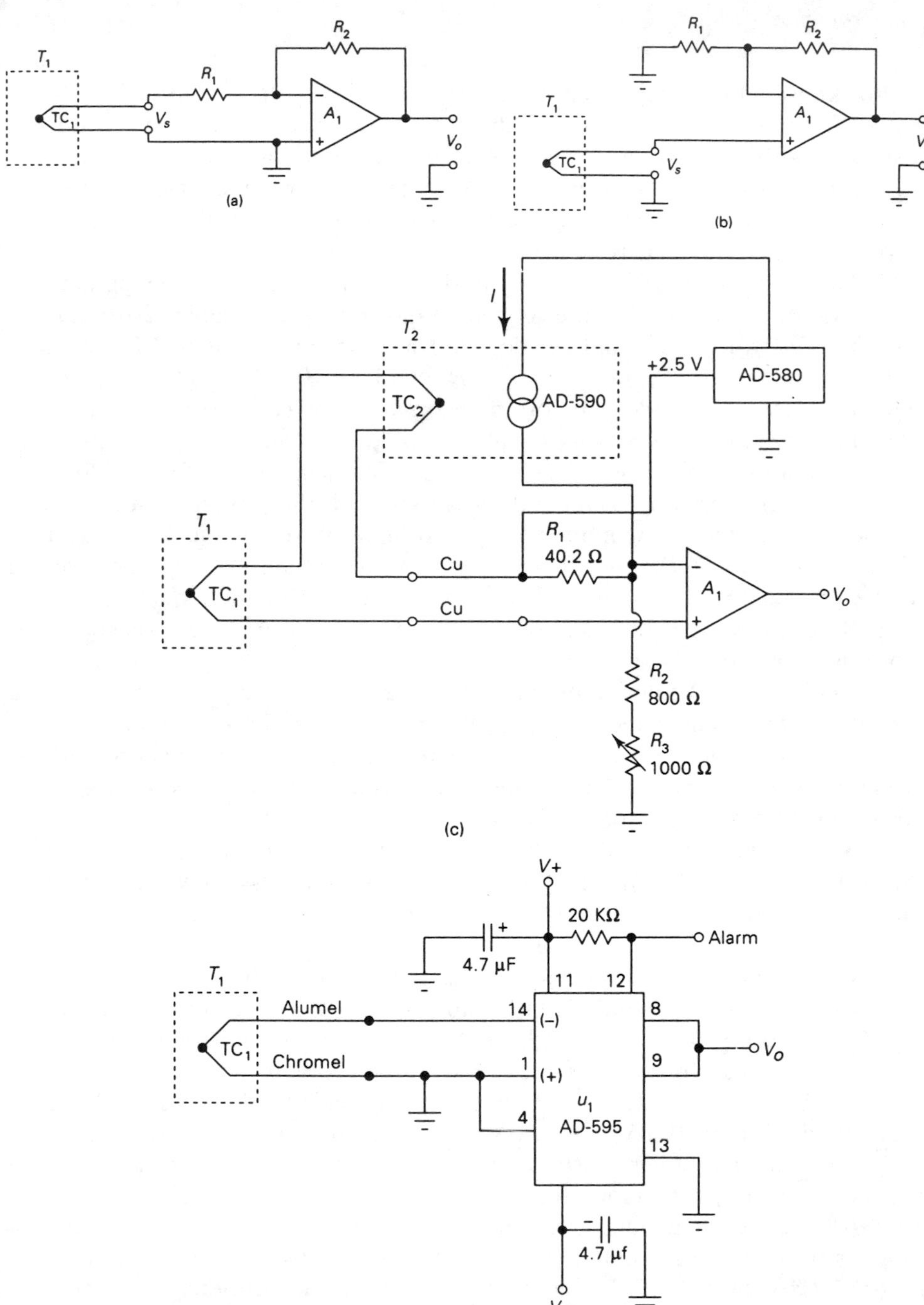

Figure 4–22 (a) Thermocouple used with inverting follower amplifier; (b) thermocouple used with noninverting follower amplifier; (c) using AD-590 as cold-junction compensation; (d) using Analog Devices, Inc., AD-595 for thermocouple processing.

would be 450 × 40 μV, or 18,000 μV (0.0018 V). If the maximum voltage allowed by the A/D converter is 10 V, then the voltage gain should be (10/0.018), or 555.6.

SOLID-STATE TEMPERATURE SENSORS

The last class of direct temperature sensor that we consider is the *solid-state PN junction*. If an ordinary solid-state rectifier diode (Fig. 4–23) is connected across an ohmmeter, the forward-biased diode resistance at room temperature can be measured. If the diode is temporarily heated with a lamp or soldering iron, the ohmmeter will show that diode resistance drops dramatically as temperature increases.

Most temperature transducers, however, use a diode-connected bipolar transistor such as shown in Fig. 4–24. We know that the base-emitter voltage V_{be} of a bipolar transistor is proportional to temperature. For a differential pair, as in Fig. 4–24, the transducer output voltage is given by

$$\Delta V_{be} = \frac{KT \ln(I_{C1}/I_{C2})}{q} \tag{4–36}$$

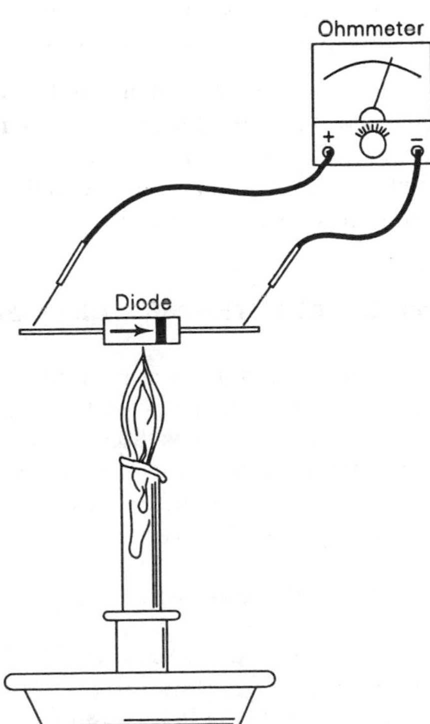

Figure 4–23 All *PN*-junction diodes show change of reverse resistance under temperature variations.

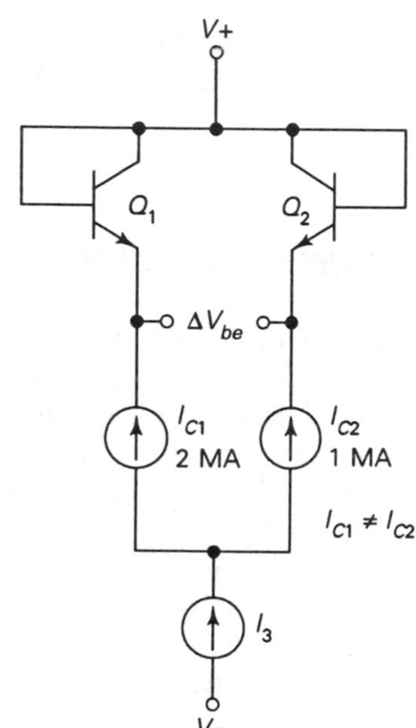

Figure 4–24 *PN*-junction transistors (diode connected) form a temperature sensor.

where K is Boltzman's constant (1.38×10^{-23} joules/K)
T is the temperature in kelvins
q is the electronic charge (1.6×10^{-19} C per electron)

The K/q ratio is constant under all circumstances. The ratio I_{C1}/I_{C2} can be held constant artificially by making I_3 a constant-current source. The only variable in the equation, therefore, is temperature. In the following sections we examine some commercial integrated circuit temperature devices based on this physical principle.

COMMERCIAL IC TEMPERATURE-MEASUREMENT SENSORS

Several semiconductor device manufacturers offer temperature-measurement/control integrated circuits (TMCIC). These devices are almost all based on the PN junction properties discussed earlier in this section, although at least one by Analog Devices, Inc. uses an external thermocouple. In this section we look at the semiconductor TMCIC devices offered by National Semiconductor, and Analog Devices, Inc.

The LM-335 Device

The National Semiconductor LM-335 device shown in Figure 4–25 is a three-terminal temperature sensor. The two main terminals are for power (and output), while the third terminal, shown coming out of the body of the

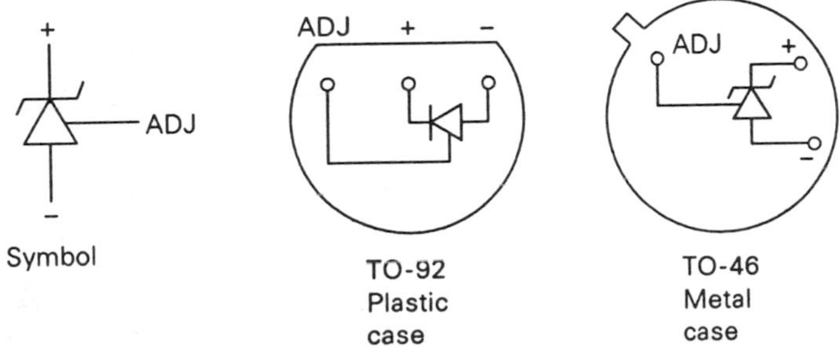

Figure 4–25 LM-335 solid-state temperature sensors (symbol and packages).

"diode" symbol is for adjustment and calibration. The LM-335 device is basically a special zener diode in which the breakdown voltage is directly proportional to the temperature, with a transfer function of close to 10 mV/K.

The LM-335 device and its wider-range cousins the LM-135 and LM-235 devices operate with a bias current set by the designer. This current is not critical but must be within the range 0.4 to 5 mA. For most applications, designers seem to prefer currents in the 1 mA range.

The accuracy of the device is relatively good and is more than sufficient for most control applications. The LM-135 version offers uncalibrated errors of 0.5 to 1°C, while the less costly LM-335 device offers errors of <3°C. Of course, good design can reduce these errors even further if they are out of tolerance for some particular application.

One difference among the three devices is the operating temperature range, as shown in the table.

Device Type No.	Temperature Range (°C)
LM-135	−55 to +150
LM-235	−40 to +125
LM-335	−10 to +100

Two packages are used for the LM-135 through LM-335 family of devices. The TO-92 is a small plastic transistor case, identified by the *Z* suffix to the part number (e.g., LM-335Z), while the TO-46 is the small metal-can transistor package (smaller than the familiar TO-5 case). This case is identified by the suffix *H* or *AH* (e.g., LM-335H or LM-335AH).

The simplest, although least accurate, method of using the LM-335 device is shown in Figure 4–26(a), where it is connected as a zener diode. The series current-limiting resistor limits the current to around 1 mA. This value of R_1 (1000 ohms) is appropriate for the +5 V power supplies that are typically found in digital electronic instruments. The resistor value can be

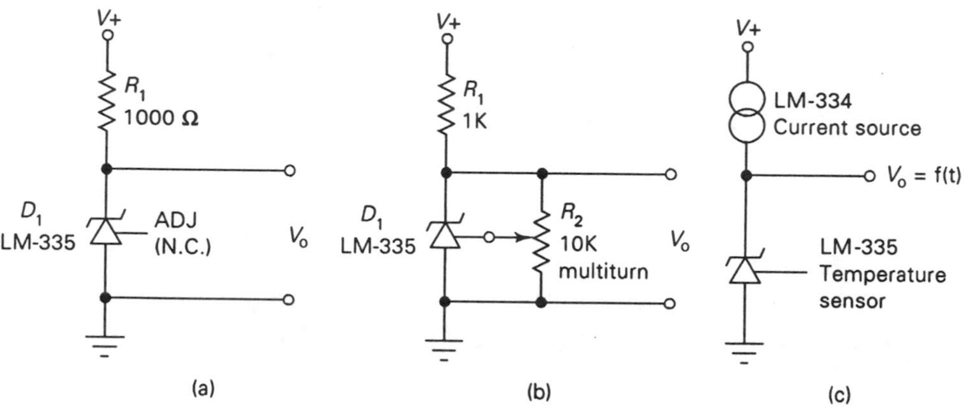

Figure 4–26 (a) Simplest LM-335 circuit; (b) calibratable LM-335 circuit; (c) current-source-driven circuit.

scaled upward for higher values of dc potential according to Ohm's law (keeping $I = 0.001$ Amperes):

$$R = \frac{(V+)}{.001} \tag{4–37}$$

The output signal in the circuit in Fig. 4–26(a) is taken across the LM-335 device. The approximate rate of voltage change is 10 mV/K. Simple math thus shows us how much voltage to expect at any given temperature. For example, suppose we want to know the output voltage at 78°C. First, we convert the temperature to kelvins, using Eq. (4–2):

$$K = °C + 273.16$$

$$K = 78 + 273.16 = 351.16$$

Next, we convert the temperature to the equivalent voltage:

$$V = \frac{10 \text{ mV}}{\text{K}} \times 351 \text{ K}$$

$$V = 3510 \text{ mV} = 3.51 \text{ V}$$

One problem with the circuit of Figure 4–26(a) is that it is not calibrated. Although that circuit works well for many applications, especially those where precision is not needed, for other cases it might be better to consider the circuit of Figure 4–26(b). This circuit allows single-point calibration of the temperature. The calibration is controlled by the 10 KΩ potentiometer in parallel with the sensor. The wiper of the potentiometer is applied to the adjustment input of the LM-335 device.

Calibration of the device is relatively simple. Only the values of the output voltage (a dc voltmeter will suffice) and the ambient temperature need to be known. In some less-than-critical cases a regular mercury-in-glass thermometer can be used to measure the air temperature after the equipment has been turned on and both the mercury thermometer and the LM-335 device have come to equilibrium. The potentiometer R_2 is then adjusted for the correct output voltage. For example, if the room temperature is 25°C (298 K), then the output voltage will be 2.98 V. The potentiometer is adjusted for 2.98 V under these conditions.

Another method uses an ice-water bath as the calibrating source. The temperature should be 0°C (273.16 K). A mercury thermometer will show the actual temperature of the bath. The potentiometer is adjusted until the output voltage is 2.73 V.

Yet another method is to use a warmed-oil bath for the calibration. The oil is heated to somewhat higher than room temperature (e.g., 40°C) and stirred slowly. Again, a mercury thermometer is used to read the actual temperature, and the potentiometer is adjusted to read the correct value. The advantage of this method is that the oil bath can be kept at a constant temperature. There are numerous laboratory vessels on the market that will keep water or oil at a constant preset temperature.

Another connection diagram for the LM-335 is shown in Fig. 4–26(c), in which a National Semiconductor LM-334 three-terminal adjustable current source is used for the bias of the LM-335 device. Again, the output voltage will be 10 mV/K.

All applications where the sensor is operated directly into its load suffer a potential problem or two, especially if the load impedance either changes or is lower than some limit. As a result, the buffered circuit of Fig. 4–27 is sometimes justified.

A buffer amplifier is used for one or both of two purposes: (1) impedance transformation or (2) isolation of the circuit from its load. Impedance transformation is important when the source impedance is high (not true of the LM-335). The isolation use is of somewhat more concern to us here for at least two reasons: (1) Changing impedances can change the output error factor, and (2) spurious oscillations on the output line can affect sensor performance. The op-amp in Fig. 4–27 places an amplifier between the sensor and its load. The gain of the amplifier in this case is unity, but a higher gain could be used if desired simply by substituting one of the gain-amplifier circuits described later.

The op-amp shown here is a GE/RCA CA-3140 device. It was chosen because of the freedom from bias currents exhibited by the BiMOS GE/RCA op-amps. The bias currents found on many other op-amps could conceivably introduce error. The CA-3140 is not the only op-amp that will work, however. Any low input bias current model will work nicely.

The noninverting input of the op-amp is connected across the LM-335. In this respect the circuit looks somewhat like the voltage reference circuits using zener diodes seen elsewhere. The bias for the LM-335 is derived from

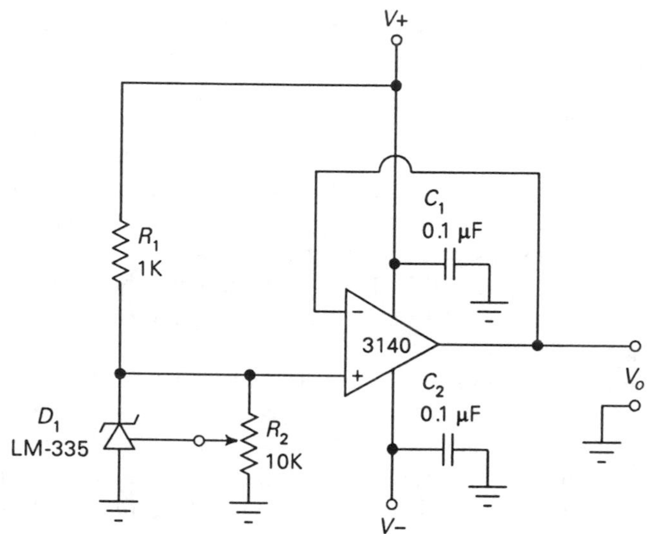

Figure 4–27 LM-335 electronic thermometer circuit.

a 12 KΩ resistor. Because there is no voltage gain in this circuit, the output voltage factor is the same as in previous designs, 10mV/K.

A circuit like the one shown in Fig. 4–27 sometimes proves useful in monitoring remote temperatures. If the op-amp is powered, a four-wire line is needed ($V-$, $V+$, ground, and temperature). The advantage is that the line losses are overcome by the higher output power of the op-amp. The LM-335 is a rugged low-impedance device, however, and in many cases such measures would not be needed.

AD-590 Devices

The Analog Devices, Inc. AD-590 [Fig. 4–28(a)] is another form of solid-state temperature sensor. This particular device is a two-electrode sensor that operates as a current source with a characteristic of 1 μA/K. The AD-590 will operate over the temperature range −55°C to +150°C. It can operate over a wide range of power supply voltages, working best in the range +4 to +30 V dc (this range is more than sufficient for most solid-state circuit applications). Selected versions are available with a linearity of ±0.3°C and a calibration accuracy of ±0.5°C.

The AD-590 comes in two different packages. The metal can (TO-52) is recognized as the small-size transistor package (smaller than TO-5). A plastic flat-pack is also available.

Being essentially a two-terminal current source, the AD-590 operates quite simply in actual circuits. Figure 4–28(b) shows the most elementary calibratable circuit for the AD-590. Because it is a current source that produces a current

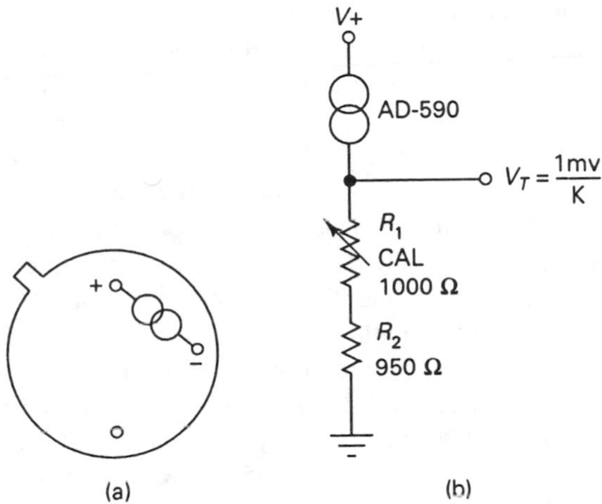

Figure 4–28 (a) AD-590 temperature sensor; (b) AD-590 circuit.

proportional to temperature, the output can be converted to a voltage by passing it through a resistor. In Fig. 4–28(b) the total resistance is approximately 1000 ohms and consists of the resistance of R_2 (950 ohms) and R_1 (a 1000 ohm potentiometer). From Ohm's law we know that 1 μA/K converts to 1 mV/K when passed through a 1000 ohm resistance. We can calculate the voltage output at any given temperature from the simple relationship

$$V_o = \frac{1 \text{ mV}}{\text{K}} T \tag{4–38}$$

Thus, if the temperature is 37°C, which is equivalent to (37 + 273) or 310 K, the output voltage will be

$$V_o = \frac{1 \text{ mV}}{\text{K}} \times 310 \text{ K} = 310 \text{ mV}$$

Potentiometer R_1 is used to calibrate this system. A rough calibration can be made with an accurate mercury thermometer (laboratory grade recommended) at room temperature. With a digital voltmeter across the output, the system is allowed to come to equilibrium (should take about 10 minutes), and the potentiometer is adjusted for the correct output voltage. For example, at a room temperature of 25°C (298 K), the output voltage will be (1 mV × 298), or 298 mV (0.298 V). A $3\frac{1}{2}$-digit voltmeter is sufficient to make this measurement.

In some cases it might be wise to delete the potentiometer and use a single 1000 ohm resistor in place of the network shown. There might be several reasons for doing this. First, the calibration accuracy might not be critical for the application at hand. Second, potentiometers are weak points in any circuit. Being mechanical devices, they are subject to stress under vibration conditions and may fail prematurely. If the temperature accuracy is not crucial, and reliability is, then a single fixed 1 percent tolerance resistor should be considered in place of the network shown in Fig. 4–28(b).

The circuit of Fig. 4–28(b) is sometimes used to make a temperature alarm. By using a voltage comparator to follow the network and biasing the comparator to the voltage that corresponds to the alarm temperature, we can create a TTL level that indicates when the temperature is over the limit. A *window comparator* will allow us to have an alarm of either under- or overtemperature conditions. Some electronic equipment designers use this method to provide an overtemperature alarm. In one application a commercial minicomputer generated a large amount of heat (it used a 65 A, +5 V dc power supply!). The specification called for an air-conditioned room for housing the computer. An AD-590 device was placed inside at a critical point. If the temperature reached a certain level (45°C), the comparator output would go LOW and create an interrupt request to the computer. The computer would then sound an alarm and display an "overtemperature warning" message on the operator's CRT screen.

The circuit of Figure 4–28(b) suffers from a problem: It allows calibration at only one temperature, which does not permit optimization of the circuit. The situation can be improved, however, by using the two-point calibration circuit of Fig. 4–29(a). The calibration curve is shown in Fig. 4–29(b). In this case an op-amp is used in the inverting follower configuration.

The *summing junction* of the amplifier (inverting input) receives two different currents. One current is the output of the AD-590 (i.e., 1 μA/K), while the other current derives from the reference voltage V_{ref} (10.000 V). Adjustment of the latter current provides the zero reference adjustment, while the overall gain of the amplifier provides the full-scale adjustment.

The op-amp selected is the LM-301 device, although almost any premium op-amp will suffice. The RCA CA-3140 BiMOS device or some of those from either Analog Devices, Inc. or National Semiconductor will also work nicely. If the LM-301 or similar device is used, then the 30 pF frequency-compensation capacitor is necessary.

Calibration of the device is simple, although two different temperature environments are required. The 0°C adjustment (R_1) can be made with the sensor in an ice-water bath (as described earlier). The upper temperature can be room temperature, provided that some means is available to measure the actual room temperature for comparison.

TEMPERATURE ALARMS

A *temperature alarm* is a circuit or device that has either a binary or three-state output that will indicate that a temperature is within limits or not. Some temperature alarms may indicate when the temperature in the measured area

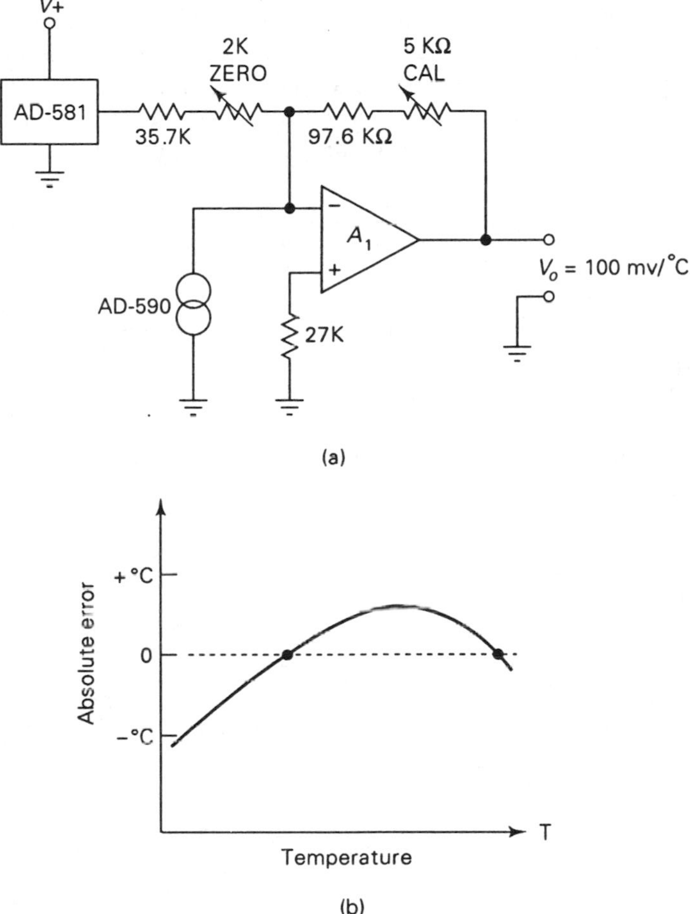

Figure 4–29 (a) AD-590 thermometer circuit; (b) calibration curve.

is above a preset level. An undertemperature alarm performs exactly the opposite function. A window alarm is one that will sound either a single alarm or unique alarms for both over- and undertemperature conditions. In other words, as long as the temperature is between two preset levels, no alarm is sent.

Bimetallic-Strip Alarms

The *bimetallic strip* is an on-off temperature sensor that permits the construction of a temperature-sensitive switch. An example of this form of temperature sensor is the water temperature sensor that plugs into the engine block of an automobile. When the water temperature reaches a certain level, the bimetallic strip closes a switch that lights up the TEMP or HOT alarm lamp on the dashboard.

Figure 4–30(a) shows the construction of a bimetallic-strip thermoswitch. The two metals are selected to have radically different thermal coefficients of expansion and are bonded together. When they are heated, the two pieces of metal try to expand at different rates, so the strip is forced into a radius of curvature R. The value of this deflection radius is

$$R = \frac{(t_1 + t_2)^3}{6\delta(T_2 - T_1)t_1 t_2} \tag{4–39}$$

where R is the radius of curvature
- t_1 and t_2 are the thicknesses of the two metal elements in the thermal strip
- T_1 is the resting temperature before curvature begins (°C)
- T_2 is the final temperature (°C)
- δ is the difference in the thermal coefficients of expansion for the two metals

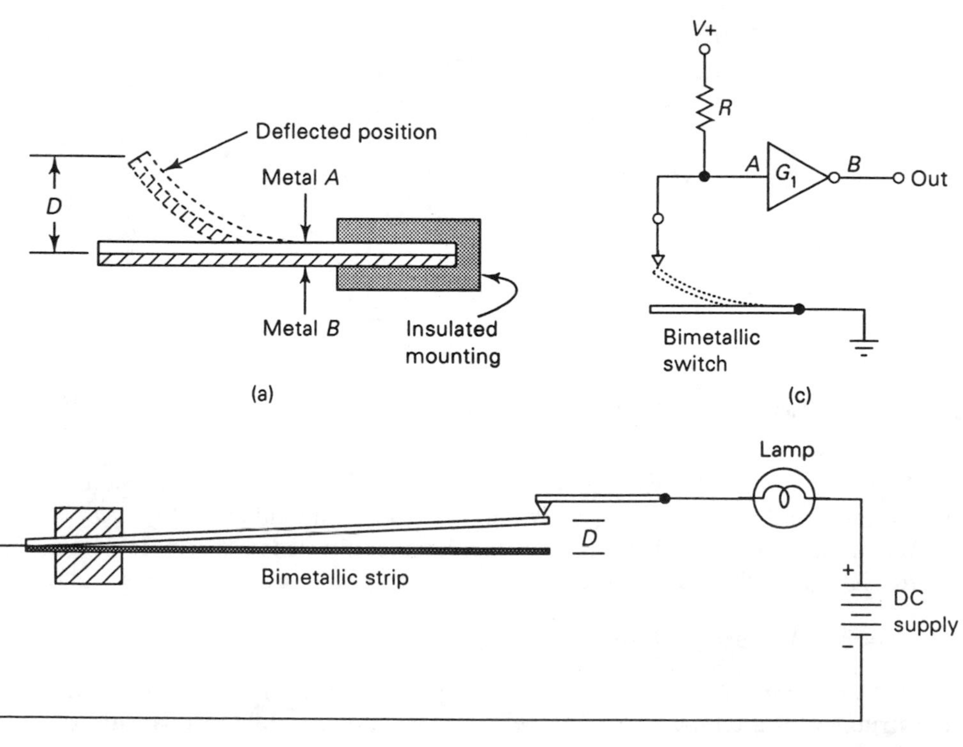

Figure 4–30 (a) Bimetallic-strip temperature sensor; (b) typical alarm circuit; (c) digital alarm circuit.

The deflection D of the end is found from

$$D = \frac{KTL^2}{t} \tag{4-40}$$

where D is the deflection in inches
L is the length of the strip in inches
T is the temperature difference $T_2 - T_1$
t is the thickness of the strip in inches
K is a constant, typically 3×10^{-6} to 7×10^{-5}

The simplest bimetallic-strip alarm circuit is shown in Fig. 4–30(b). A lamp is lit by closure of the switch. This is the circuit of the overheating warning light on automobiles. Fig. 4–30(c) shows a typical electronic alarm circuit based on the bimetallic strip. A digital inverter, G_1, is used as the sensor electronics. The rules of this device are simple: When the input (point A) is HIGH (near $V+$), the output (point B) is LOW (near ground), and when the input is LOW, the output is HIGH. Under normal conditions, below the alarm threshold, the bimetallic switch is open, so the input of the gate is held HIGH by resistor R connected to the $V+$ source. Under this condition the output of the gate is LOW. But when the temperature passes a critical threshold, the bimetallic switch closes, the gate input is shorted to ground— so it is forced to the LOW level—and the output snaps HIGH to indicate an overtemperature condition.

Heat is the greatest killer of electronic equipment. Whenever electronic gear meets premature death or experiences a large number of failures per unit of time, then it is highly probable that one of the root causes is overtemperature. Nothing is 100 percent efficient. Thus, it is necessary to know when the temperature exceeds a certain safe limit. For example, the equipment in a two-way radio repeater site may have its own internal blowers to keep equipment cool, but these may fail (or the ambient temperature may exceed normal expectations and make them ineffective). The control operator needs to know when the repeater is at risk of failure from overtemperature. A similar situation exists for computer bulletin board operators: a defective fan on the computer power supply will cause overheating and subsequent failure of the expensive unattended computer.

Warning-Alarm Circuits

Temperature sensors can be used in any of a number of ways. If they are used only for measurement, then the dc voltage output of the circuit will suffice for the required data. The thermistor can be used in a Wheatstone bridge (with dc output amplifier), whereas the thermocouple and solid-state sensors are used with dc amplifiers only. A good hint to remember is to make the scaling function *numerically the same* as the output voltage, so that a simple voltmeter can be used without special calibration. For example, if a sensor with a 10 mV/K output is to read degrees Celsius, the dc output is offset by 2.732 V dc (i.e., V_o of the buffer amplifier is zero when the sensor input is 2.732 V), and the voltage will then read 10 mV/°C. Thus, a temperature of 25°C will produce an output of (25°C × 10 mV/°C), or 250 mV.

Alarm circuits are a little different. In this case, the requirement is to produce an output that can be used to monitor a temperature. Figure 4–31 shows a simple circuit for use with a thermistor. The CMOS inverter will change state when the input voltage crosses a threshold that is half the algebraic difference between $V+$ and $V-$, which in this case is +2.5 V ($V+ = 5$ V dc, $V- = 0$). The thermistor R_2 and a fixed resistor R_1 form a resistor voltage divider with an output voltage V_1 of

$$V_1 = \frac{(5.00 \text{ V dc})(R_2)}{R_1 + R_2} \tag{4-41}$$

The values of R_1 and R_2 are designed to set the trip point (+2.5 V) when the temperature applied to R_2 exceeds the design temperature maximum. Resistor R_1 can be made variable to set the trip point and compensate for thermistor tolerances.

The output voltage of the CMOS inverter can be applied to a computer input and so serve as a flag bit for an overtemperature condition, if that is the application. Alternatively, the output can be used to drive a relay amplifier.

Figure 4–31 shows a voltage comparator used to produce an overtemperature output bit V_o when the trip-point temperature is exceeded. The circuit will work with any voltage-output temperature sensor. The output of the sensor V_1 is applied to the noninverting input of the comparator, while the voltage output from a potentiometer V_2 is applied to the inverting input. As long as $V_2 > V_1$, the comparator produces a LOW output; when $V_2 = V_1$, the output drops to zero; when $V_2 < V_1$, the output snaps HIGH.

A *temperature-to-frequency converter* (*TFC*) circuit is shown in Fig.

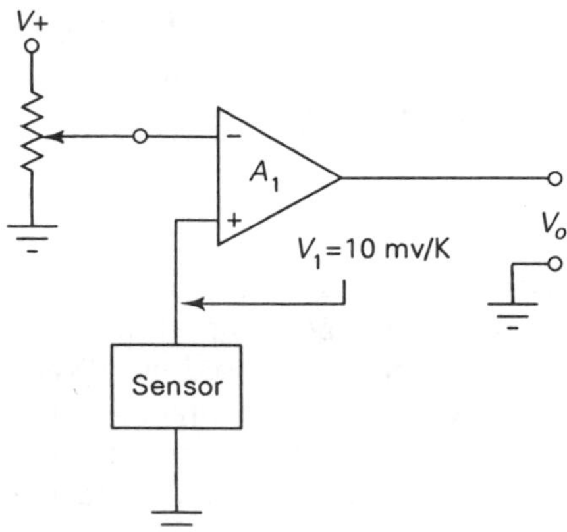

Figure 4–31 Alarm circuit using temperature sensor and a comparator A_1.

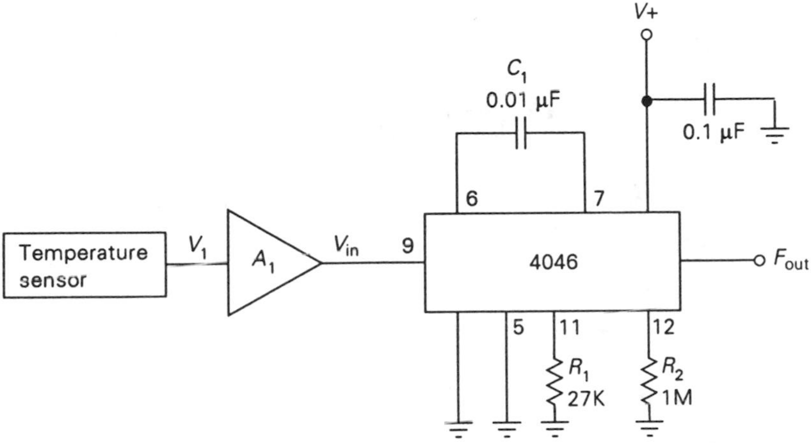

Figure 4–32 Temperature-to-frequency converter.

4–32. This circuit can be used to transmit a tone that indicates temperature from a remote location over either radio link or telephone link to a control operator who monitors its state. An LM-2907 frequency-to-voltage converter can be used at the receiving end to produce a dc voltage that is proportional to the temperature.

The TFC circuit is based on the CMOS 4046 chip, which contains both a voltage-controlled oscillator and a phase detector (not used here), which can be used to make the 4046 into a near-dc to 1 MHz phase-locked loop (PLL). With the values shown, the 4046 will oscillate at approximately 1000 Hz when the input voltage V_{in} applied to pin 9 is $(V+)/2$. Changing R_1 and C_1 will change the operating frequency.[3] The input voltage is supplied by a temperature sensor and any scaling dc amplifier (A_1) that may be needed to match the output range of the sensor to the input range of the 4046.

REFERENCES AND NOTES

1. P. A. Kinzie. *Thermocouple Temperature Measurement.* New York: John Wiley, Wiley-Interscience Series, 1973.
2. *Manual on the Use of Thermocouples in Temperature Measurement*, ASTP Publication STP-470A. American Society for Testing and Materials.
3. See *CMOS Cookbook, 2nd Ed.* by Don Lancaster and Howard M. Berlin (Sams 22459, $18.95) for details on how to use this chip. The 4046 can be used to make a dandy frequency meter for measuring subaudio frequencies on a regular digital frequency counter.

5 Position and Displacement Sensors

Position and displacement are two related parameters that are the basis of a large number of different measurements. The physical concepts underlying position and displacement are very simple and are usually covered in any first course on physics (even in high school). For the sake of establishing the frame of reference and terminology, however, let's review a few of these concepts.

Position is merely the specification of a place in the universe, whether that universe is of cosmic scale or limited to a few centimeters. Figure 5–1(a) shows the concept graphically. A ball is placed alongside a ruler or measurer of some sort so that we can establish a reference point from which to distinguish one position from another. In the case shown, the reference point is the zero end of the ruler. This universe is unidimensional, so the ball can be only at some point along the ruler, from 0 to 7 distance units. In Fig. 5–1(a), the ball is at a position 3 units from zero.

In Fig. 5–1(a) the position is referenced to an "anchor point" (called *zero*) on the ruler. The position of the ball is understood only along the x-axis from zero to the maximum value of 7 units. Where it is in relation to other objects in another frame of reference (off the ruler) is not important. If we had chosen to do so, we could have located the zero reference point any place along the ruler. Distances to the right of the zero are (by convention) called "$+x$," while distances to the left are designated "$-x$."

But we can relate the position of the ball to other objects in the same frame of reference. For example, a little hexagonal device is also shown in Fig. 5–1(a). The ball is at position 3, while the hexagon is at position 5. We can say that the hexagon is at 5, or at $(5 - 3)$, or $+2$ units from the ball

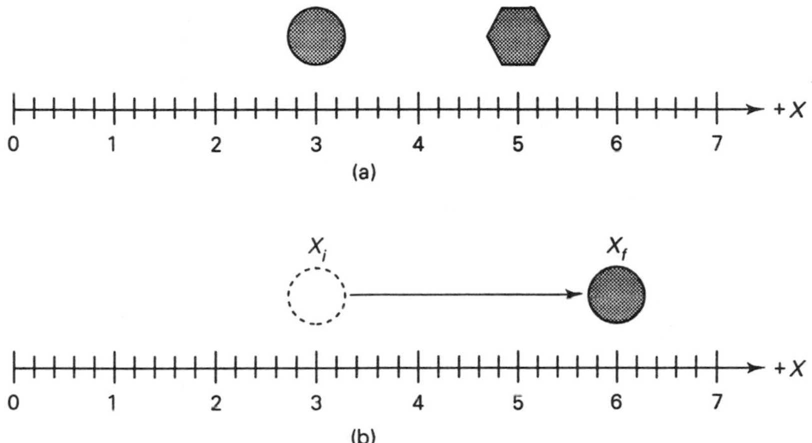

Figure 5–1 (a) Positions marked at 3 and 5 on x-axis; (b) displacement from 3 to 6.

(or, alternatively, saying the ball is -2 units from the hexagon is just as valid—there is no preferred frame of reference).

The concept of *displacement* is shown in Fig. 5–1(b). In this figure the ball is *moved* from position X_i at $+3$ on the x-axis to a new position X_f at $+6$ on the x-axis. Thus, the ball moved (i.e., was *displaced* $[(+6) - (+3)]$, or $+3$ units. Thus, displacement is said to be the change in position x, or Δx. *Velocity* is defined as the *displacement per unit of time*, or $\Delta x/\Delta t$.

When we remove ourselves from the unidimensional universe and travel to flatland, that is, a two-dimensional world, then it is necessary to use two dimensions to specify position. In Fig. 5–2(a) we see the *Cartesian coordinate* system (after René Descartes, 1596–1650), also called the *rectangular* or *x-y* coordinate system. In this system there are two axes. The horizontal axis is $-x \leq 0 \leq +x$, while the vertical axis extends from $-y \leq 0 \leq +y$. A point has to be designated in relation to both the X and Y axes to be located in this universe. According to convention, the points are usually specified as an ordered pair (x, y), such as the point (3, 5) in Fig. 5–2(a).

Displacement likewise requires ordered pairs of points, one pair each for the origin and the destination as the ball is displaced from (7, 5) to (–3, 5). Note that two paths are shown in Fig. 5–2(b). The direct path A is the shortest distance between the two points. But the displacement is also identical for path B, for the ball will end up at the same point. In position and displacement instrumentation systems it becomes important to know the path of displacement. For although the physicist will recognize that path A and path B create identical displacement, the two paths are not the same when measured by real-world sensors. The final result is, of course, the same, but the sensor is sensitive to path as well as displacement.

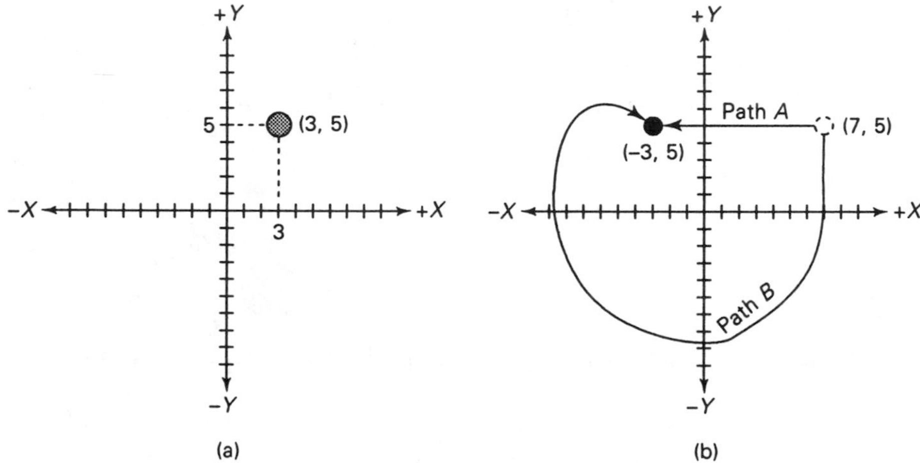

Figure 5–2 (a) Position described in two-dimensional space requires two coordinate points: one on the x-axis, one on the y-axis. (b) Displacement on two-dimensional plane is independent of the path. The end point remains the same for path A and path B.

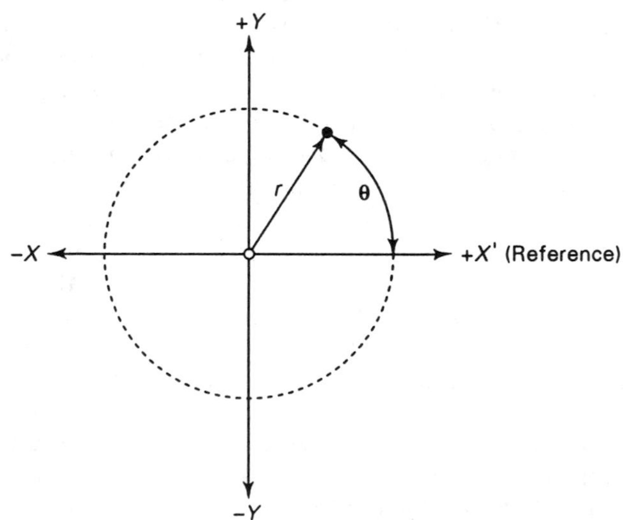

Figure 5–3 Polar coordinate notation.

Another system for specifying displacement and position is the *polar coordinate* system shown in Fig. 5–3. In this case the point of interest P is specified by a vector measured relative to a zero origin (which serves as the magnitude reference point) and a reference direction (which serves as the 0° angle reference). The vector is specified by a length OP and an angle θ with respect to the reference direction. The position is sometimes written $r\theta$ (e.g., $8 \angle 30$, which means a magnitude r of 8 units and an angle of 30°).

In Fig. 5–3 the polar coordinate system is superimposed on a Cartesian system (which is by no means necessary) so that the 0° reference direction is along the $+x$-axis. Converting between the two systems necessitates a little trigonometry. To convert from polar to rectangular coordinates, we use the equations

$$x = r \cos \theta \tag{5–1}$$

$$y = r \sin \theta \tag{5–2}$$

or, when converting from rectangular to polar coordinates, we use the equations

$$r = \sqrt{x^2 + y^2} \tag{5–3}$$

$$\theta = \tan^{-1}\left(\frac{y}{x}\right) \tag{5–4}$$

for $-180° \leq \theta \leq +180°$.

In general, it is customary to use Cartesian coordinates when dealing with linear, that is, straight-line, displacement and polar coordinates when dealing with angular displacement.

POTENTIOMETER POSITION/DISPLACEMENT SENSORS

A *potentiometer* is a variable resistor that has three terminals: two are fixed, and one is variable. Figure 5–4 shows a *linear potentiometer*, that is, one in which the resistance is a function of the position of the wiper along a resistance element. The element may be a carbon composition layer, a metal oxide film (more modern), or a wire-wound assembly. The *total resistance* R_t is measured between the fixed end terminals, A and B. The wiper C is placed at a point along the element, creating two segments (AC and CB), each of which has a fraction of the total resistance (R_{AC} and R_{CB}). The ratio of these two resistances is determined by the position of the wiper along element AB.

A *rotary potentiometer* (not shown) is made by bending the element into a circle. The most common form of rotary potentiometer uses a 270° element, although 360° and angles other than 270° are sometimes seen.

Figure 5–5(a) shows the simplest form of potentiometer position/displacement sensor. The potentiometer is connected so that point A is grounded, and point B is excited by a reference voltage V_{ref}. The output voltage (V_o) will vary from zero to $+V_{ref}$ [Fig. 5–5(b)] as the position of the wiper varies from zero to L.

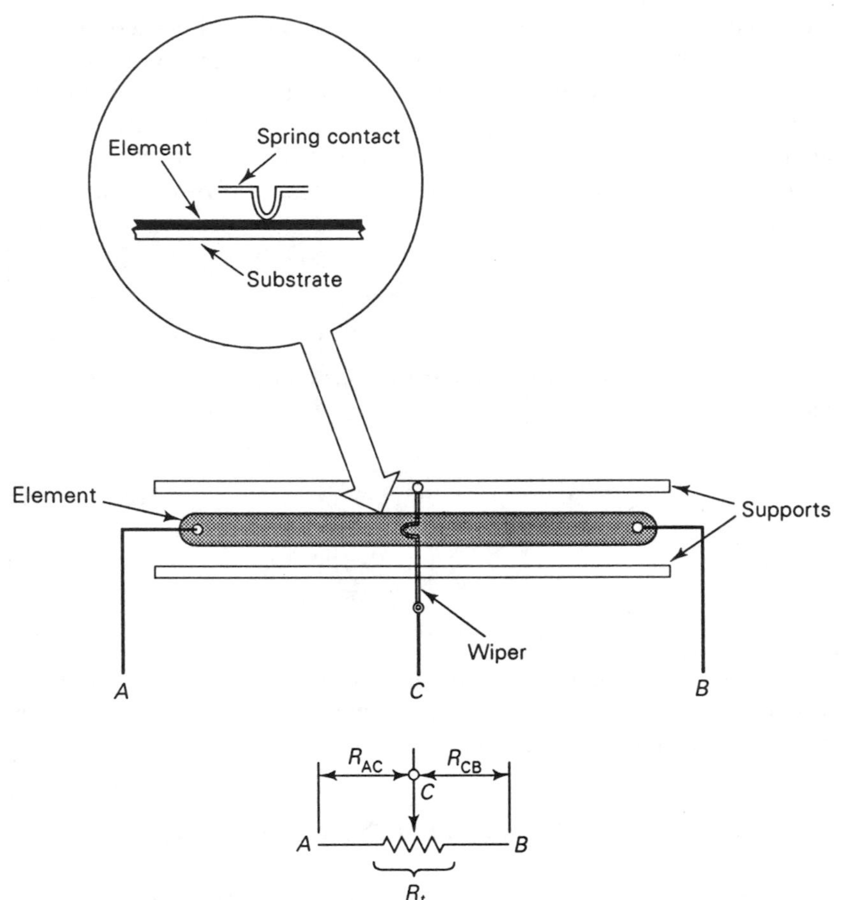

Figure 5–4 Potentiometer position/displacement sensor.

It is important that this excitation voltage be stable and accurate so that the voltage at point C will reflect the position of the wiper. The wiper is connected to an actuator or some other mechanical device that will cause its position to change with changes in position.

If the motion is in only one quadrant, then the system of Fig. 5–5(a) works nicely, but when the motion forces the displacement to either side of the zero reference point [Fig. 5–6(a)], then it is prudent to connect the potentiometer to both $-V_{ref}$ and $+V_{ref}$ excitation potentials. The *polarity* of the output potential indicates the *direction* of the displacement. The graph of output voltage versus position is shown in Fig. 5–6(b).

Figure 5–7(a) shows the basic method of connecting the potentiometer R_2 in practical circuits. There may or may not be a series resistor R_1, which is sometimes used for calibrating the potentiometer. If the load resistance

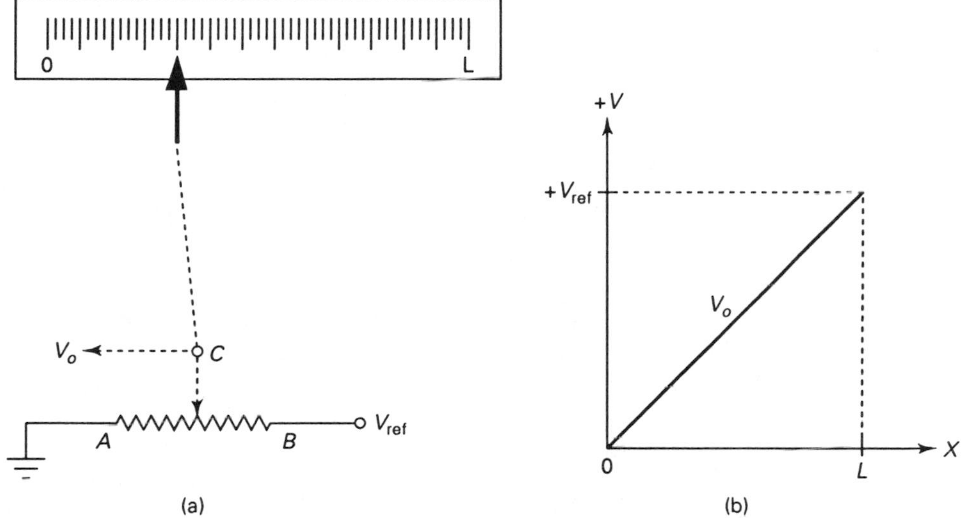

Figure 5–5 (a) Single-quadrant position/displacement sensor; (b) characteristic curve.

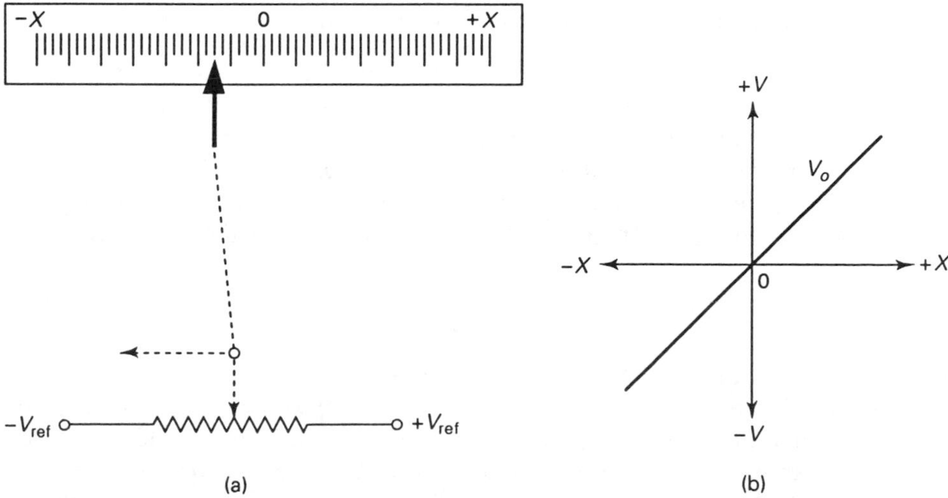

Figure 5–6 (a) Two-quadrant position/displacement sensor; (b) characteristic curve.

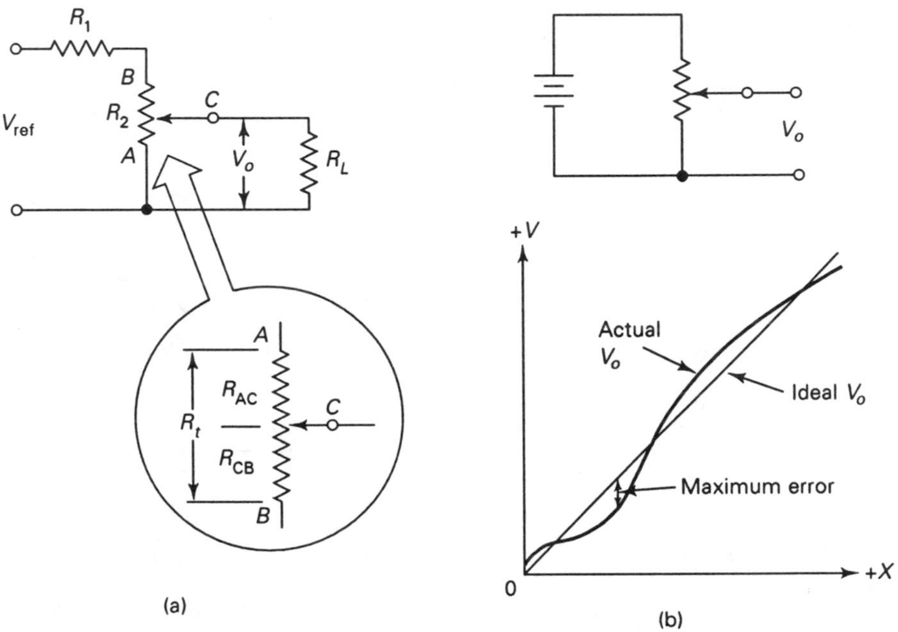

Figure 5–7 (a) Potentiometer sensor circuit; (b) Actual curve versus ideal curve.

R_L is infinite, then the output voltage V_o is found from the equation

$$V_o = \frac{V_o R_{CB}}{R_1 + R_{AC} + R_{CB}} \tag{5–5}$$

where R_{AC} and R_{CB} are functions of the position of the wiper.

Unfortunately, because the circuit is not ideal, there are problems with it. When the load resistance is not infinite (which is all the time) and is not very large with respect to the potentiometer resistance, then it will load the potentiometer by different amounts as a function of position [Fig. 5–7(b)], causing nonlinearity. The reason for this is that the actual resistance from point C to point A is the parallel resistance defined by the equation

$$R_{CB'} = \frac{R_L R_{CB}}{R_L + R_{CB}} \tag{5–6}$$

One of the traditional ways to solve this problem is either to use trimmer potentiometers at the ends of the measurement sensor or to place a tap on the sensor and then connect a compensating resistor from B to the tap. But there is a better way that will work in most cases. Figure 5–8 shows the use of op-amps to buffer the sensor potentiometer. In Fig. 5–8(a) the op-amp is used in the inverting follower configuration. The output voltage is a function of position

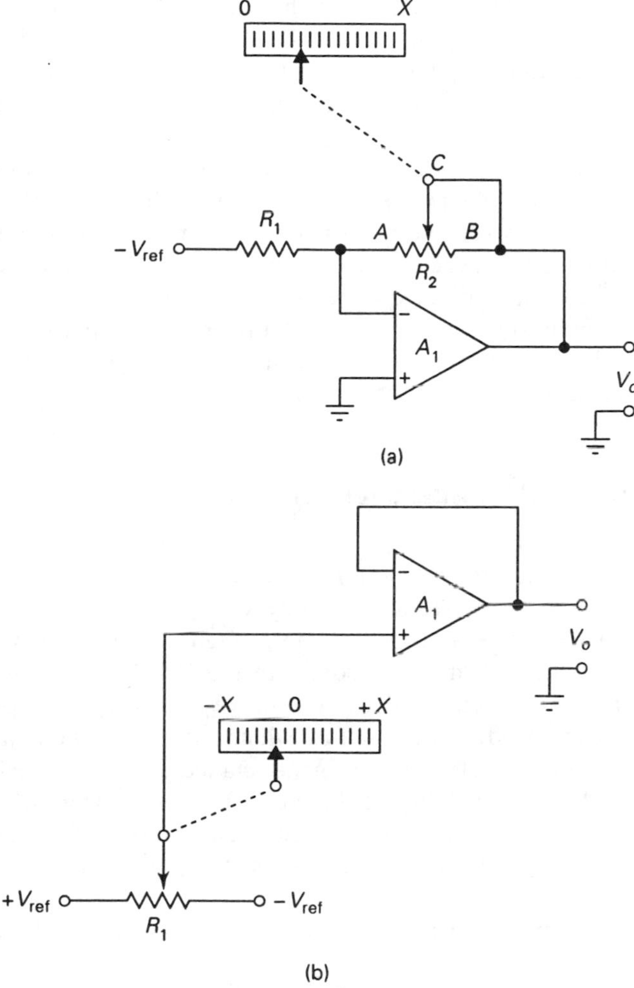

Figure 5–8 (a) Inverting amplifier for position sensor; (b) noninverting amplifier for position sensor.

X, and is found from the equation

$$V_o = -V_{ref}\frac{R_2}{R_1} \tag{5–7}$$

Where R_2 is the potentiometer position sensor resistance. Note that the potentiometer is connected as a *rheostat*, that is, with the wiper connected to one end of the resistance element (BC in this case).

A disadvantage of the system of Fig. 5–8(a) is that the measurement is limited to one quadrant. If two-quadrant measurement is desired, then a circuit such as Fig. 5–8(b) is used. In this circuit the op-amp is connected in the noninverting follower configuration. The gain of this amplifier is unity, but there are also gain versions. The output voltage in this circuit is equal to the input voltage V_1, and V_1 is a function of the position of the potentiometer wiper on the element.

A limitation sometimes encountered on linear potentiometer position sensors is that the range of travel is limited. That is, the length of the element is relatively short. To use the sensor for larger ranges of positions or for larger displacements it is necessary to use either a sensor with a longer range of travel or a mechanical linkage to reduce the travel.

A rotary potentiometer is used to measure angular displacement. These devices are perhaps even more common than linear potentiometers, for they are used as gain and position controls on instruments.

OTHER POSITION/DISPLACEMENT SENSORS

Although potentiometers are among the most commonplace position/ displacement sensors, they are not the only kind. There are also *inductive*, *capacitive* (see Chapter 3), and *optical* types. Figure 5–9 shows two different types of inductive displacement sensors. In Fig. 5–9(a) the inductor has a core element that moves when the actuator moves. The actuator places more or less of the core inside the inductor, which affects its inductance. The displacement Δx translates to a change in inductance Δl because the inductance is a function of the permeability of the coil. In most sensors there are two coils, or a single coil that is center-tapped. The ratio method can be used with these devices, or the sensor can be used in an inductive Wheatstone bridge.

A *linear differential variable transformer* (LDVT) is shown in Fig.

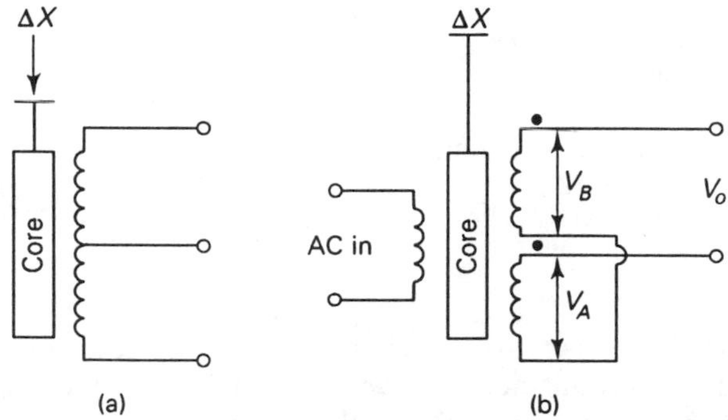

Figure 5–9 (a) Inductive-bridge position sensor; (b) LVDT position sensor.

5–9(b). This is a three-coil inductive position sensor, connected so that one coil is connected to a source of an ac excitation potential. The other two coils are used as secondary windings connected in the series-opposing configuration. The output voltage V_o is the sum of the two coil voltages V_A and V_B, or

$$V_o = V_A - V_B \tag{5-8}$$

When the core is at the zero position, it rests equally inside coil A and coil B, so the inductive reactances of these two coils are equal to each other, and the voltages across the coil are equal. But because the coils are connected out of phase with each other ("opposing"), the output voltage will be zero under that condition. If the core moves in one direction, then a positive output voltage is created, and if in the other direction, a negative voltage is created.

A capacitive position sensor is shown in Fig. 5–10. The capacitance of a parallel-plate capacitor is proportional to the area of the plates facing each other and the dielectric constant k between the plates ($k = 1.006$ for air) and inversely proportional to the spacing between the plates. If any of these factors is varied, then the capacitance of the assembly will also change. A capacitive position sensor is designed to alter one of the three parameters under changes in position.

An optical position sensor is shown in Fig. 5–11. This version is an angular displacement sensor in which a mirror is rotated by the actuator shaft. A light source is deflected to a position-sensitive sensor array, a film emulsion, or a phosphorous screen.

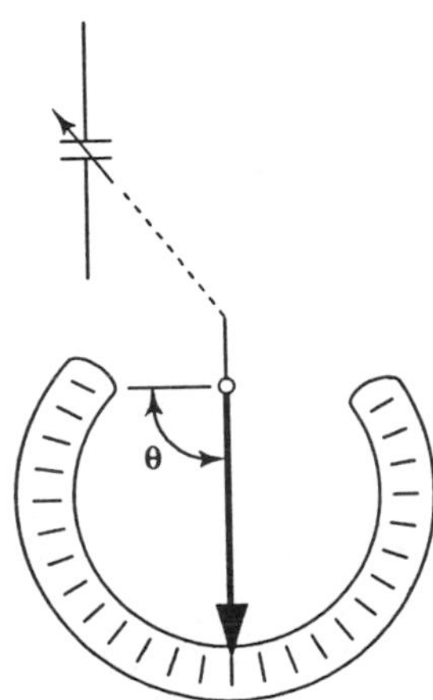

Figure 5–10 Capacitor angular position sensor.

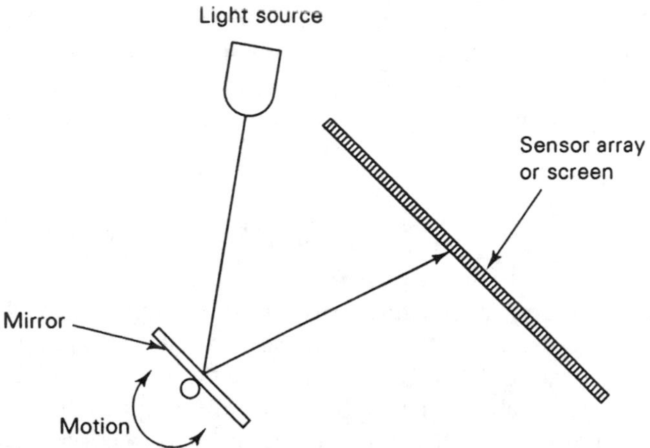

Figure 5–11 Optical reflective position sensor.

Position/displacement sensors are used not only in simple position or displacement applications but also in vibration, acceleration, and velocity measurements. The reason for their use in velocity and acceleration measurements is found in the simple relationships among the displacement Δx, velocity v, and acceleration a: Velocity is the first derivative of x with respect to time, while acceleration is the second derivative.

6 Force and Pressure Sensors

Force and pressure are related concepts, so the sensors used to measure these parameters can be grouped together in a single discussion. Other authors opt for separate discussions, and that approach is also valid, but for our purposes the discussions are combined.

Force F is a vector quantity, so it has both a magnitude and a direction of application. It is not rigorously proper to say that a force of n units is applied without also stating in what direction and where.

The concept of force is well rooted in classical Newtonian physics as that physical quantity that will cause a change in momentum of a body that possesses mass m. Thus, when enough force is applied to a body at rest, the body will be set in motion; if a body is in motion when force is applied, the motion will change one or both aspects of its velocity vector (direction or speed).

Mass is a measure of the amount of mattter in an object as well as its resistance to a change in motion. *Weight*, on the other hand, is a measure of the gravitational force of attraction between an object of mass m and the Earth. Newton's equation governs force:

$$F = MA \tag{6–1}$$

where F is the force
M is the mass (kilograms)
A is the acceleration

In the case of weight, a is replaced by g, the gravitational acceleration.

UNITS OF FORCE

The *Système International* (SI) unit of force is the *newton* (N), which is defined in MKS terms as 1 *kilogram · meter per second squared* (kg · m/s^2). In CGS terminology the unit is the *dyne*, or 1 *gram · centimeter per second squared* (g · cm/s^2). In speaking of weight, the pound (lb) is used. To convert a pound-force to a Newton force, multiply by 4.4468.

In some fields, units of mass are commonly used to indicate a force. For example, a physiologist may use the *gram-force* in measuring the contractile properties of a rodent heart. The argument is made that grams are mass, not force, but the terminology is correct if the unit is properly labeled a gram-force, because the force represented is the force of gravity on a mass of 1 g, or about 980 dynes.

SENSING FORCE

A number of different sensors are used to measure force, but they can be roughly classified according to the transduction method they employ. Some, for example, are compression types of sensors, in which the density of a packed material is changed by the application of a force. Others depend on strain at a specified point, while others use the Δx deflection to directly measure the applied force.

Examples of force sensors include cantilever and suspended ("bridge") beams, proving rings and columns, and diaphragms on which a strain-gage element is mounted. If the deflection is small, then the strain-gage approach is viable, but for larger deflections it is necessary to use a gross deflection or change-of-position sensor. In general, the basic methods can be summarized as follows:

Acceleration methods measure the acceleration of a known mass on which the unknown force operates. The acceleration method is a dynamic one in which the force is applied to a pendulum consisting of a lever arm and a standard mass.

Gravity-balance methods compare the unknown force with the action of the gravitational force on a known mass. The gravity-balance method (Fig. 6–1) is based on the same principle as the ordinary balance scale used in science laboratories. If mass M is on one side of the platform (or beam arm), then the point will return to zero only when a counterbalancing force F is applied to the other end of the platform. The gravitational force acting on the mass is MG, where G is the gravitational constant 6.673×10^{-11} N · m^2/kg^2 in MKS units. The force that restores the pointer to zero balance is equal to mg.

Either the zero-balance or the deflection method can be used. If an

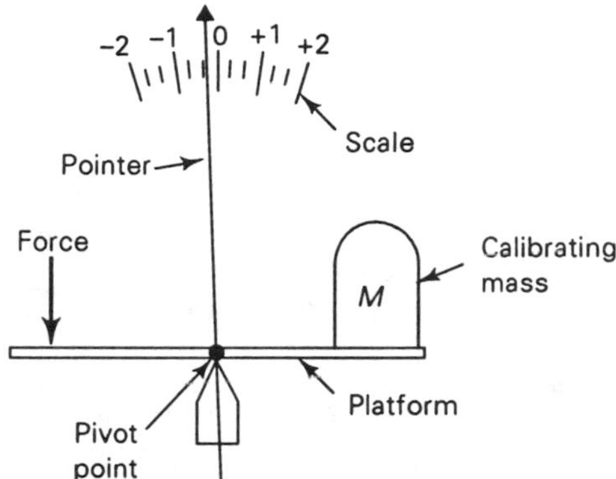

Figure 6–1 Force measurement by balance scale.

unknown force is applied to one end of the platform and sufficient known mass is added to the other end to restore balance, then the product of the gravitational constant and the sum of the calibrated mass elements is the value of the force. Alternatively, a fixed mass is placed on one end of the platform, and the scale of the pointer is calibrated in units of force as a function of angular deflection.

Pressure-sensing methods convert the force to a fluid pressure, which is measured using a pressure transducer. Pressure can be sensed in several different ways (some of which are discussed in this chapter). A method formerly used placed a piston into a cylinder filled with either loosely packed carbon granules or a viscous fluid. The so-called carbon-cell method, which was also the basis for the carbon microphones used in communications many years ago, utilizes the change in the electrical resistance of the carbon segment when compressed [Fig. 6–2(a)]. An external dc power supply is used to pass an electrical current through the carbon element and a series load resistor R_L. When force F is applied to the piston contact, the carbon granules are packed more tightly, reducing the bulk electrical resistance and increasing the current flow. By Ohm's law ($V = IR$), an increase in the current flow increases the voltage drop across the load resistance.

The fluid-filled cylinder version of the method is shown in Fig. 6–2(b). In this type of sensor the pressure change produced by the applied force is registered on a pressure gauge and is a measure of the level of the force. Such load cells are filled with oil or other viscous substances.

Electromagnetic balance methods compare the unknown force with the action of an electromagnet on a standardized permanent magnet. Figure 6–3 shows an electromagnet force sensor. The permanent magnet generates a fixed, calibrated field. When a dc current passes through the windings of the electromagnet, another magnetic field is generated that either aids or opposes the field of the permanent magnet. As a result, the rod on which the coil is mounted will move in a direction that is determined by the polarity

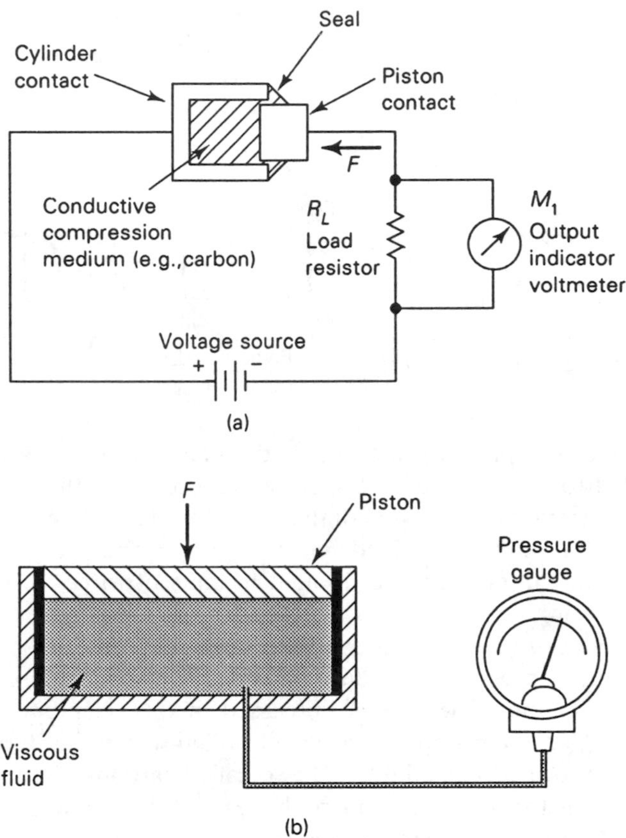

Figure 6–2 (a) Compression load cell; (b) fluid-filled load cell.

of the magnetic field and by an amount that is proportional to the current flowing in the electromagnet's coil. The current level is controlled by an adjustable dc power supply. Again there are two ways to use this sensor. Either the magnitude of the deflection (or change of position δx) measured with a deflection sensor (e.g., LDVT), or the current level needed to restore the rod to its original position when the force is applied can be used to determine the magnitude of the force.

Strain-gage methods use either piezoresistive or piezoelectric elements to measure the strain caused by the force as it deflects a beam, diaphragm, or other object. Many of the force sensors on the market are based on deflection measurements, and the strain gage is well suited to this application if the expected deflections are small. Some force transducers use a ceramic *piezoelectric bimorph* (Fig. 6–4) to measure deflection. This sensor consists of two piezoelectric elements sandwiched together. The voltage appearing

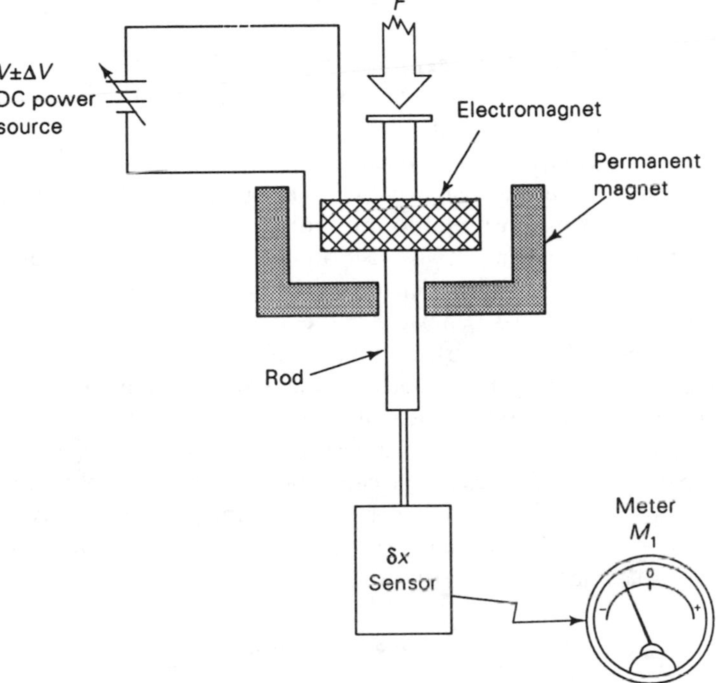

Figure 6–3 Magnetic force sensor.

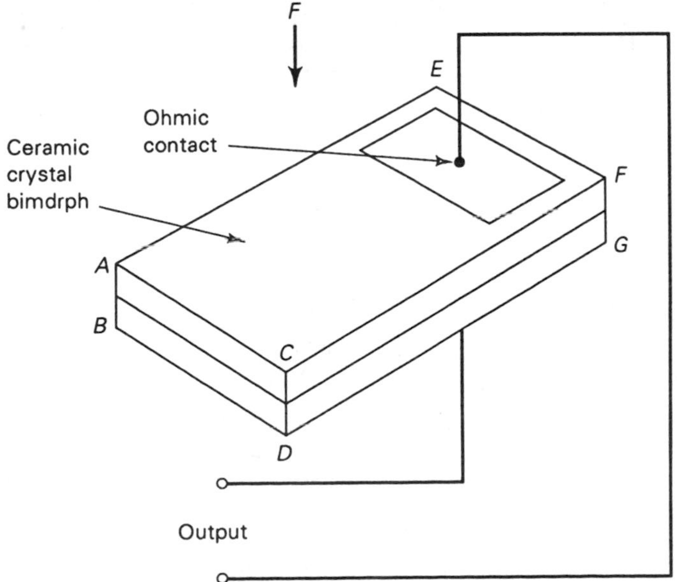

Figure 6–4 Piezoelectric bimorph force sensor.

across the output wires is a function of the deflection, which is proportional to the applied force.

A *beam-deflection* force sensor is shown in Fig. 6–5. The sensor beam is shown at rest in Fig. 6–5(a). Strain-gage elements are placed on the top and bottom of the beam so that a deflection will place one of them in tension and the other in compression. When a force F is applied to the beam [Fig. 6–5(b)], it will deflect a distance δx, causing the strain-gage elements to produce an output proportional to the force.

Deflection force sensors also use other forms of sensing elements. In Fig. 6–6(a) the force sensor is a cantilever beam that deflects δx when a force F is applied. In this case, however, the sensing element can be either a variable inductor [Fig. 6–6(b)] or a variable capacitor [Fig. 6–6(c)]. The inductance of the coil in Fig. 6–6(b) changes when the ferrite or powdered iron core is moved in or out of the coil. The position of the core is a measure of δx, so the inductance is also a measure of deflection. In the capacitive sensor the two factors that affect the capacitance of the parallel-plate capacitor are the cross-sectional area of the plates and the spacing s between them. The spacing is altered by changes in the position of the actuating rod, which alters the capacitance in proportion to δx.

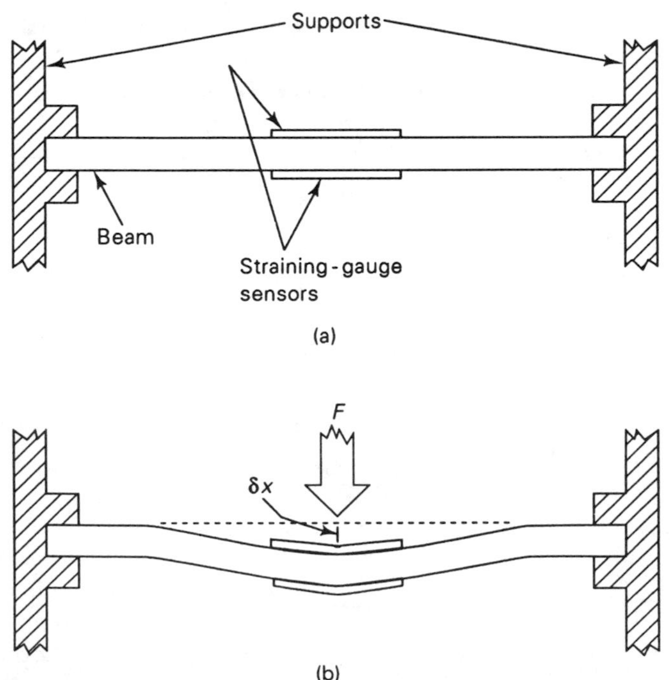

Figure 6–5 Double-ended cantilever-beam force sensor; (a) at rest; (b) deflected.

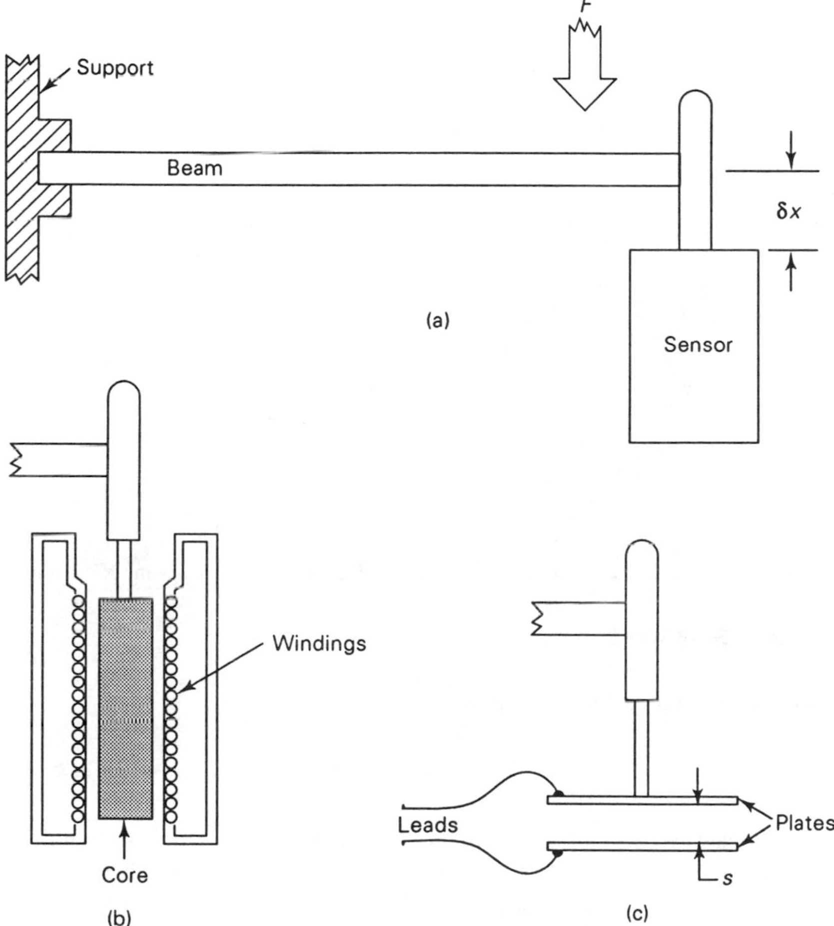

Figure 6–6 (a) Single-ended cantilever force sensor; (b) inductive pickup; (c) capacitive pickup.

Another type of force-sensing element is the *proving ring* (which may also be a solid box or column), shown in Fig. 6–7. One point of the ring is mounted on a bearing point, and an unknown force is applied along the axis that passes through this point and the center of the ring. When the force is applied, the ring shape is deformed from a circle to an oblate circle, or near-ellipse. In some ring sensors, a δx sensor is mounted interiorly to the ring and measures the decreased diameter between A and B. In others, a pair of strain-gage sensors are mounted at A and B, and another pair at C and D (or C' and D') to measure net deflection.

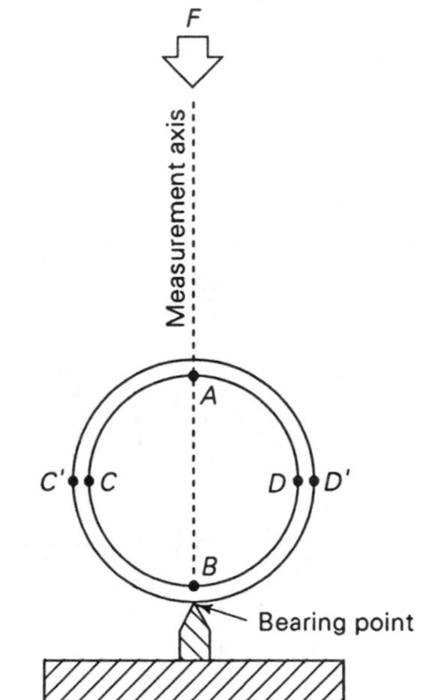

Figure 6–7 Proving-ring force sensor.

PRESSURE SENSORS

Pressure is defined as *force per unit of area*, or

$$P = \frac{F}{A} \tag{6–2}$$

The basic SI unit of pressure is the *pascal* (Pa), which is equal to 1 *newton per square meter* (N/m^2). The pascal is a rather small unit, so the units *kilopascal* (kPa) and *megapascal* (MPa) are also used. Other units of pressure include the CGS *dynes per square centimeter* ($dyne/cm^2$); the British Engineering System unit *pounds per square inch* (psi); and *millimeters of mercury* (mm Hg), or *torr* (1 torr = 1 mm Hg—a special case that we shall discuss shortly). The last unit refers to the height of a column of mercury that is supported by the applied pressure. In some low-pressure measurements, the column height might be measured in terms of *centimeters of water* (cm H_2O). The density of water is about 13.6 times less than the density of mercury, so the same pressure will create a much higher column of water, permitting greater resolution in the measurement.

The pressure in a system can be increased either by increasing the applied force or by reducing the cross-sectional area over which the force operates.

When the force in any system is constant (that is, nonvarying), that pressure is said to be *static* or *hydrostatic*; if the force is varying, the force is said to be *dynamic* or *hydrodynamic*. Physiological pressures (e.g., human arterial blood pressure) are examples of hydrodynamic pressures; the pressure head in a stoppered keg of beer or in a commode tank is a hydrostatic pressure—at least until the keg is tapped or the commode is flushed.

PASCAL'S PRINCIPLE

Pascal's principle (after French scientist/theologian Blaisc Pascal, 1623–1662) governs pressures in a closed system. This physical law states that *pressure applied to an enclosed fluid is transmitted undiminished to every portion of the fluid and the walls of the containing vessel.* Let's consider an example. If a pressure is applied to a stoppered system (e.g., the syringe and vessel in Fig. 6–8), then the same pressure is felt throughout the interior of both the syringe and the vessel. Changing the applied pressure at the rear of the plunger causes the same change to be reflected at every point throughout the interior of the system.

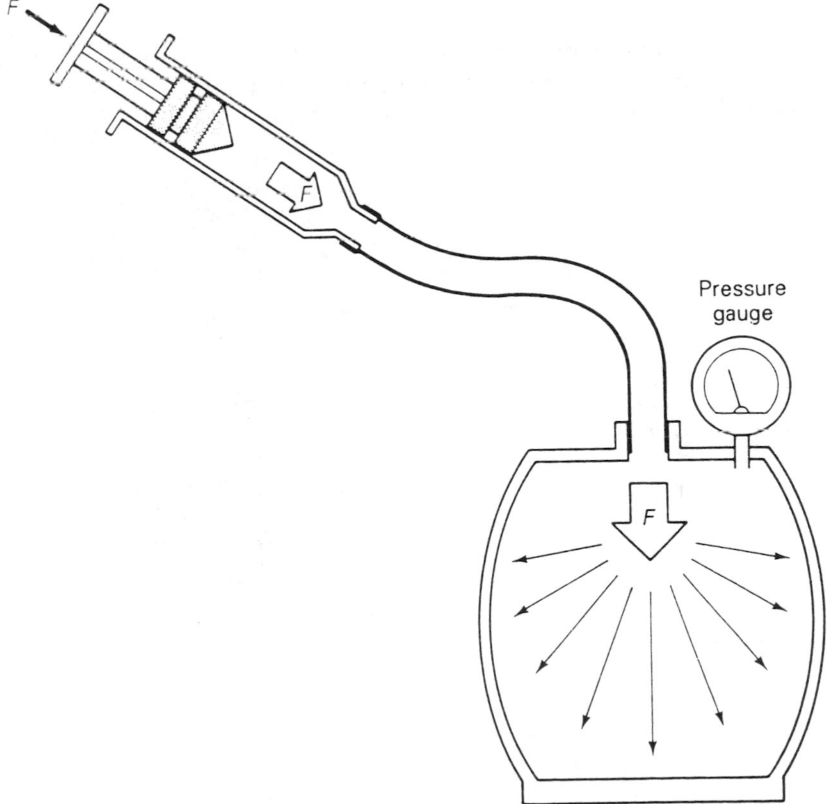

Figure 6–8 Illustration of Pascal's principle.

Pascal's principle always holds true in hydrostatic systems. In hydrodynamic systems it holds true only for quasi-static changes—that is, when a very small change is made, and the turbulence is allowed to die down before subsequent measurements are made. Pascal's principle holds approximately true for those hydrodynamic systems where the flow is reasonably nonturbulent (no true nonturbulent flow exists) and the pipe orifice is small compared with its length. The simple model holds true in those cases, however, only in the center of the flow mass but not at the pipe-wall boundaries.

BASIC PRESSURE SENSORS

Because a pressure is a force applied to a specified area, it is possible to use most of the same sensors for both force and pressure measurements. Indeed, versions of the spring-loaded force sensor are seen frequently in pressure measurements. All that is needed to convert the force output data to pressure output data is the area of the actuator over which the applied pressure-force operates.

Another form of pressure sensor is the *Bourdon tube* shown in Fig. 6–9. The classical circular Bourdon tube is shown in Fig. 6–9(a). It consists of a loop of flexible, hollow spring tubing. When a pressure is applied to

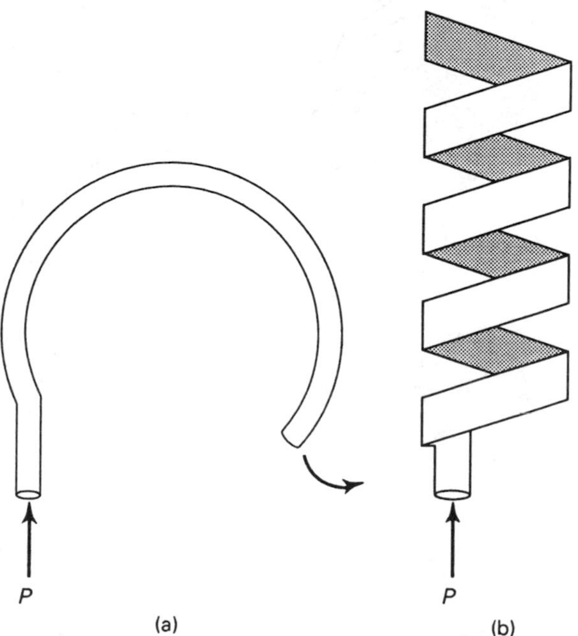

Figure 6–9 (a) Classical Bourdon tube pressure sensor; (b) helical Bourdon tube pressure sensor.

the input venturi, the tube attempts to straighten out. The amount of deflection of the end is proportional to the applied pressure. A helical version is shown in Fig. 6–9(b). The flexible spring tubing is coiled into a solenoidal helix that attempts to unwind when the pressure is applied. Coiled and twisted forms of Bourdon tube are also available (not illustrated). It is possible to attach a point and scale to any of these devices, although in most cases it is more practical to use the Bourdon tube in conjunction with a displacement sensor such as a potentiometer or LDVT.

Perhaps the most common form of pressure sensor is the diaphragm type. A *diaphragm* is a thin metallic plate that is deflected in the center when a pressure is applied. The diaphragm sensor is well suited to several different measuring methods; strain gage, LDVT, and inductive Wheatstone bridge are commonly found.

Figure 6–10 shows several different configurations of a diaphragm pressure sensor. In all three cases shown there are two chambers in the sensor, one on each side of the diaphragm. An *absolute pressure sensor* is shown in Fig. 6–10(a). In this device, one side of the diaphragm is ported to the unknown pressure P_x. The other chamber, on the opposite side of the diaphragm, is evacuated to as near a vacuum as can be maintained. Thus, the deflection of the diaphragm is proportional to the absolute pressure. A *gauge pressure sensor* is shown in Fig. 6–10(b). One side of the diaphragm is ported to the atmosphere, while the other is ported to the unknown pressure. A *differential pressure sensor* is shown in Fig. 6–10(c). In this device one port is for one pressure P_1, while the other is for another pressure P_2. The differential pressure ΔP is the difference between P_1 and P_2:

$$\Delta P = P_1 - P_2 \tag{6–3}$$

In some differential pressure sensors one side is sealed at a specific pressure, called a *reference pressure*, whereas in others the port is open so that a pressure drop in a system can be measured (see the discussion on flow measurements in Chapter 13).

We now consider examples from medical, physiological, and life sciences

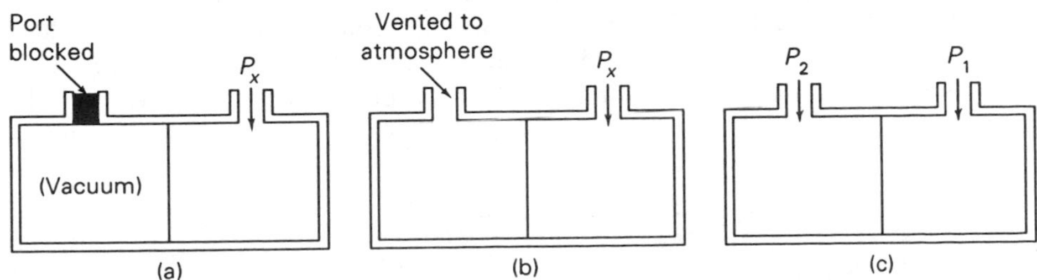

Figure 6–10 Types of pressure sensors: (a) absolute pressure; (b) gauge pressure; (c) differential pressure.

applications. The principles are the same, so the information can easily be converted from one system to another.

MEDICAL PRESSURE-MEASUREMENT SYSTEMS

The most common medical pressure measurement is of blood pressure, and the most common blood pressure measurement is the arterial blood pressure. Also of interest in special cases are venous pressures, central venous pressure (CVP), intracardiac blood pressure, pulmonary artery pressure, spinal fluid pressures, and intraventricular brain pressures. The principal difference among these pressures is primarily in the range of measurement; oftentimes the same instruments are used for all these different pressures.

A medical pressure-measurement system needs to be the least invasive possible (it is, however, invasive) and sterile (to prevent infection). In addition, electrical isolation from the ac power mains needs to be maintained at a higher level for patient safety.

In physiological systems the situation is somewhat more complicated. Unfortunately, many students in life sciences lack the mathematics background to perform proper analysis of blood flow systems. Many medical students, for example, are taught a simplistic model that fails to take into account four complicating factors: (a) blood has a particulate nature (it is not a strict fluid, but contains red cells and other material), (b) blood vessel walls are distensible and not rigid, (c) blood viscosity changes under certain influences, and (d) blood vessel walls are not smooth, especially in older subjects. This model is analogous to Ohm's law and states that:

$$P = R \times F \tag{6–4}$$

where P is the pressure difference in torr
F is the flow rate in milliliters/second
R is the blood vessel resistance in peripheral resistance units (PRU), where 1 PRU allows a flow of 1 ml/s under 1 torr pressure

The actual situation in physiological blood flow systems is a lot more complex because of the factors listed. The vessel diameter changes (which is one way the body regulates blood pressure) both from systemic readjustments and because the beating heart forms a pulsatile pressure wave. The flow rate (and hence the other parameters) is thus better (but still imperfectly) given by *Poiseuille's law*;

$$F = \frac{P\pi R^4}{8\eta L} \tag{6–5}$$

where F is the flow in cubic centimeters per second
P is the pressure in dynes per square centimeter

η is the coefficient of viscosity in dyne · seconds per square centimeter
R is the vessel radius in centimeters
L is the vessel length in centimeters

The study of pressure in turbulent or large lumen systems, or in the boundary area close to the pipe/vessel wall, is the subject of advanced engineering mechanics and physics courses. For our purposes, we may assume that either Pascal's principle holds true absolutely or the system can be made quasi-static for measurement, discussion, or analysis purposes.

Pulsatile waves result from a cardiac pumping action that is not constant. In physiological systems the heart of the subject animal or human beats in a manner that produces a pulse flow (which can be felt with the finger tips where arteries run close to the surface—in the wrist for example). Figure 6–11 shows the human arterial blood pressure waveform, here used as an example of pulsatile systems. Several values can be measured in this system, namely,

1. Peak pressure (systolic)
2. Minimum pressure (diastolic)

$$P_{sys} = P_{max}$$

$$P_{dias} = P_{min}$$

$$\bar{P} = \frac{1}{T_2 - T_1} \int_{T_1}^{T_2} P\,dt = \text{MAP}$$

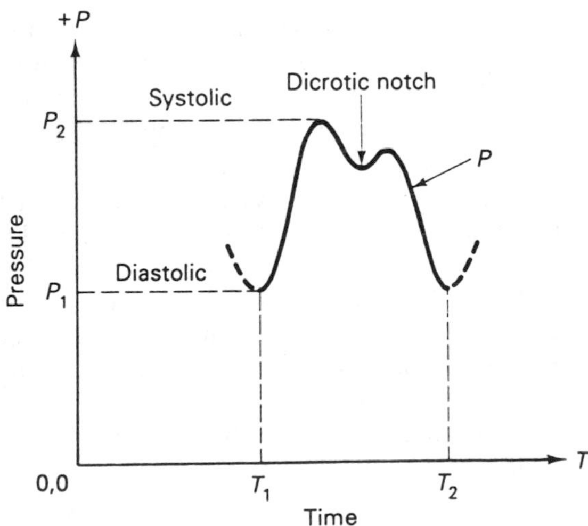

Figure 6–11 Human arterial blood pressure waveform.

3. Dynamic average (one-half peak minus minimum)
4. Average pressure (called mean arterial pressure)

Thus, in a discussion of "pressure" in a pulsatile system, the type of pressure must be specified. In a later section we discuss the methods used for measuring these pressures electronically. Engineers have an ongoing communication problem with both clinicians and medical researchers regarding the average pressure readings on electronic blood pressure monitors. The problem lies in each group's definition of "mean arterial pressure" (i.e., the time average). The correct definition, which is used in the design of the typical instrument, is written as a calculus expression:

$$\overline{P} = \frac{1}{T_2 - T_1} \int_{T_i}^{T_2} P \, dT \tag{6–6}$$

In medical and nursing schools, and in typical Intensive Care Unit nursing courses, however, a synthetic (and often incorrect) definition is used (referring to Fig. 6–11) because rigorous mathematics is not stressed. The approximation used is

$$\overline{P} = P_1 + \frac{P_2 - P_1}{3} \tag{6–7}$$

In medical terminology, this definition states that the mean arterial pressure (or MAP) is equal to the diastolic (P_1) plus one-third the difference between systolic and diastolic ($P_2 - P_1$).

The problem faced by the biomedical engineer (and more often either the biomedical equipment technician who repairs the equipment or monitoring technicians who operate the equipment) is that the synthetic definition is merely an approximation of the functional calculus definition for healthy people! In many sick people, however, the portion of the waveform between the dicrotic notch and time T_2 (Fig. 6–8) is very heavily damped, so the actual MAP is less than the functional MAP measured by the electronic instrument.

A simple test, which is revealed by inserting values into both equations, is to place a constant pressure on the system and see what happens to the readings. In that case, $P_1 = P_2 =$ MAP, so all three digital readouts should be the same. In other words, a test of a pressure system when the displayed MAP and the calculated MAP are different is to pump a constant pressure into the system (or press the CAL button) and see if all three pressures are the same: diastolic, systolic, and MAP. If they differ when the pressure is constant, then there is a system fault.

The CAL button alternative works only to test the electronics. A fault in the transducer will not be revealed by this test unless a real pressure is applied.

In a later section we deal with electronic measurement of pressures, so we will return to Figure 6–11 to see the relationships among various pressures.

BASIC PRESSURE MEASUREMENTS

The air forming our atmosphere exerts a pressure on the surface of the earth and all objects on the surface (or above it). This pressure is usually expressed in atmospheres (atm), pounds per square inch (lb/in.^{2} or psi), or other pressure units. The magnitude of 1 atm is approximately 14.7 lb/in.^{2} at mean sea level.

If a pressure is measured with respect to a perfect vacuum (defined as 0 atm), then it is called *absolute pressure*; and if against 1 atm ("open air") it is called a *gauge pressure*. Two gauge pressures, or a gauge pressure and an absolute pressure, can be measured relative to each other to form a single measurement called *relative pressure* or *differential pressure*. Pressures in fluid pipelines, storage tanks, and the human circulatory system are usually gauge pressures (if measured at a point) or differential pressures (if measured between two points along a length).

Figure 6–12 shows the Torricelli manometer, named after Evangelista Torricelli (Italian scientist, 1608–1647), which is used to measure atmospheric pressure. An evacuated, small-lumen glass tube stands vertically in a pool of mercury. The end that is inside the mercury pool is open, while the other end is closed. The pressure exerted by the atmosphere on the surface of the mercury pool forces mercury into the tube, forming a column. The mercury column rises in the tube until its weight (i.e., gravitational force) exactly balances the force of the atmospheric pressure. Torricelli found that a 760 mm column of mercury can be supported by atmospheric pressure at sea level. Thus, 1 atm is equivalent to 760 mm Hg (also sometimes given in weather reports and aviation in inches, i.e., 1 atm = 760 mm Hg = 29.92 in. Hg).

The proper unit of atmospheric pressure, as established by scientists in international agreement and adopted in the United States by the National Institute for Standards and Technology (National Bureau of Standards), is the *torr* (named after Torricelli), where 1 torr = 1 mm Hg. In medicine and medical science (e.g., physiology) millimeters of mercury is still used instead of the correct torr.

Gauge pressures are usually given in millimeters of mercury (or inches) above or below atmospheric pressure. A *manometer* is any device that measures gauge pressure, positive or negative. By convention, pressures above atmospheric pressure are signed positive, and those below atmospheric pressure are signed negative. Also by convention, negative gauge pressures are called vacuums, and negative-reading manometers are called vacuum gauges (positive-reading manometers are also called pressure gauges). Both instruments are nonetheless properly called manometers, and in this chapter we discuss both mercury and electronic manometers.

All measurements require a reference point, and for gauge pressures the zero reference is a pressure of 1 atm. Although the absolute value of the atmospheric pressure varies from one place to another, and even in the

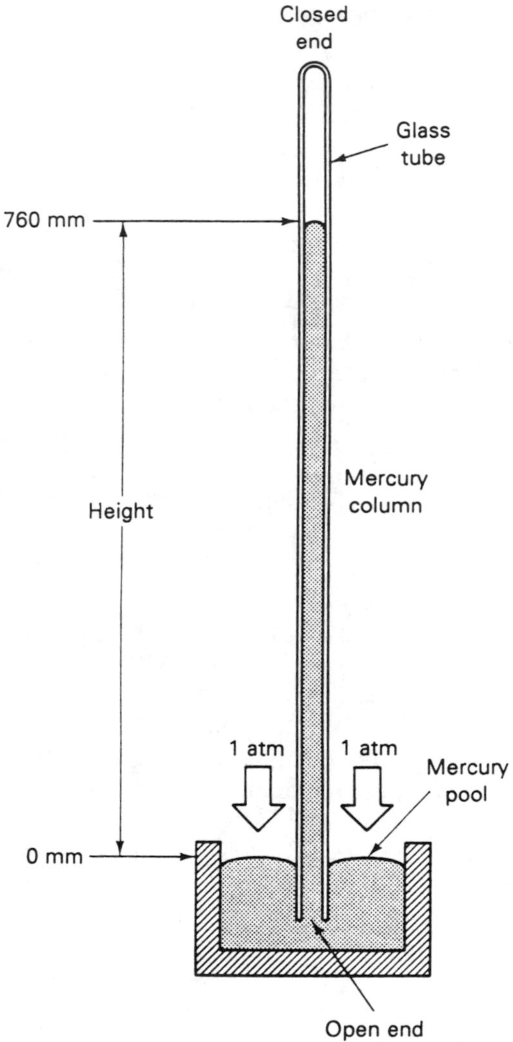

Figure 6–12 Torricelli manometer.

same location over the space of a few hours, the zero point can be established by setting the indicator to zero on the scale and opening the manometer to the atmosphere.

Figure 6–13 shows a mercury manometer that is similar to those used to measure pressures and calibrate electronic pressure manometers. The open tube is connected to a mercury reservoir that is fitted with a rubber squeeze ball pump that can be used to increase pressure in the system. A valve is used to either open the chamber to the atmosphere or close it off.

If the valve is open to the atmosphere, then the pressure on the chamber

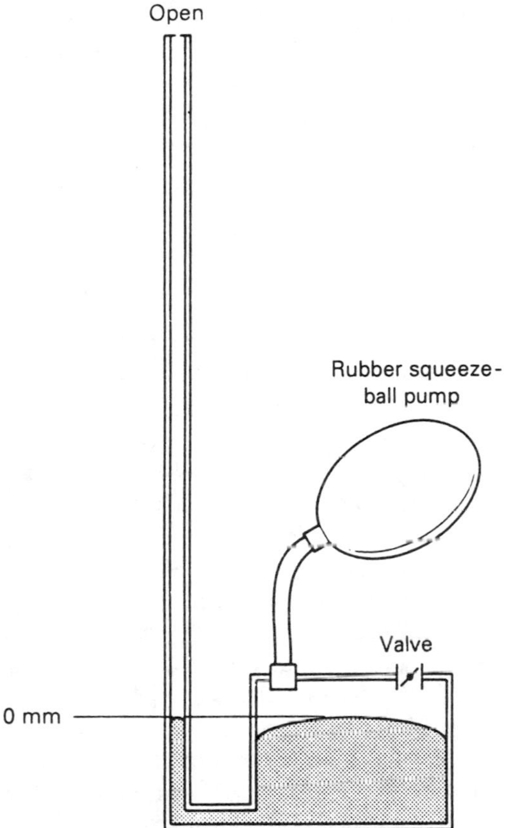

Figure 6–13 Pressure manometer.

is equal to the pressure on the column, that is, 1 atm. Under this condition the mercury column in the tube is at the same height as the mercury in the chamber. This point is defined as 0 mmHg. If the valve is closed, and the pressure inside the chamber is increased by operating the pump, then the mercury in the column will rise to a level proportional to the new pressure above atmospheric pressure.

If the rubber ball in Fig. 6–13 is replaced with a connection to a closed pressure system other than the squeeze ball, then the mercury will rise to a level proportional to the pressure in that system. Medical personnel can use this manometer as a calibrating device by adding a tee-connector in the line between the rubber squeeze ball and the chamber. One port of the tee goes to the rubber ball, one port goes to the chamber, and the third port goes to the transducer or other instrument being calibrated.

Gauge pressure is used for measurements because it is easy to establish the reference or zero point (open the manometer to the atmosphere) and because it can easily be recalibrated for each use (no matter where in the world the measurement is made). In addition, for most practical applications the absolute pressure conveys no additional information over gauge pressure. Should absolute pressure be needed, it is relatively easy to measure the atmospheric pressure (with a device like that in Fig. 6–12) and then add that value to the pressure measured with the manometer.

BLOOD PRESSURE MEASUREMENT

Until 1905 the only method available for the measurement of blood pressure was to insert a Torricelli manometer tube into the vein or artery being measured and note how far up the tube the blood flowed after a few pulses. This method is still used for some spinal fluid and central venous pressure measurements. It was pioneered in 1773 by English physician Stephen Hales, who used an open-ended glass tube in the neck artery of an unanesthetized horse.

Although direct pressure measurement is more accurate than indirect methods, it is also more difficult and more dangerous to the patient, so it is unsuited for routine use. Today, although direct (water manometer) measurements of some blood pressures are performed in special care situations (e.g., intensive care unit, operating room), the usual medical direct pressure method uses an electronic pressure sensor. Most blood pressure measurements taken outside of intensive care settings use an indirect method.

Sphygmomanometry is an indirect method of measuring pressure. An inflatable cuff is placed over the patient's arm and inflated to a point where it occludes the underlying artery, so no blood can flow. A stethoscope placed downstream from the occlusion is used to monitor the onset of blood flow. The operator slowly releases the pressure in the cuff (3 mm Hg/s is optimum) until a series of sharp, snapping Korotkoff sounds are heard. These are produced by the turbulence of blood under pressure as it breaks through the occlusion into the downstream artery. This event occurs when the cuff pressure equals the systolic pressure. The operator continues to monitor the pressure until the Korotkoff turbulence dies out, an event that occurs when the cuff pressure equals the diastolic pressure.

This method of measuring blood pressure was pioneered in 1905 by Nicolas Korotkoff but was not used widely until after 1935, when the Korotkoff readings were finally correlated with the Hales method values. It wasn't until well after World War II that nurses were permitted to take blood pressure readings, as that was considered "doctor's work" up until that time.

Modern electronic blood pressure measurements consist of two types. One is an electronic version of the 1773 Hales method, in which a thin, hollow catheter is introduced into the artery and then connected to a transducer (Figure 6–14). The transducer outputs an electrical analogue of the pressure waveform that can be directly calibrated in millimeters of mercury. The

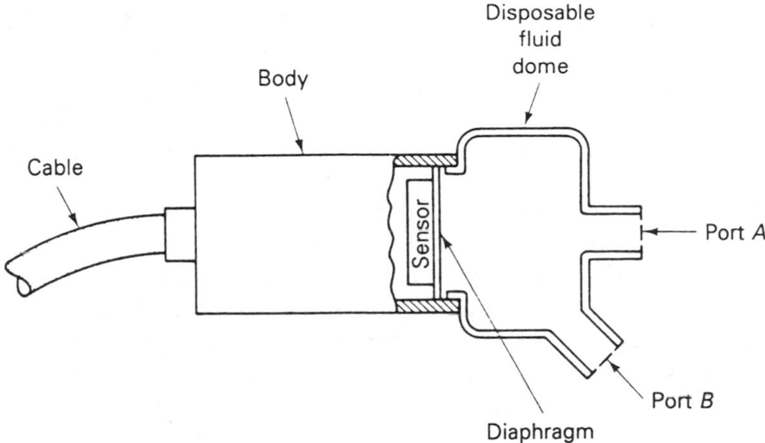

Figure 6–14 Pressure sensor cutaway view.

second type of electronic measurement is used by many all-electronic home-type blood pressure kits. In these instruments a microphone (optimized for low-frequency sounds) is placed under the cuff, and the pressure is released. When the Korotkoff sounds appear, the internal circuitry records the cuff pressure as the systolic; when they disappear, it records the diastolic pressure.

Another (and related) method for indirectly measuring blood pressure is used in certain automatic bedside instruments. It was observed that the pressure inside the blood pressure cuff begins to oscillate as the decreasing cuff pressure approaches the mean arterial pressure (MAP). This phenomenon is thought by some to be an example of chaos in a physical system ("Chaos" is a new science that is just emerging. Its first widespread impacts were in cardiology and the type of turbulent dynamic systems represented by the blood pressure measuring system.) The diastolic and systolic pressures can be calculated from a formula derived from either Eq. (6–6) or (6–7) that utilizes this pressure variation and its peaks (at the MAP).

Other automatic blood pressure cuffs rely on either a low-frequency microphone or an ultrasonic transducer embedded in the cuff. These transducer elements are used to automatically sense the Korotkoff sounds. These instruments are relatively sensitive to the positioning of the microphone or transducer over the brachial artery. An incorrect pressure reading could be the result of improper adjustment of the cuff position.

FLUID-PRESSURE TRANSDUCERS

Figure 6–14 shows a cutaway view of a typical blood pressure transducer. The body of the transducer contains the circuitry, which is separated from the fluid dome by the transducer's pressure-sensitive diaphragm. This thin metallic membrane feels the pressure in the dome and is distended in proportion to the applied pressure. The other side of the diaphragm is connected to either the core of an LVDT transformer or a piezoresistive Wheatstone bridge transduction element. The dome is used to contain the fluid and in most medical applications is disposable (to prevent cross infection between patients).

There are at least two ports on the transducer dome. These ports are controlled by stopcocks, either built-in or added-on. When the stopcock is opened, the fluid flows into or out of the transducer, but when the stopcock is closed, the transducer is basically a closed system. In normal operation, one port is connected to the system being measured, while the other port is initially open to the atmosphere. With the atmospheric port open, the transducer is at zero gauge pressure, so the electronic instrument it drives can be "zeroed."

Figure 6–15 shows the calibration setup for an electronic blood pressure measurement instrument. The transducer is connected to an electronic manometer. In this test setup the atmospheric port is connected to a regular mercury (or in low-pressure cases, water) manometer. When the relief valve is open to the atmosphere, the pressure monitor is set to zero. The valve is then closed, and a pressure is pumped onto the system by the squeeze-ball

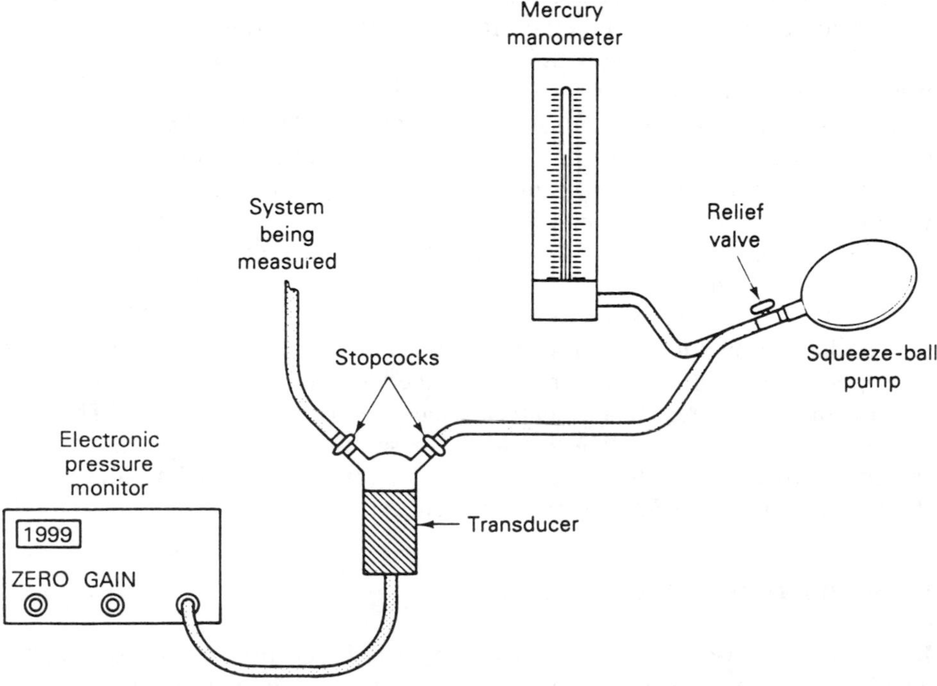

Figure 6–15 Pressure sensor calibration system.

pump. In most cases it is a good idea to set the pressure (as read on the mercury manometer) to some standard value such as 100 torr or 200 torr. The SPAN or GAIN control on the electronic monitor can then be adjusted to the same reading as obtained on the mercury manometer.

In biomedical applications it is critical that the operator be instructed in proper technique. One reason is to prevent contamination of the patient's lines. Another is that incorrect settings of sometimes complicated stopcock arrangements can cause the system to be pumped with air, which could enter the patient as a potentially fatal air embolism. Instruction from a medically qualified person is essential to preventing these disasters.

FLUID-MEASUREMENT SYSTEM FAULTS

The measurement of fluid pressures with an electronic apparatus is not always as simple as it appears, unless certain precautions are observed. Several problems can affect the data acquired. For example, although it seems a trivial observation, in multibranch or varying-diameter systems the pressure being measured must be specified. One problem with pressure measurements is that the pressure measured by the instrument is not the same pressure measured by the blood pressure cuff. Typically, the blood pressure cuff measures a pressure in the upper arm, whereas the electronic instrument catheter is placed distal (downstream) from that point, typically in the radial artery in the wrist. The problem is that these two pressures are normally different, so an accurately taken manual blood pressure (albeit accuracy is a problem with the manual method) will normally be a slightly higher reading than the downstream reading—taken after the artery has branched. Also, height of the measurement site relative to the heart affects the reading.

Two other problems frequently encountered are resonance and damping of the pressure in nonstatic systems. Of course, if the pressure is a dead pressure, that is, one that does not vary or varies quasi-statically, then the problem does not exist. But if the pressure has a dynamic waveform, as in a blood pressure system, then certain problems with the system plumbing can cause errors.

There are several causes of damping in blood pressure systems. One is clogging of the tubing to and from the transducer. This is especially likely when blood enters the tubing and clots. Other types of fluids will exhibit damping due either to phenomena like clotting or blockage of the sampling catheter by particulate matter. Damping adds a certain degree of inertia to the system, with a resultant loss of frequency response and peak data that are of medical significance.

Another cause of damping is purely a procedural problem. Sometimes the wrong form of tubing is used for the transducer plumbing system. The correct tubing is stiff-walled. If the tubing is rubber, neoprene, or some other distensible material (like intravenous IV tubing!), then there will be a problem. The cause of the problem is that increases and decreases of the pressure waveform cause the tubing diameter to increase or decrease in response. Unfortunately, changing the diameter of the measurement system also changes the pressure, so the measurement interferes with the system being investigated. This problem is especially common in medical or life sciences areas because users will press various forms of medical tubing into service. Both surgical tubing and IV set tubing are sometimes used—erroneously—and will result in a damped waveform. The results are incorrect waveshape and readings.

Resonance or "ringing" [Fig. 6–16(a)] in the system is caused mainly by two phenomena. The first is the presence of air or other gas bubbles in the system. It is almost impossible to set up a pressure system without having air enter the plumbing. It is therefore necessary to purge the system of air bubbles by placing the transducer and plumbing in an attitude where the air can rise to the top—near a port or stopcock—and be removed.

The second cause of resonance is improper length of tubing. Like any dynamic mechanical or electrical system, the transducer plumbing has a certain resonant frequency. If the length and diameter of the tubing are such that the system is resonant for the applied waveform, then ringing will result. Typically, a ringing system produces a jagged waveform or one in which the peak pressure indication is substantially larger—and sharper in shape—than can be justified by the application.

All waveshapes can be represented mathematically by a Fourier series of alternating current frequencies consisting of a fundamental frequency and a collection of harmonics (integer multiples: 2, 3, . . .) or subharmonics (reciprocals of integer multiples: 1/2, 1/3, . . .). For this reason an ECG amplifier must have a frequency response of 0.05 to 100 Hz, and a pressure amplifier a response of 0.05 to 85 Hz (1 Hz = 1 cycle per second). In the case of a perfect square wave, the Fourier series consists of a fundamental and an infinite series of odd-order harmonics (3, 5, 7, 9, . . .). In medical applications such as pressure monitoring, however, the "ideal" square wave has harmonics only to about 85 Hz, so the edges will appear rounded [see the ideal square wave in Fig. 6–16(b)]. This is the wave that should result from a quick, single-stroke flushing of a blood pressure transducer with a syringe.

Note in Fig. 6–16(b) the ringing square wave. In this case the plumbing or some other factor has caused the high frequencies to be accentuated out of balance. This is an example of an underdamped system. The damping factor is a measure of the system response and is approximated by the equation

$$\text{D.F.} = \frac{-\ln(BA)}{\sqrt{\pi^2 + \ln(B/A)}} \tag{6–8}$$

(see Fig. 6–16 for definitions of A and B).

The opposite problem, a severely overdamped system is shown in Fig.

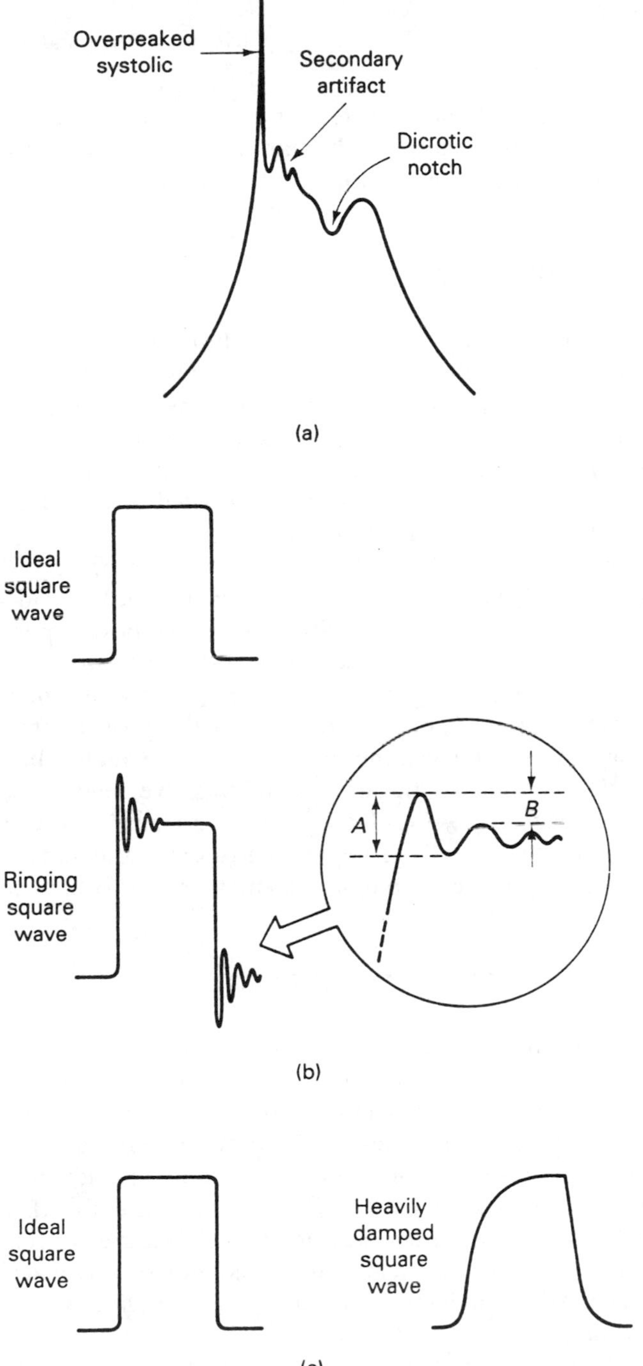

Figure 6–16 Pressure system artifacts: (a) ringing or peaking on an arterial pressure signal; (b) ideal square wave; (c) ringing square wave; (d) rounded-off (poor high-frequency response) square wave.

6–16(c). In this case, the higher frequencies are not passed, so the pressure waveform changes much more slowly than the actual applied pressure. This problem is sometimes caused by fouled pressure lines.

STATIC PRESSURE HEAD

Another measurement factor to consider is transducer placement when the transducer is not physically part of the system being measured. For example, when the pressure at the bottom of a tank is being measured, the transducer should be placed at the same height as the bottom of the tank. This line is called a *pressure datum reference line*. In the human blood pressure application, the transducer must be placed at the level of the patient's heart to prevent hydrostatic pressure head differences.

Another example of hydrostatic pressure head problems is shown in Figure 6–17. The fluid in the tubing has weight, so it will add a pressure of its own due to the force of gravity. Figure 6–17(a) shows a positive pressure head, in which the tubing approaches the transducer from above. Similarly, a negative pressure head [Fig. 6–17(b)] is produced when the plumbing approaches from below. Figure 6–17(c) shows the proper scheme (where it can be accomplished) in which the tubing is routed equally above and below the datum reference line, so the positive and negative head cancel each other.

In the event that the transducer cannot be placed to prevent a positive or negative head, the positive or negative offset can sometimes be tolerated, or electronic corrections can be made in the amplifier or processor circuits.

PRESSURE MEASUREMENT CIRCUITS

There are several basic forms of pressure amplifier circuits. Some of them are so simple that it is easy to build them from op-amps or other linear integrated circuit devices. Common types include dc, isolated dc, pulsed excitation, and ac carrier amplifiers. The dc amplifiers work only with resistance strain-gage transducers or those newer forms of inductive transducers that output a rectified dc signal. AC carrier amplifiers work with both resistive strain gages and inductive transformers (inductive Wheatstone bridges and LDVTs). The pulsed excitation works with resistive strain gages but with only some of the inductive transducers (depending on the inductance, duration of the pulse, and other factors).

Regardless of the design, however, certain features are common to all forms of pressure amplifiers. Some devices are narrowly limited in range for special purposes, whereas others are capable of a wider range for more general applications. In cases where moderate accuracy is needed, internal calibration methods are sufficient, but where superior accuracy is a require-

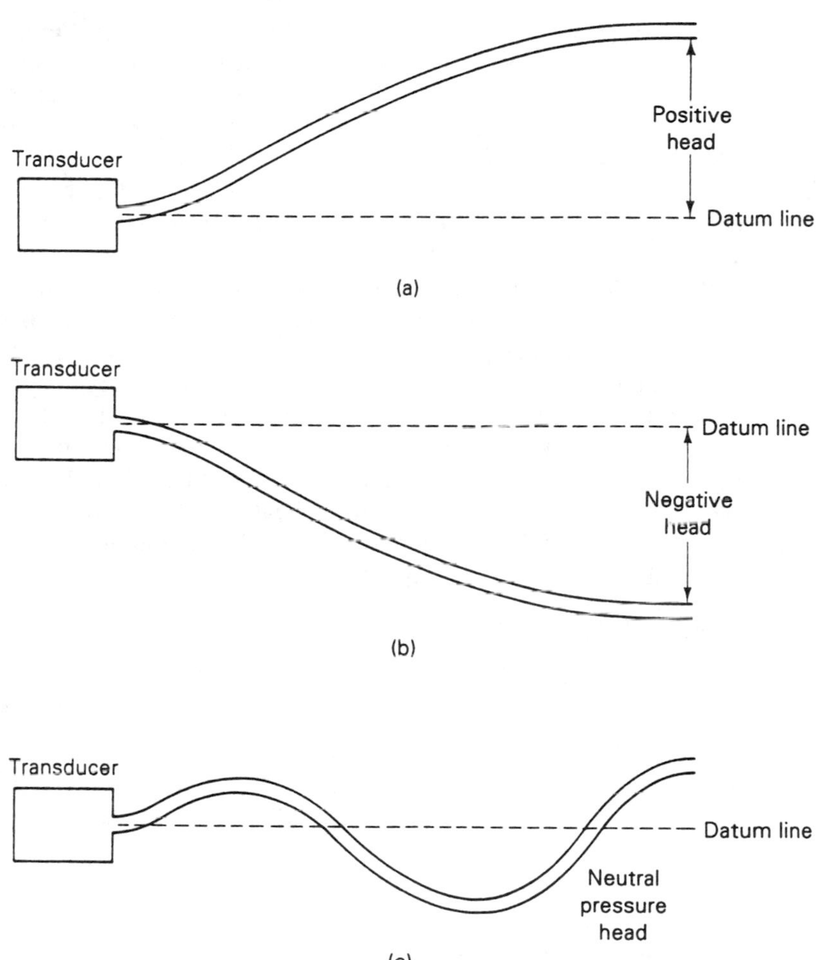

Figure 6–17 Plumbing defects leading to pressure errors: (a) positive head; (b) negative head; (c) properly balanced system.

ment, the system should be calibrated against a mercury manometer. When designing pressure amplifier/display systems, manufacturers provide both zero control and gain control. There are no perfect transducers on the market. Although some transducers contain internal circuitry (sometimes in the connector), all will exhibit offset and gain problems to some extent. The offset is caused by errors in the strain gages, errors in the other circuitry, and distention of the diaphragm. The gain error is caused by variations in the sensitivity of the transducer. For example, for one set of 12 Statham transducers (a good brand), nominally rated at a sensitivity of 5 μV/V/torr, the factory calibration certificates showed the real sensitivity figures to be 37 to 65 μV/V/torr.

There are three ways to standardize the transducer for medical applications. One is to tightly specify the offset and sensitivity to the suppliers and force them to hand select or high-grade their product. Another is to provide an internal balance and sensitivity adjustment. Figure 6–18 shows a transducer with both zero offset and sensitivity trimmer potentiometers installed. The final method is to provide a calibration factor for the transducer (which is printed on a label on the body of the transducer) that is used to adjust the external amplifier circuitry.

The calibration resistor R_{cal} is used to provide a standardized offset to the transducer bridge in order to mimic a specified pressure level. If the resistor is carefully chosen, then pressing the CAL button will cause the output voltage to shift an amount equal to the shift of a standard pressure (usually 100 mm Hg for arterial transducers, and 10 mm Hg for venous models).

Figure 6–19 shows the simplified circuit of a dc pressure amplifier that uses the calibration factor method. The pressure amplifier A_1 is a dc amplifier, so the pressure transducer is a resistive Wheatstone bridge strain gage. Diode D_1 provides the 7.5 V dc excitation to the transducer, and the potentials for the BALANCE and CAL FACTOR controls. The calibration factor for the transducer will sometimes change, so it is necessary to provide a procedure for measuring the new factor:

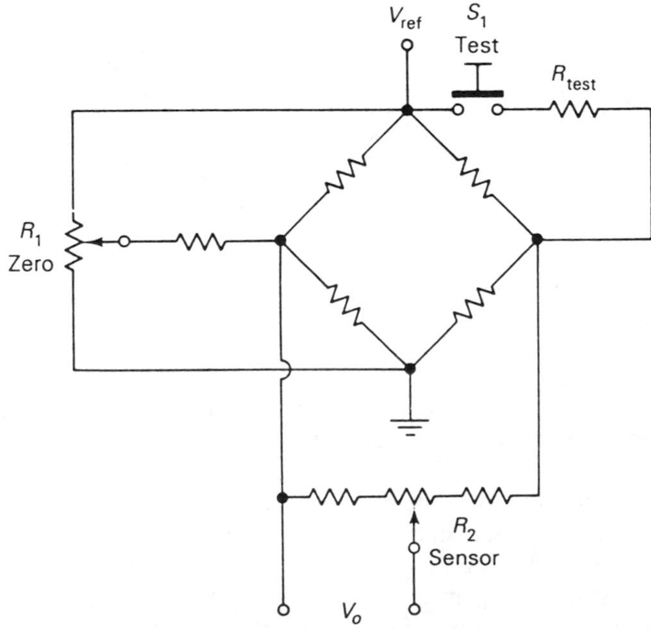

Figure 6–18 Wheatstone bridge pressure sensor.

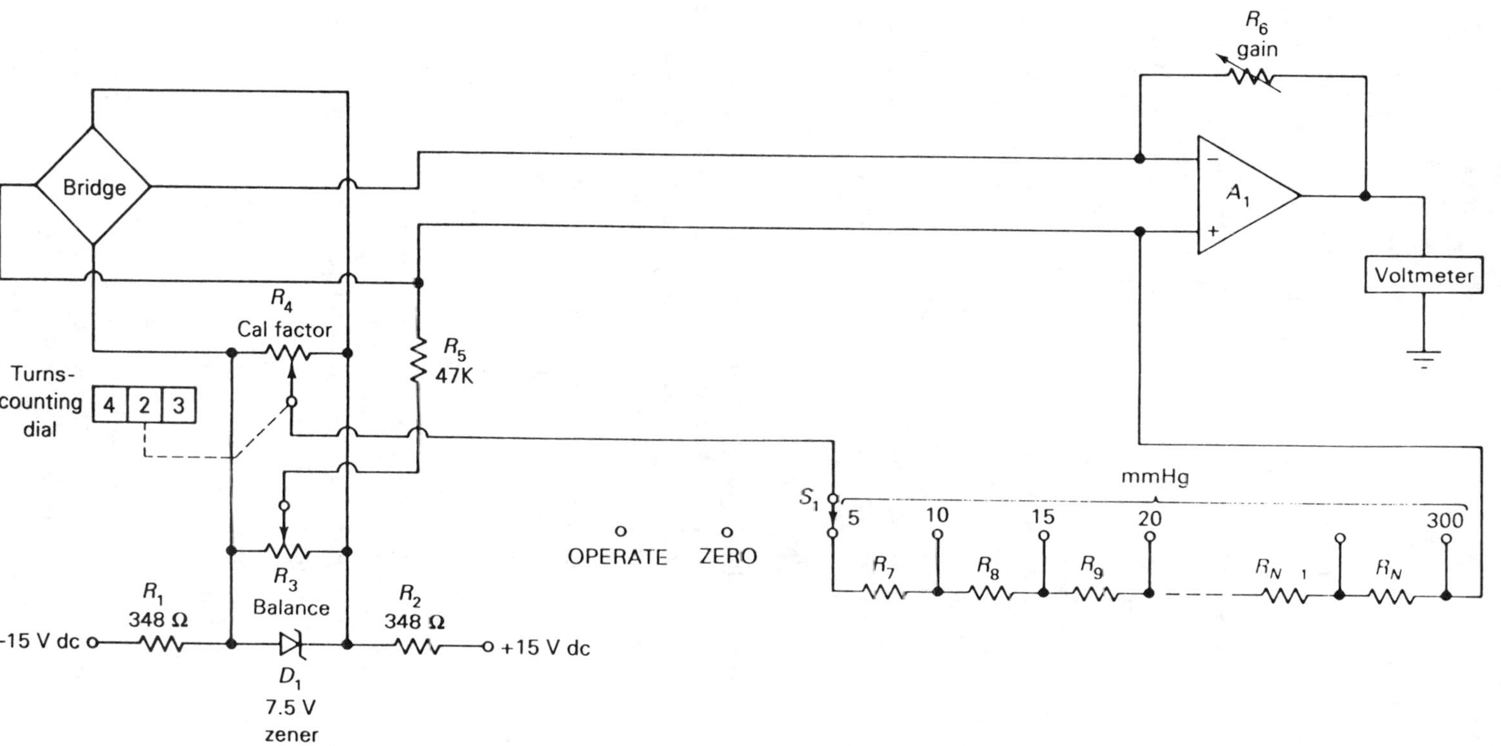

Figure 6–19 DC pressure amplifier schematic.

1. Calibrate the dc pressure amplifier with an accurate mercury manometer.
2. Set switch S_1 to the OPERATE position, open the transducer stopcock to the atmosphere, and adjust R_3 for 0 V output (i.e., 0 torr reading on the display).
3. Close the transducer stopcock and then pump a standard pressure (e.g., 100 torr, or at least half-scale). Adjust gain control R_6 until the meter reads the correct (standard) pressure. Check for agreement between the meter and the manometer at several standard pressures throughout the range (e.g., 50, 100, 150, 200, 250, and 300 torr). This step is needed to ensure that the transducer is reasonably linear, that is, that the diaphragm was not strained by out-of-range pressures or vacuums.
4. Turn switch S_1 to the position corresponding to the applied standard pressure, and then adjust the CAL FACTOR control R_4 until the same standard pressure is obtained on the meter. The CAL FACTOR control is ganged to a turns-counting dial. The number appearing on the dial at the position of R_4 that creates the same standard pressure signal is the calibration factor for that transducer. Record the turns-counter reading for future reference.

For a period of time (usually 6 months) the calibration factor will not need to be redetermined unless damage to the transducer occurs. The calibration factor is entered into the amplifier by turning R_4 (or digitally in modern instruments). The following procedure is normally used:

1. Open the transducer stopcock to the atmosphere, place switch S_1 in the 0 torr (0 mm Hg) position, and adjust R_3 for a 0 V output indication (0 torr on the display).
2. Set switch S_1 to a position that is most convenient for the range of pressures to be measured. In general, select a scale that places the reading midscale or higher. Set the CAL FACTOR knob to the figure recorded previously.
3. Adjust the gain control R_6 until the meter reads the standard pressure used in the original calibration.

Amplifier A_1 is the input amplifier, and it should be a low-drift, premium model. Both gain and zero controls are provided, so the amplifier will work with a wide variety of transducers.

The excitation voltage of the transducer is determined (as a maximum) by the transducer manufacturer, with a value of 10 V being common. In general, it is best to operate the transducer at a voltage lower than the maximum to prevent drift due to self-heating. Pressure amplifier manufacturers typically specify either 5 V or 7.5 V for a 10 V (max) transducer.

The required amplifier gain can be calculated from the required output voltage that is used to represent any given pressure. Because digital voltmeters are used extensively for readout displays, the common practice is to use an output voltage scale factor that is numerically the same as the full-scale pressure. For example, 1 mV or 10 mV/mm Hg is common. Thus, with a maximum pressure range of 400 mm Hg (common on arterial monitors) and a scale factor of 1 mV/mm Hg, the output voltage will be 400 mV, or 0.40 V, for a pressure of 400 mm Hg. No further scaling of the meter output is needed.

The sensitivity and the excitation potential give the transducer output voltage at full scale. For example, let's assume that a 400 mm Hg pressure amplifier, scaled at 1 mV/mm Hg is desired. What is the output voltage of the transducer if a +5 V dc excitation is applied, and the sensitivity is 50 μV/V/mm Hg?

$$V_t = \frac{50\ \mu\text{V}}{\text{V} \cdot \text{mm Hg}} \times (5\ \text{V}) \times (400\ \text{mm Hg}) \tag{6-9}$$

$$V_t = (50\ \mu\text{V}) \times (5) \times (400) = 100{,}000\ \mu\text{V} = 100\ \text{mV} \tag{6-10}$$

The output of the transducer at full scale will thus be 100 mV. This potential is the amplifier input voltage, so the gain can be calculated:

$$A_v = \frac{V_o}{V_{\text{in}}} \tag{6-11}$$

$$A_v = \frac{400\ \text{mV}}{100\ \text{mV}} = 4 \tag{6-12}$$

A variation on the dc amplifier scheme is the isolated dc amplifier, which provides a very high impedance (10^{12} ohms or more) between the input and the dc power supply terminals. This electrical isolation is required in medical applications for patient safety.

An example of a pulsed excitation amplifier is shown in Fig. 6–20(a). The Wheatstone bridge strain gage is excited with a short-duration biphasic pulse instead of dc. This method allows the transducer to be excited to a voltage high enough to make a measurement possible without a constant flow of current to aggravate the self-heating problem inherent in dc transducers. The pulse typically has a 1 ms duration and a 25 percent duty cycle (which translates to a 4 ms period, or a frequency of 250 Hz). An advantage of the short duty cycle is that operations like amplifier drift cancellation can be incorporated.

Amplifier A_1 is a *DC* pressure amplifier, while A_2 is a unity-gain summation stage. The output signal indicator is a digital voltmeter that will update the display only when the strobe (STR) line is HIGH. Switches S_1 through S_3 are CMOS electronic switches that close when the control line C is HIGH. All circuit action is controlled by a four-phase clock. Phases P_1 and P_2 excite the transducer and operate the amplifier drift cancellation cir-

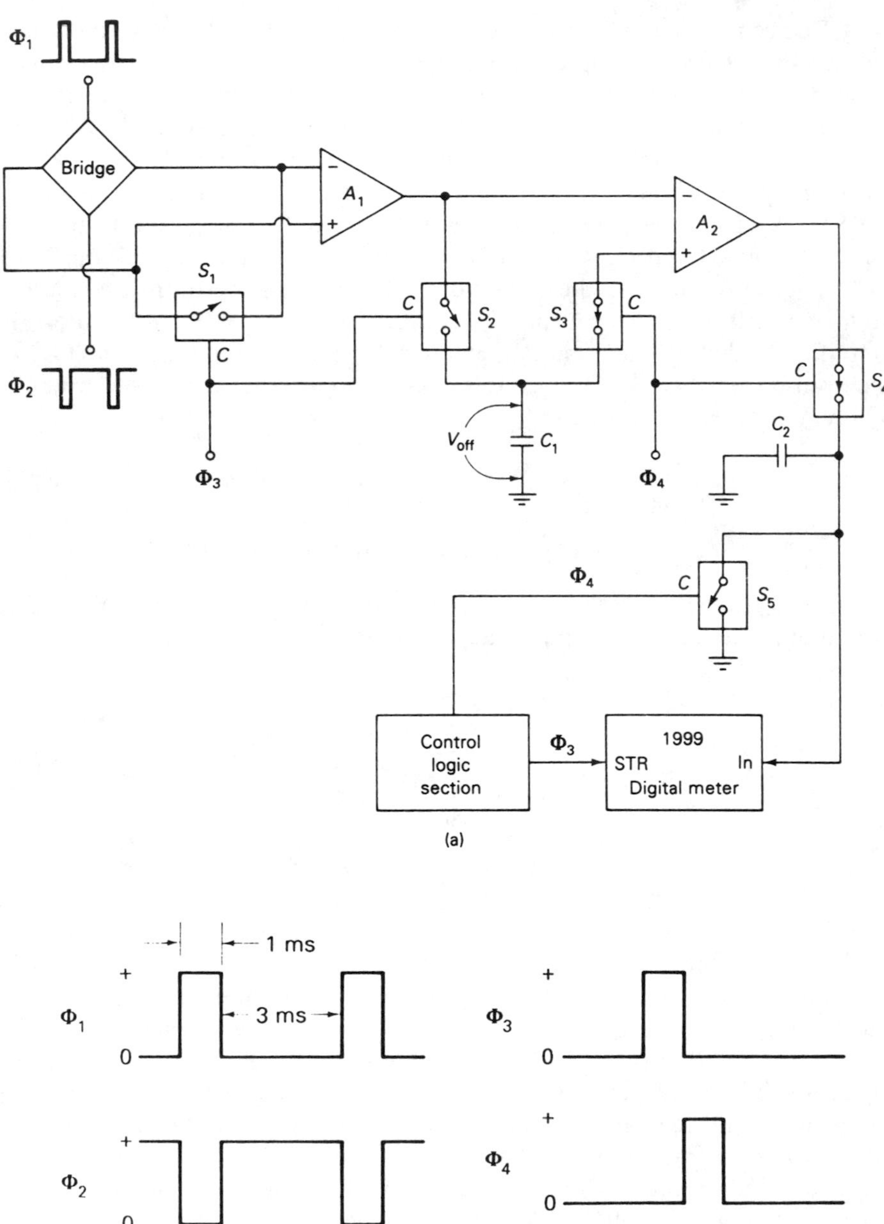

Figure 6–20 (a) Pulsed pressure amplifier circuit; (b) timing waveforms.

cuit. Phase P_3 updates the display meter, and phase P_4 resets the circuit following the update.

All dc amplifiers tend to drift (although modern premium IC op-amps have sharply reduced drift), that is, they create output voltage offset voltages due to thermal changes. Capacitor C_1 and amplifier A_2 serve as a drift cancellation circuit [Fig. 6–20(a)].

The transducer is excited only when P_1 is HIGH positive and P_2 is HIGH negative. At all other times the transducer is not excited, which keeps transducer self-heating to a minimum. Amplifier A_1 will drift, however, because of its high gain and inherent offset voltages. The purpose of this circuit is to charge capacitor C_1 during the nonexcited period (during which A_1 input is shorted by S_1) and then add the capacitor voltage algebraically to the signal, thereby removing the amplifier drift component.

Ac carrier amplifiers (Fig. 6–21) use an ac signal for transducer excitation, so they will operate equally well with resistive strain gages and inductive transducers. The carrier frequencies are typically 200 to 5000 Hz, with 400 and 2400 Hz being the most common. The Hewlett-Packard 8800-series pressure carrier amplifiers, for example, produce a 2400 Hz, a 5 V_{rms} signal. Carrier amplifiers are probably the most stable on the market because of their narrow bandwidth and heavy feedback design. Some inexpensive carrier amplifiers use simple envelope detectors to extract the pressure waveform, but all proper instruments use a variant of the circuit shown in Fig. 6–21. This circuit uses a quadrature, or phase-sensitive, detector to extract the signal information.

PRESSURE PROCESSING

Only rarely is a simple pressure amplifier necessary for dynamic measurements (the same is not true where static pressures are involved). Such a system is shown in Figure 6–22. The pressure waveform P is the analog output of a pressure amplifier, and it is fed to four different circuits: a peak detector (maximum pressure), inverted peak detector (minimum pressure), a time integrator, and a differentiator (dP/dt).

The peak detectors and integrator can be calibrated by applying a constant value of pressure P. But since the differentiator measures dP/dt, a varying signal is necessary. Typically, a sawtooth or ramp of the same amplitude as a standard value of the P-signal is applied, as in Figure 6-23. This signal produces an output from the differentiator (as shown) that is a constant voltage level (i.e., dP/dt = constant) that can be measured.

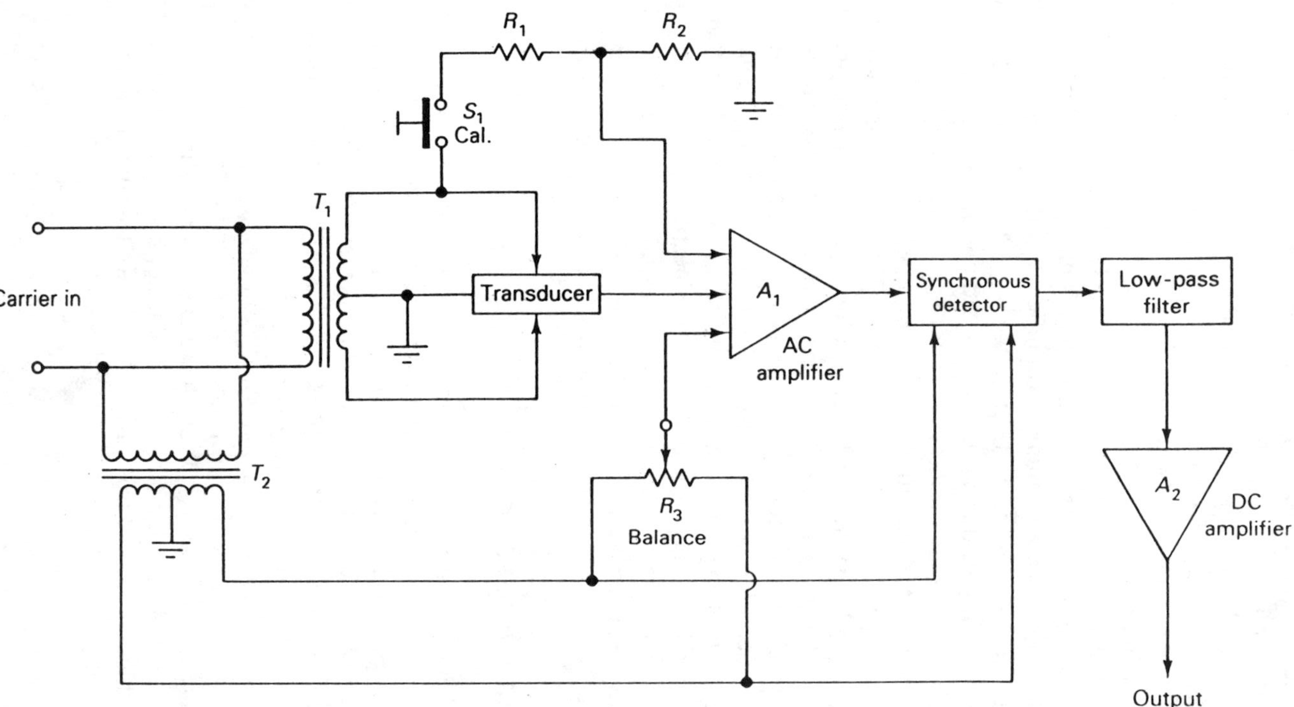

Figure 6–21 AC carrier amplifier circuit.

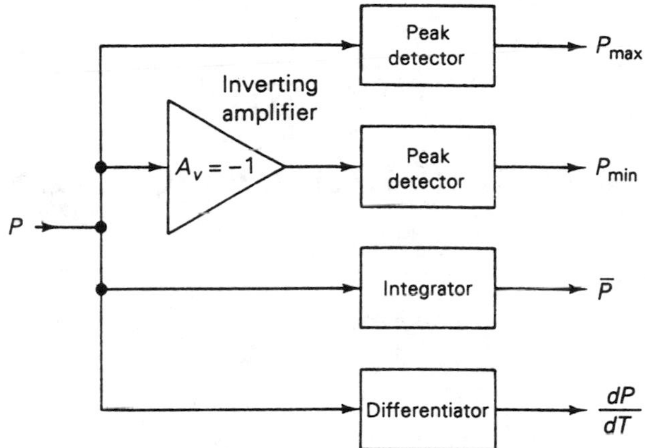

Figure 6–22 Postprocessing for pressure amplifier.

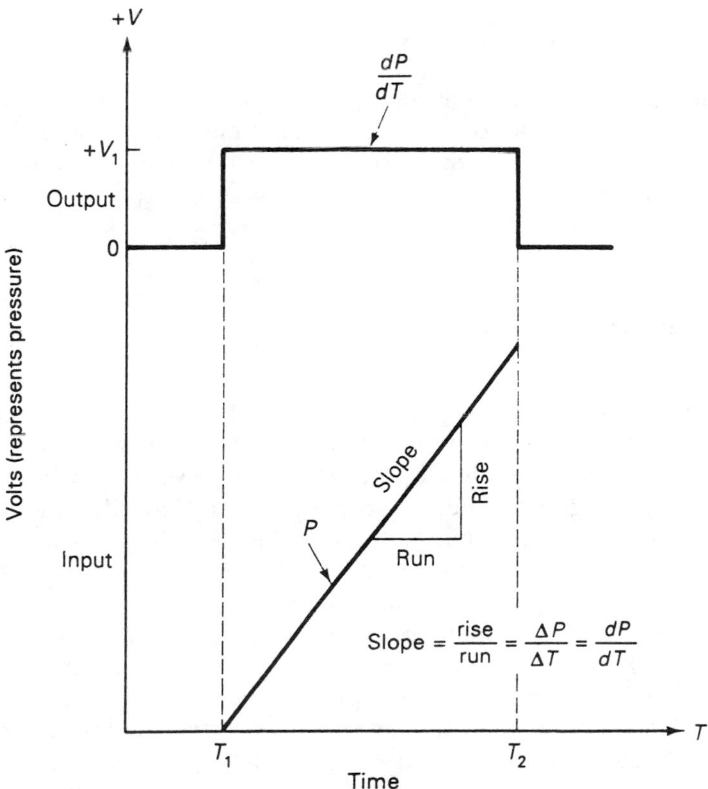

Figure 6–23 Calibrating the *dP/dt* circuit.

7 Vibration and Acceleration Sensors

Acceleration and vibration sensors are often the same basic instrument, although there are examples of each that will not perform the other job. Because of the considerable overlap, however, it is reasonable to consider them in the same discussion.

ACCELERATION

The principle of acceleration is firmly ensconced in Newtonian physics and, although replaced early in the twentieth century by the New Physics, remains viable today. *Acceleration* is the rate of change of velocity with respect to time. The forcc of gravity produces acceleration, and indeed Einstein insisted on an *equivalence principle* that maintained that gravity is an acceleration. In Newton's famous equation, $F = Ma$ the acceleration term a is replaced by the gravitational constant G in appropriate problems.

Acceleration is a vector quantity, so it has both magnitude and direction. One cannot talk properly of an acceleration of, say, 1 g, without specifying in which direction the acceleration operates.

The units of acceleration are meters per second squared (m/s^2) or radians per second squared (rad/s^2). In the English system the units are often expressed in terms of inches per second squared ($in./s^2$) or feet per second squared (ft/s^2). Another unit used for acceleration is g, 9.806.65 m/s^2, which

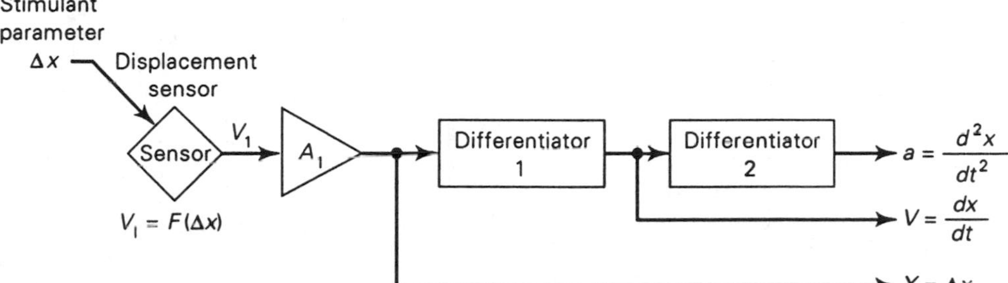

Figure 7–1 Generating velocity and acceleration signals from a displacement transducer.

is often reduced to 9.81 m/s^2 where the greater precision is not needed. The English system equivalent is 386.087 in./s^2 (or 386 in./s^2), or 32.2 ft/s^2.

From Newtonian physics we know that the following relationships hold true:

$$v = \frac{dx}{dt} \tag{7–1}$$

and

$$a = \frac{dv}{dt} = \frac{d^2x}{dt^2} \tag{7–2}$$

where x represents position

We can therefore consider generating an acceleration signal from the output of a displacement sensor. Figure 7–1 shows the block diagram for a system that accomplishes this job. A displacement sensor is affected by a stimulant and creates an output signal V_1 that is proportional to the displacement Δx. This Δx signal is amplified or otherwise processed in an amplifier stage A_1 to become the $X = \Delta x$ signal. Under the right circumstances a velocity or displacement sensor can be used to measure acceleration.

SEISMIC ACCELEROMETERS

One of the most common forms of accelerometer is the *damped-mass system* of Fig. 7–2. This device consists of a mass M suspended between a spring and a damping device such as a viscous-fluid-filled dashpot. In the system described in Fig. 7–2, X_m is the position of the mass M, while X_c is the position of the accelerometer case. The spring obeys Hooke's law, so it has a restorative force kx, while for the damping factor the force expression is

$$F = \beta \dot{X}_m \tag{7–3}$$

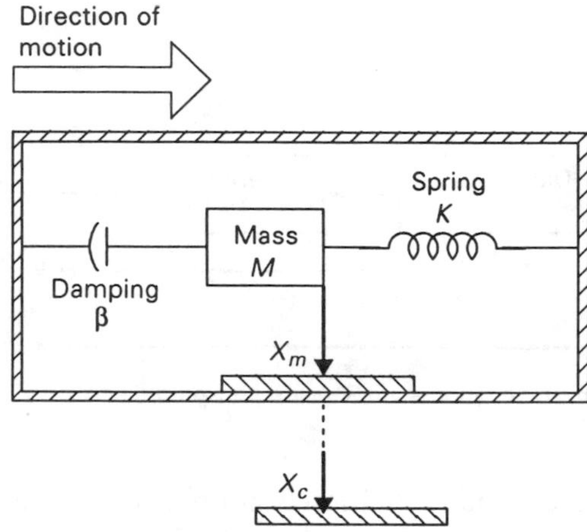

Figure 7–2 Mass accelerometer measurement.

The equations of the system in Fig. 7–2 are

$$M\left(\frac{d^2(X_c - X_m)}{dt^2}\right) = \beta\dot{X}_m + kX_m \tag{7–4}$$

$$M\ddot{X}_c = M\ddot{X}_m + \beta\dot{X}_m + KX_m \tag{7–5}$$

$$\frac{X_m}{\ddot{X}} = \frac{M/K}{\left[(m/K)\,\dfrac{d^2}{dt^2}\right] + \left[(\beta/K)\,\dfrac{d}{dt}\right] + 1} \tag{7–6}$$

But the second derivative of position is *acceleration*, so we can rewrite Eq. (7–6) in the form

$$\frac{X_m}{a} = \frac{M/K}{\left[(M/K)\,\dfrac{d^2}{dt^2}\right] + \left[(\beta/K)\,\dfrac{d}{dt}\right] + 1} \tag{7–7}$$

Thus, the expression involves the positions of the accelerometer mass and case, and the acceleration a. The mass will tend to oscillate with a frequency F (in hertz or in radians per second $2\pi F$, often written ω). For a system that is critically damped, the sensor will provide reasonably accurate data from zero to just below ω. Equation (7–6) can be rewritten in a form that takes this oscillation into effect:

$$\frac{X_m}{a} = \frac{M/K}{\left(\dfrac{1}{\omega}\right)^2 \dfrac{d^2}{dt^2} + \left(\dfrac{2Z}{\omega}\right)\dfrac{d}{dt} + 1} \tag{7–8}$$

Z is the mechanical impedance of the system, usually called a *damping constant.* The value of Z gives some information about the characteristics of the sensor:

- $Z = 1$: Critical damping; denominators of the equations contain a pair of repeated roots.
- $Z < 1$: Undercritically damped; denominators contain two complex roots.
- $Z > 1$: Overcritically damped; denominators contain two real roots.

Some means of transduction is needed to produce an output signal from the bouncing mass. In Fig. 7–3 the electrical signal is produced by a potentiometer connected to electrical source potentials. When the mass is at rest, the wiper on the potentiometer is at the midpoint, so the output potential V_o is zero. When the mass moves upward, the potential goes positive, and when it moves down, the potential goes negative.

Another form of accelerometer tranduction is shown in Fig. 7–4. The seismic mass M is mounted to the case through a pair of wires that are stretched taut. These wires are considered electrical resistances, so the sensor can be used in a piezoresistive strain-gage type of circuit, such as the Wheatstone bridge. When the mass accelerates along the sensor axis, one wire is placed under greater tension, while the other wire is placed under less tension. The oscillating frequency of the mass is proportional to the tension:

$$T = 4\pi L^2 r^2 \rho F^2 \qquad (7\text{–}9)$$

An inductive sensor placed close to the mass will provide the frequency information.

A *dual-spring accelerometer* is shown in Fig. 7–5. In this case the mass is a permanent magnet balanced between two springs. An inductive coil is placed adjacent to the coil, so that when the magnet moves, an electrical

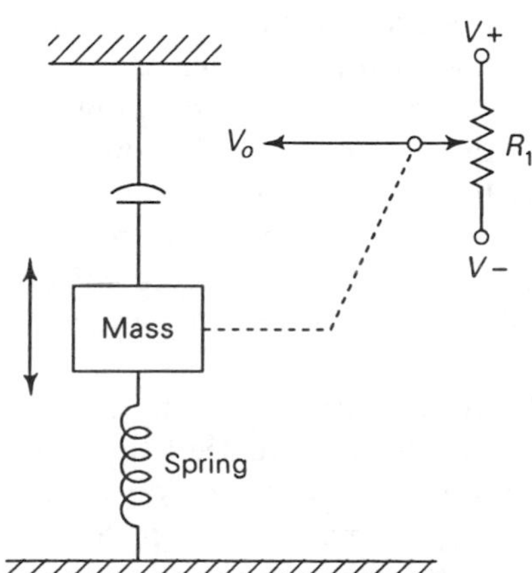

Figure 7–3 Deriving an electrical output signal.

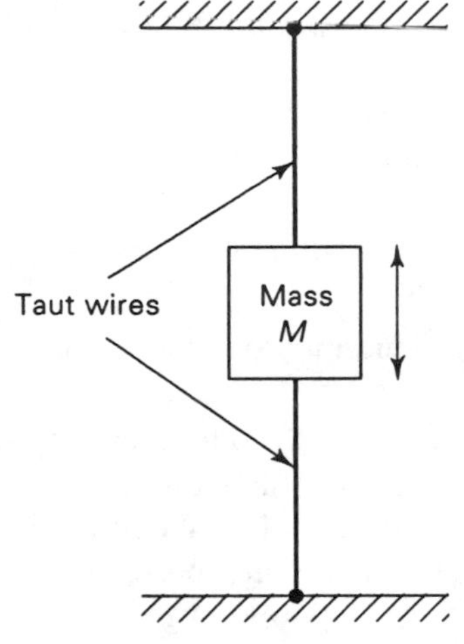

Figure 7–4 Taut-wire accelerometer.

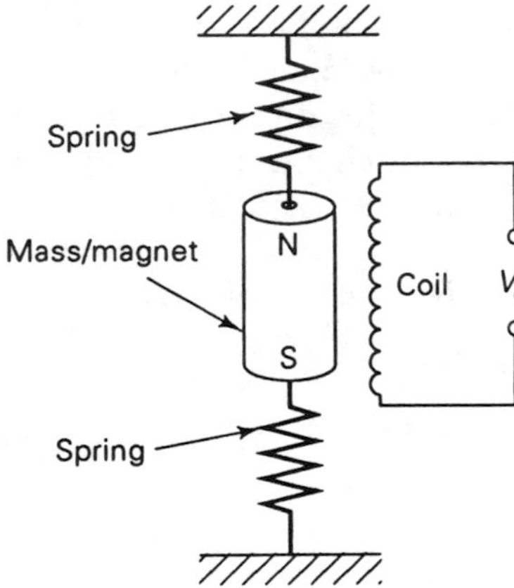

Figure 7–5 Inductive pickup for mass accelerometer.

current is induced in the coil. When the mass vibrates, therefore, an oscillating voltage will appear across the ends of the coil. Another application of this same principle is shown in Fig. 7–6. The mass/magnet is placed inside the coil form. The springs shown in Fig. 7–6 are cantilever leaf springs attached to a housing or container. In simple vibration sensors the mass inside the coil is a magnetically permeable core that changes the inductance of the coil, rather than inducing current in the coil. Such sensors are used with oscillator circuits where the inductance of the coil is part of the frequency-setting network.

Strain gages (discussed in Chapter 6) are used for a wide variety of sensors. In Fig. 7–7 we see a strain gage used in an accelerometer sensor. A seismic mass is suspended on a cantilever spring attached to the sensor wall. When the mass moves, it flexes the spring, placing one strain gage in tension and the other in compression. The strain-gage elements are usually

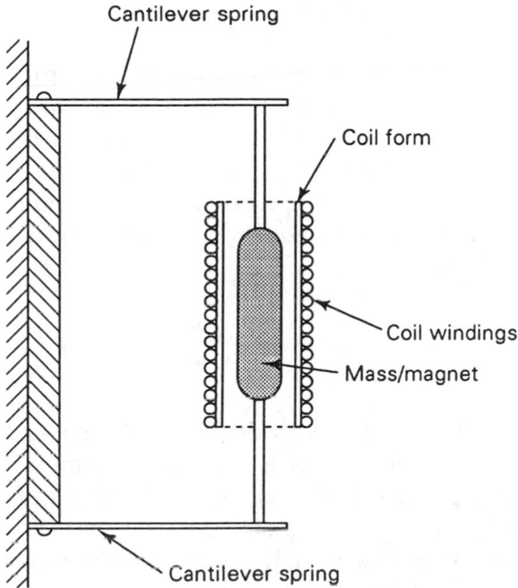

Figure 7–6 Alternative inductive pickup method.

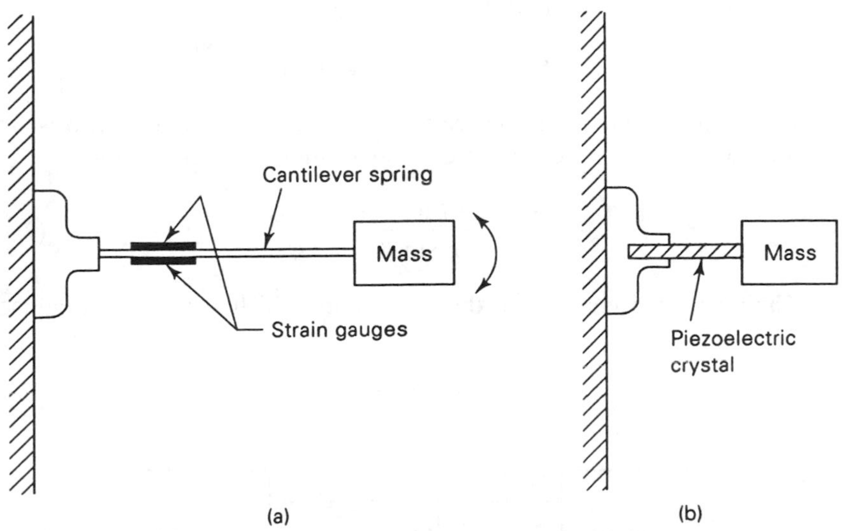

Figure 7–7 Cantilever-beam accelerometer using (a) piezoresistive sensors; (b) piezoelectric sensor element.

used in Wheatstone bridge circuits, so they produce an output that is proportional to the acceleration of the mass.

A *double cantilever* or *bridged cantilever* accelerometer is shown in Fig. 7–8. Some of the more modern accelerometers based on this method are piezoresistive silicon (IC) based.[1] In these sensors the seismic mass is suspended from supports through a pair of strain-gage elements. These devices have good sensitivity at frequencies to 5 kHz and provide very good off-axis rejection. The resonant frequency of the seismic harmonic oscillating mass can be found from

$$F = \frac{1}{2\pi}\sqrt{\frac{k}{M}} \tag{7–10}$$

The spring constant k is force divided by displacement Δx. A parameter called the *transduction efficiency* β is used to determine the sensitivity ψ. Factors in determining β include the position and number of the piezoresistive elements, the doping profile, doping elements, and aspect ratio (L/W) of the elements. Typical values of sensitivity, expressed in microvolts per gram of acceleration per volt of excitation potential, range from 5 μV/g/V to 5 mV/g/V. The sensitivity is

$$\psi = \frac{\beta M}{k} = \frac{(2\pi)^2\beta}{F^2} \tag{7–11}$$

A pendulum accelerometer is shown in Fig. 7–9. In this form of sensor a mass is hanging from a pendulum of length L.

$$F = Ma = -mg \sin \theta \tag{7–12}$$

For a simple pendulum, where the displacement from center S is small compared with length L, the ratio of displacement to acceleration is

$$\frac{S}{a} = \frac{L\theta}{-g \sin \theta} = -\frac{L}{g} \tag{7–13}$$

The resonant frequency of the pendulum (which accounts for its use in

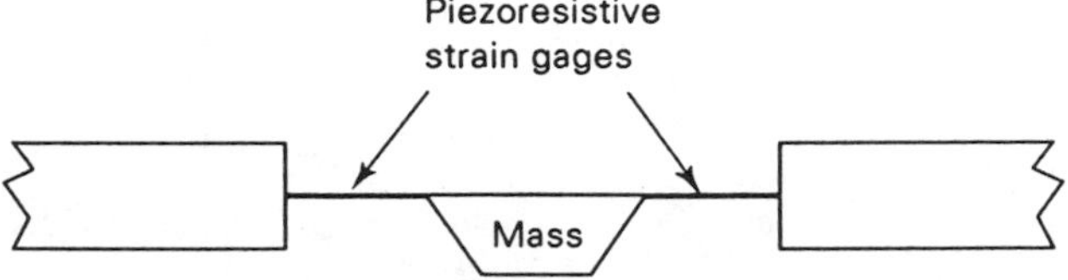

Figure 7–8 Bridge cantilever accelerometer.

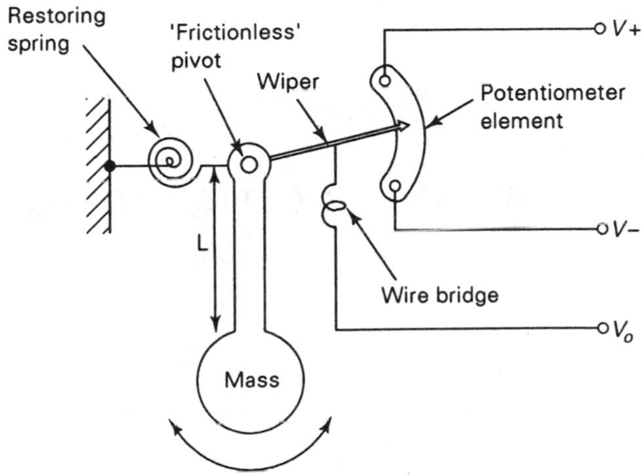

Figure 7–9 Pendulum accelerometer.

clocks since the time of Galileo) is

$$T = 2\pi\sqrt{\frac{L}{g}} \tag{7–14}$$

for small angular deflections.

The electrical transduction element in the pendulum acceleration sensor is a potentiometer element packaged with the pendulum assembly. When the pendulum is at rest, the potentiometer wiper is at the midpoint, and the output voltage is zero. Motion of the pendulum causes the voltage to vary from positive to negative in a nearly sinusoidal waveform.

REFERENCE

1. Allen, Henry V., Stephen C. Terry, and James W. Knutti. IC Sensors. "Understanding Silicon Accelerometers." *Sensors* (September 1989) 17ff.

8 Electrodes for Bioelectric Sensing

Bioelectricity is a naturally occurring phenomenon that arises because living organisms are composed of positive and negative ions in various quantities and concentrations. Ionic electrical conduction is different from *electronic conduction*, which is perhaps more familiar to engineers and technologists. *Ionic conduction* involves the migration of ions—positively or negatively charged atoms or molecules—through a region, whereas *electronic conduction* involves the flow of electrons under the influence of an electrical field. In an *electrolytic solution*, ions are easily available. Electrical potential differences occur when the concentration of ions is different between two points.

Ionic conduction is a very complex, nonlinear phenomenon. But for small-signal applications, where there is only a very small—indeed, minuscule—current flowing, especially in the context of sensor electrodes, modeling it as a flow of ordinary current between points of potential difference is a reasonable first-order approximation. Chemists would find this model wanting, however, except in the most elementary classes, but their need for understanding is deeper than that of the typical instrumentation specialist or end user. Keep in mind, however, that situations where a more substantial current flows across an electrode-to-tissue interface change the situation entirely, and a more sophisticated higher-order model is needed.

Bioelectrodes are a class of sensors that transduce ionic conduction into electronic conduction so that the signal can be processed in electronic circuits. The usual purpose of bioelectrodes is to acquire medically significant bioelectrical signals such as *electrocardiograms* (ECGs), *electroencephalograms* (EEGs), and *electromyograms* (EMGs). Both clinical and research examples are easily found, although in many cases the two are indistinguishable from one other. Most such bioelectrical signals are acquired from one of three forms of electrodes: *surface macroelectrodes, indwelling macroelectrodes*, and *microelectrodes*. Of these, the first two are generally used in vivo, while the latter are used in vitro.

In this chapter we discuss the acquisition of biopotentials by the types of electrodes commonly used in biomedical and other life sciences application. Again, it should be recognized that this discussion is generic and representative, not exhaustive, for the subject can be quite complex.

SOURCE OF BIOPOTENTIALS

Biopotentials exist in organisms because the body is made up of cells resting in an electrolytic fluid consisting mostly of sodium, potassium, and chloride ions. The differential concentrations of these ions inside and outside each cell create a potential gradient across the membrane that makes up the cell wall. For a typical human cell, the transmembrane potential is between 70 and 90 mV, with the inside of the cell being negative with respect to the outside environment.

ELECTRODE POTENTIALS

The skin and other tissues of higher-order organisms, such as humans, are electrolytic, so they can be modeled as an electrolytic solution. In some models the solution is shown as saline, reflecting the fact that we are very similar to salt water in our bodily composition. Imagine a metallic electrode immersed in an electrolytic solution [Fig. 8–1(a)]. Almost immediately after immersion, the electrode begins to discharge some metallic ions into the solution, and some of the ions in the solution start combining with the metallic electrode. (Incidentally, this is the phenomenon at the heart of the electroplating and anodizing processes.)

After a short while, a *charge gradient* builds up, creating a potential difference, or *electrode potential* [V_e in Fig. 8–1(a)], or *half-cell potential*. Keep in mind that this potential difference can be due to either differences in concentration of a single ion type or of two dissimilar ions. For example, if there are two positive ions (+ +) in location A, and three positive ions (+ + +) in location B, then there will be a net difference of 3 − 2, or 1, with point B being more positive than point A. Two basic reactions can take place at the electrode/electrolyte interface: an *oxidizing reaction*, in which a metal gives up electrons to form metal ions; and a *reduction reaction* in which electrons combine with metal ions to form the metal.

A complex phenomenon is seen at the interface between the metallic electrode and the electrolyte. Ions migrate toward one side of the region or

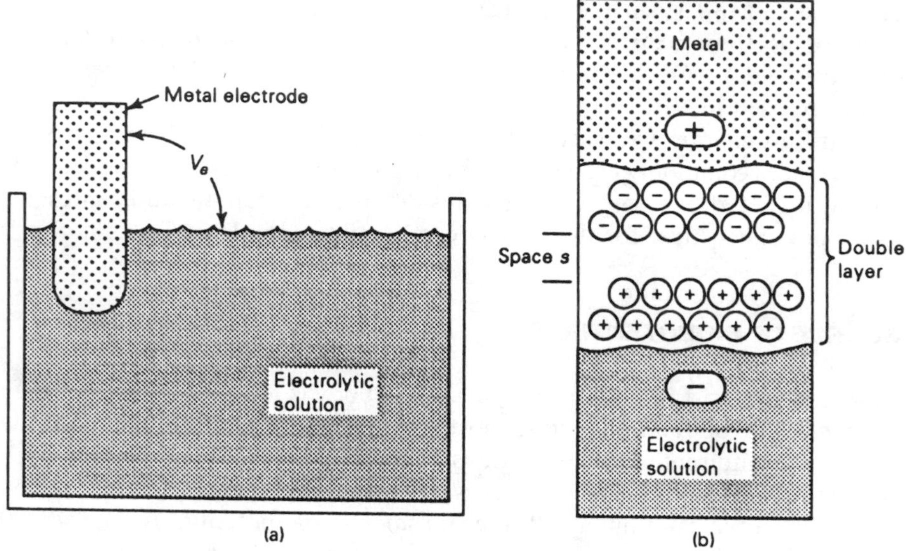

Figure 8–1 (a) A potential develops between a metal electrode and the electrolytic solution in which it is immersed; (b) ionic migration that generates the half-cell potential.

TABLE 8–1 Half-Cell Potentials of Common Elements

Material	Half-Cell Potential (V)
Aluminum (Al^{3+})	−1.66
Zinc (Zn^{2+})	−0.76
Iron (Fe^{2+})	−0.44
Lead (Pb^{2+})	−0.12
Hydrogen (H^+)	0
Copper	+0.34
Silver (Ag^+)	+0.80
Platinum (Pt^+)	+0.86
Gold (Au^+)	+1.50

the other [Fig. 8–1(b)], forming two parallel layers of ions of opposite charge. This region is called the *electrode double layer*, and its ionic differences are the source of the electrode or half-cell potential V_e. Different materials exhibit different half-cell potentials, as shown in Table 8–1.

By international scientific agreement, the standard electrode against which half-cell potentials are measured is the *hydrogen electrode* (Pt, H_2; H^+), which is assigned a half-cell potential of 0 V by convention. The standard hydrogen electrode consists of gaseous hydrogen bubbled over a piece of platinized platinum immersed in a hydrochloric acid solution. The half-cell potentials cited for any given electrode are actually the differential potential between the electrode and the standard hydrogen reference electrode.

Now let's consider what happens when two electrodes (call them *A* and *B*), made of *dissimilar metals*, are immersed in the same electrolytic solution (Fig. 8–2). Each electrode will exhibit its own half-cell potential, V_{ea} and V_{eb}, respectively. Because the two half-cell potentials are different, there is a net potential difference V_{ed} between them, which causes an electronic current I_e to flow through an external circuit. The differential potential, sometimes called an *electrode offset potential*, is only a first-order approximation for the small-signal case and is defined as

$$V_{ed} = V_{ea} - V_{eb} \tag{8–1}$$

For example, in the case where a gold (Au^+) electrode and a silver (Ag^+) electrode are immersed in the same electrolyte,

$$V_{ed} = V_{e(\mathrm{Au})} - V_{e(\mathrm{Ag})} \tag{8–2}$$
$$V_{ed} = (+1.50 \text{ V}) - (+0.80 \text{ V}) = +0.70 \text{ V}$$

Or, in the frequently seen case of copper (Cu^{2+}) and silver (Ag^+), which can exist erroneously in electronic circuits that use copper connecting wires:

$$V_{ed} = V_{e(\mathrm{Ag})} - V_{e(\mathrm{Cu})} \tag{8–3}$$

$$V_{ed} = (+0.80 \text{ V}) - (+0.34 \text{ V}) = 0.46 \text{ V} \tag{8–4}$$

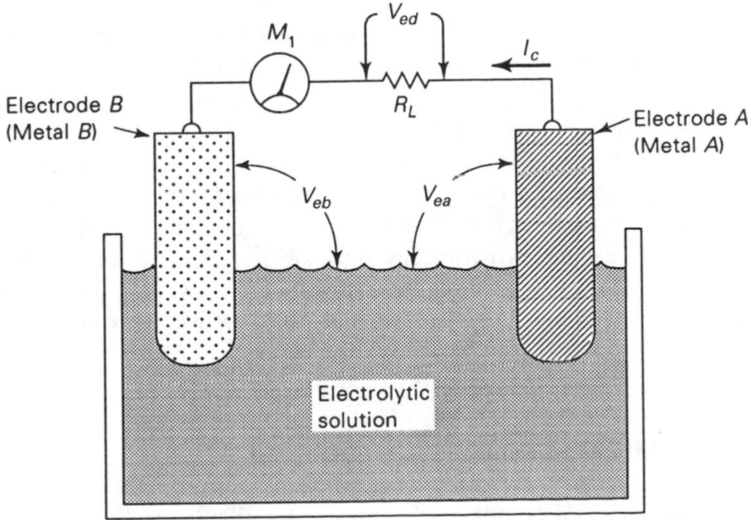

Figure 8–2 Measuring the half-cell potential between two different electrode materials.

The electrode offset potential will be zero when the two electrodes are made of identical materials, which is the usual case in bioelectric sensing.

Care must be given to the selection of materials when designing electrodes for bioelectric sensing. The choice of materials, as noted above, will affect the half-cell and offset potentials. Besides having an initial materials dependency, the actual half-cell potential exhibited by any given electrode may change slowly with time (hours to days). Some candidate materials look good initially but change so much with time and chemical environment that they are rendered almost useless in practical applications. Such a phenomenon is called *long-term drift*.

There are two general categories of electrode material combinations. A *perfectly polarized* or *perfectly nonreversible* electrode is one in which there is no net transfer of charge across the metal/electrolyte interface; in these only one of the two possible types of chemical reactions can occur. A *perfectly nonpolarizable* or *perfectly reversible* electrode is one in which there is an unhindered transfer of charge between the metal of the electrode and the electrode. Although these idealized situations are not obtained in reality, care must be given to selecting the right electrode. In general, a reversible electrode such as *silver–silver chloride* (Ag–AgCl) is used.

Body fluids are very corrosive, so not all materials are acceptable for bioelectric sensing. In addition, some materials that form reversible electrodes (e.g., zinc–zinc sulfate) are toxic to living tissue. For these reasons, materials such as the noble metals (e.g., gold and platinum), some tungsten alloys, silver–silver chloride (Ag–AgCl), and a material called platinum–platinum black are used to make practical biopotentials electrodes. In general medical use for simple surface recording of biopotentials, the Ag–AgCl electrode is most often used. Unless otherwise specified, it can generally be assumed that this material is used in clinical electrodes.

Figure 8–3 shows why the Ag–AgCl electrode is so popular with medical instrument designers. This electrode consists of a body of silver onto which a thin layer of silver chloride is deposited. The AgCl provides a free two-way exchange of Ag^+ and Cl^- ions, so that (ideally) no double layer forms. Thus, the electrode potential is very small. In the manufacture of Ag–AgCl electrodes it is necessary to use spectroscopically pure silver. Such silver is 0.99999 fine (i.e., 99.999 percent pure), compared with ordinary jeweler's and silversmith's silver, which is 0.999 fine (i.e., 99.9 percent pure) (Note: Sterling silver is 92.5 percent 0.999-fine silver, and 7.5 percent copper).

ELECTRODE MODEL CIRCUIT

Figure 8–4(a) shows a circuit model of a biomedical surface electrode; Fig. 8–4(b) shows the equivalent physical situation. This model more or less matches the equivalent circuit of ECG and EEG electrodes. In this circuit

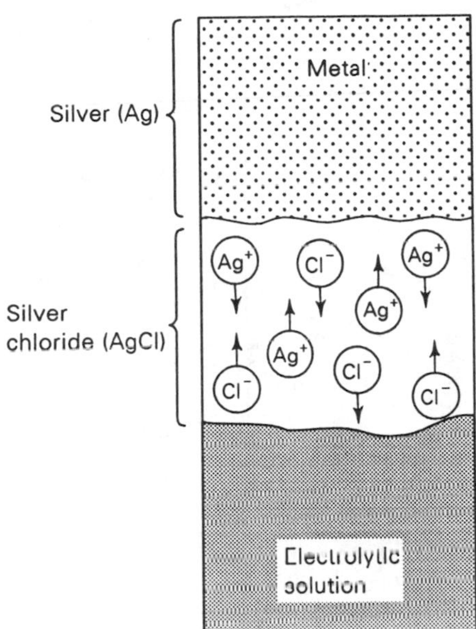

Figure 8–3 Ionic structure of Ag–AgCl electrodes.

a differential amplifier is used for signal processing, so it cancels the effects of electrode half-cell potentials V_1 and V_2. Resistance R_c represents the internal resistances of the body, which are typically quite low. The biopotentials signal is represented as a differential voltage V_d. The other resistances in the circuit represent the resistances at the electrode/skin contact interface. The surprising aspect of Fig. 8–4(a) is the usual values associated with capacitors C_1 and C_2. Although some capacitance is normally expected, it usually surprises people to learn that these contact capacitances can attain values of several microfarads. The reason is that the two layers of the double layer are equivalent to the two plates of a parallel-plate capacitor, and the spacing between them is very small. The small spacing creates a large capacitance.

Electrode Potentials and Recording Problems

Electrode half-cell potentials become a significant problem in bioelectric signal acquisition and recording because of the great difference between these dc potentials and normal biopotentials. For example, a typical half-cell potential for a biomedical electrode is 1 V or so, whereas biopotentials are less than 1/1000 the half-cell potential! The surface manifestation of the ECG signal is 1 to 2 mV, while EEG scalp potentials are on the order of 50 μV. Thus, the half-cell electrode voltage is 1000 times greater than the peak ECG potential, and 20,000 times greater than the EEG signal!

The instrument designer must provide a strategy for overcoming the effects of the very large difference between biopotentials and half-cell potential offset. Because the half-cell potential forms a large dc component of

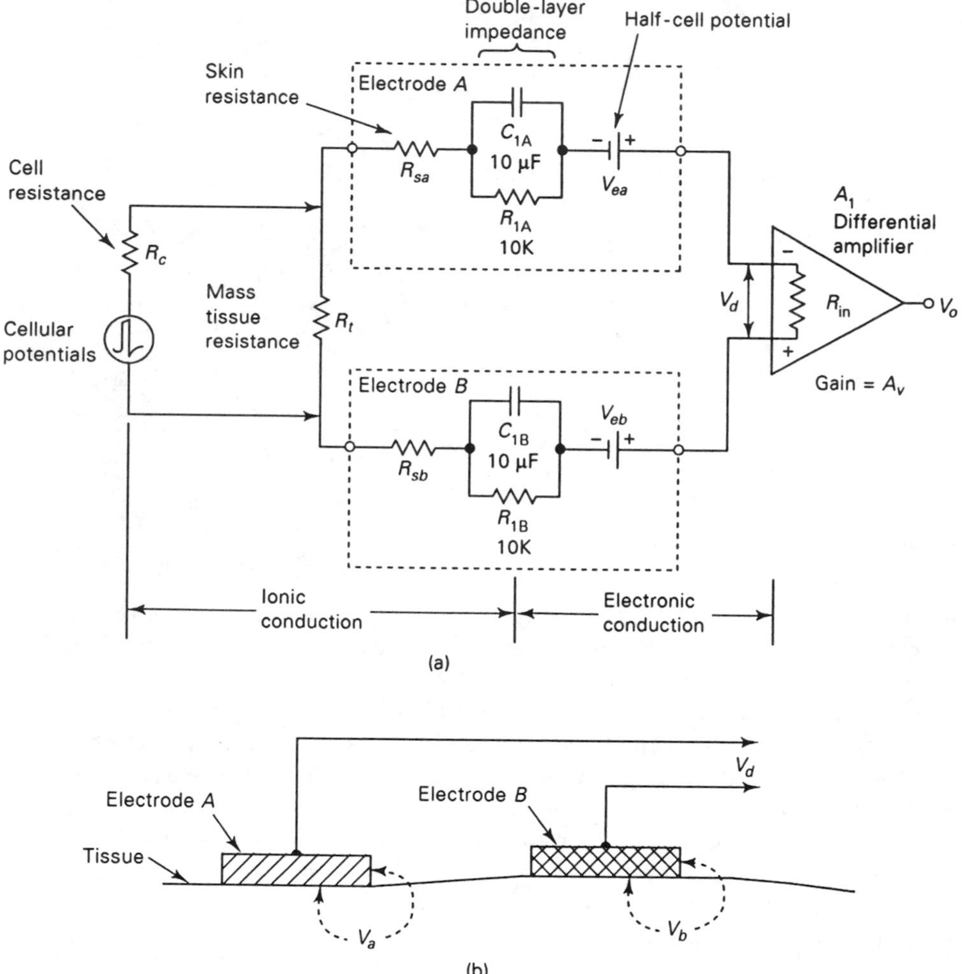

Figure 8–4 (a) Equivalent circuit model for a biomedical electrode. (b) If two electrodes are used, the differential potential is the algebraic sum of V_a and V_b.

the minute signal voltage, it is necessary to find an appropriate strategy that uses a combination of the following approaches.

First, a differential dc amplifier can be used to acquire the signal. If the electrodes are identical, then the two half-cell potentials should be the same. Theoretically, at least, the equal potentials would be seen by the differential amplifier as a single common-mode potential and thus would cancel in the amplifier output. A limitation on this approach is that the gains required to process low-level signals also act on tiny differences between the two half-cell potentials. A difference of 1 mV—which is only 0.1 percent of the total—looks like any other 1 mV dc signal to the gain of 1000 ECG amplifier.

Second, the signal acquisition circuit may be designed to provide a counteroffset voltage to cancel the half-cell potential of the electrode. Although this approach has certain initial appeal, it is limited by the fact that the half-cell potential changes with time and with the relative motion between skin and electrode. Electrode motion can cause a wildly varying dc baseline.

Third, the input amplifier can be ac-coupled. This approach permits removal of the signal component from the dc offset. This option is, perhaps, the most appealing, especially where variations of the dc offset are of substantially lower frequency than the signal frequency components. In that case the normal −3 dB frequency response limit can be used to tailor the attentuation of variations in the dc offset.

In some biomedical applications, however, signal components are near-dc. For example, the frequency content of the ECG signal is 0.05 to 100 Hz. In medical ECG equipment, therefore, the baseline can be expected to shift every time the patient moves around in bed and disturbs the electrodes.

In most cases the first and third options are selected for biopotential amplifiers. An ac-coupled differential input amplifier is required for signal acquisition.

SURFACE ELECTRODES

Surface electrodes are those that are placed in contact with the skin of the subject. Also in this category are certain needle electrodes of such a size that they cannot be inserted inside a single cell (the criterion that defines a microelectrode). There is some case for including needle electrodes in the category of "indwelling" electrodes, but that is not generally the practice in biomedical engineering.

Surface electrodes (other than needle electrodes) vary in diameter from 0.3 to 5 cm in diameter, with most being in the 1 cm range. Human skin tends to have a very high impedance compared with other voltage sources. Typically, normal skin impedance as seen by the electrode varies from 0.5K ohm for sweaty skin surfaces to more than 20K ohms for dry skin surfaces. Problem skin, especially dry, scaly, or diseased skin, may reach impedances in the 500K ohms range. In any event, surface electrodes must be treated as a very high impedance voltage source—a fact that seriously influences the design of biopotential amplifier input circuitry. In most cases the rule of thumb for a voltage amplifier is to make the input impedance of the amplifier at least 10 times the source impedance. For biopotential amplifiers this requirement means 5M ohms or greater input impedance—a value easily achieved using either premium bipolar, BiFET, or BiMOS op-amps.

Typical Surface Electrodes

A variety of electrodes have been designed for surface acquisition of biomedical signals. Perhaps the oldest form of ECG electrodes in clinical use are the strap-on variety [Fig. 8–5(a)]. These electrodes are 1 to 2 $in.^2$ brass plates that are held in place by rubber straps. A conductive gel or paste is used to reduce the impedance between electrode and skin.

A related form of ECG electrode is the suction cup electrode shown in Fig. 8–5(b). This device is used as a chest electrode in short-term ECG recording. For longer-term recording or monitoring, such as continuous monitoring of a hospitalized patient in a coronary or intensive care unit, the paste-on column electrode is used instead.

A typical column electrode is shown schematically in Fig. 8–5(c). The electrode consists of a Ag–AgCl metal contact button at the top of a hollow column that is filled with a conductive gel or paste. This assembly is held in place by an adhesive-coated foam rubber disk.

The use of a gel- or paste-filled column that holds the actual metallic electrode off the surface reduces movement artifact. For this reason (among several others), the electrode of Fig. 8–5(c) is preferred for the monitoring of hospitalized patients.

There are, however, several problems associated with this type of elec-

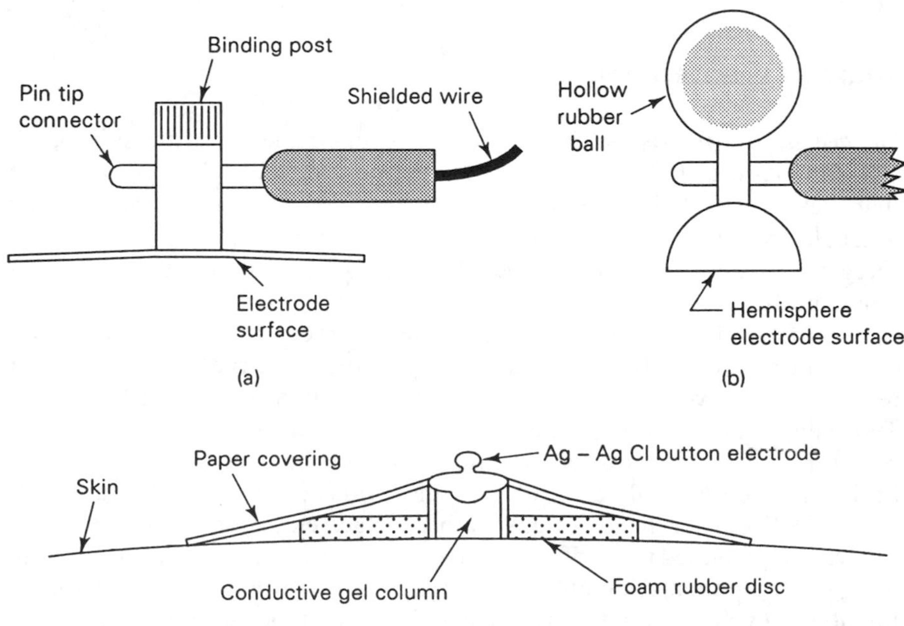

Figure 8–5 (a) Strap-on electrodes; (b) suction cup electrode; (c) gel-filled column electrode.

trode. One of the problems is the inability of the adhesive to stick for long on sweaty or clammy skin surfaces. The electrode also must not be placed over bony prominences. Usually the fleshy portions of the chest and abdomen are selected as electrode sites. Various hospitals have different protocols for changing the electrodes, but in general, the electrode is changed at least every 24 hours—and often more frequently, as few last as long as 24 hours. In some hospitals the electrode sites are moved and electrodes changed once every 8-hour nursing shift to prevent ischemia of the skin at the site.

The surface electrodes that we have discussed thus far are noninvasive types. That is, they adhere to the skin without puncturing it. Figure 8–6 shows a needle electrode. This type of ECG electrode is inserted into the tissue immediately beneath the skin by puncturing the skin at a large oblique angle (i.e., close to horizontal with respect to the skin surface). The needle electrode is used only for exceptionally poor skin, especially on anesthetized patients, and in veterinary situations. Of course, infection is an issue in these cases, so needle electrodes are either disposed of (after one use) or resterilized in ethylene oxide gas.

INDWELLING ELECTRODES

Indwelling electrodes are those that are intended to be inserted into the body. These are not to be confused with needle electrodes, which are intended for insertion into the layers beneath the skin. The indwelling electrode is typically a tiny exposed metallic contact at the end of a long insulated catheter (Fig. 8–7). In one application the electrode is threaded through the patient's veins (usually in the right arm) to the right side of the heart in order to measure the intracardiac ECG waveform. Certain low-amplitude, high-

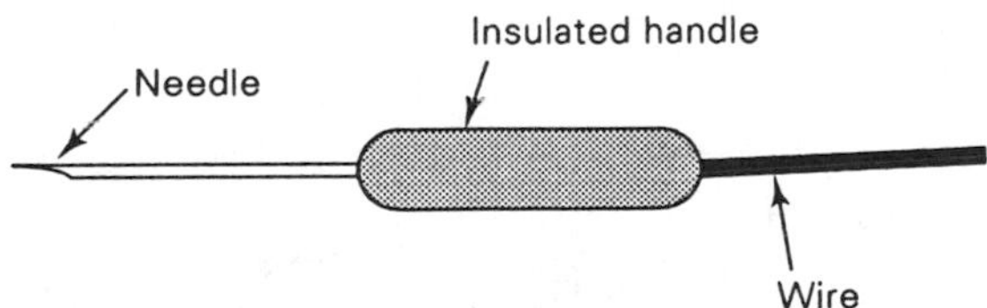

Figure 8–6 Needle or pin electrode.

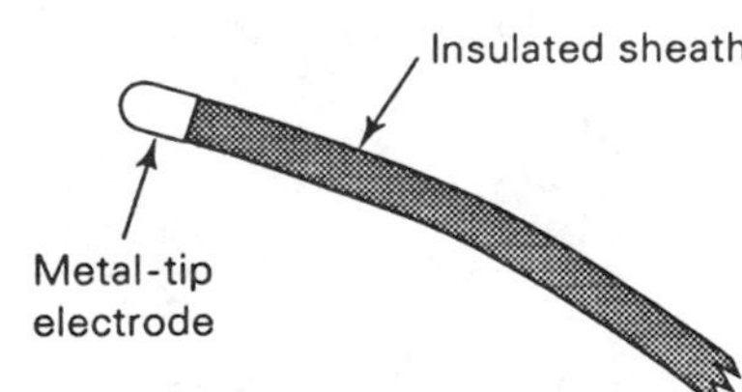

Figure 8–7 Indwelling electrode.

frequency features (such as the bundle of His element) become visible only when an indwelling electrode is used.

MICROELECTRODES

A *microelectrode* is an ultrafine device that is used to measure biopotentials at the cellular level (Fig. 8–8). In practice the microelectrode penetrates a cell that is immersed in an "infinite" fluid (such as physiological saline) that is in turn connected to a reference electrode. Although several types of microelectrodes exist, most of them are of one of two basic forms: metallic contact or fluid filled. In both cases, an exposed contact surface of about 1 to 2 μm is in contact with the cell. As might be expected, this fact makes microelectrodes very high impedance devices.

Figure 8–9 shows the construction of a typical glass-metal microelectrode. A very fine platinum or tungsten wire is slip-fit through a 1.5 to 2 mm glass pipette. The tip is etched and then fire-formed into the shallow-angle taper shown. The electrode can then be connected to one input of the signal amplifier.

There are two subcategories of this type of electrode. In one type the metallic tip is flush with the end of the pipette taper, while in the other a thin layer of glass covers the metal point. This glass layer is so thin as to require measurement in angstroms. It drastically increases the impedance of the device.

The fluid-filled microelectrode is shown in Fig. 8–10. In this type the glass pipette is filled with a 3 *M* solution of potassium chloride (KCl), and

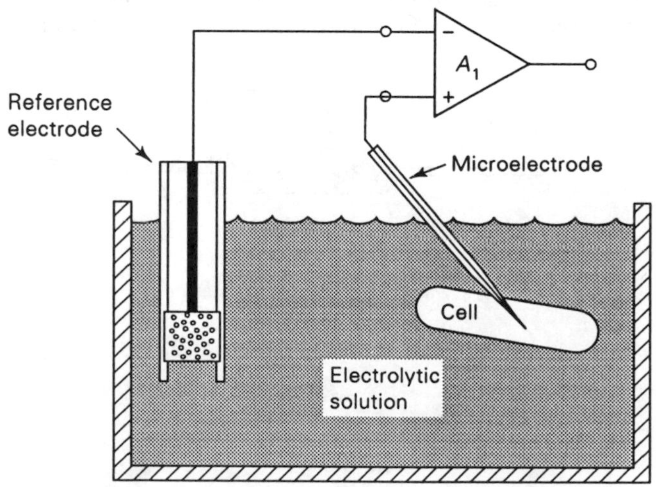

Figure 8–8 Microelectrode system for measuring cell potential.

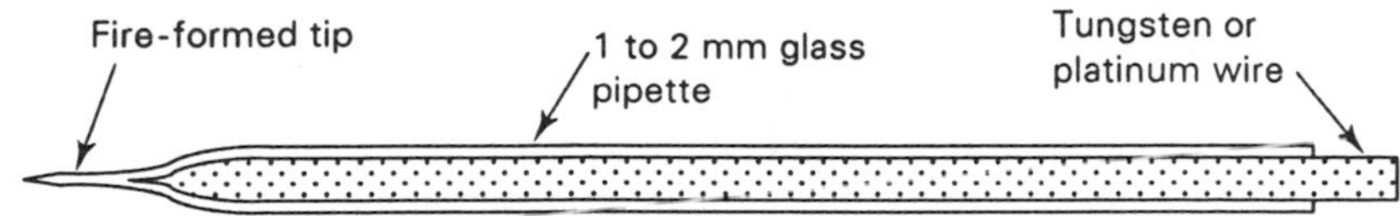

Figure 8–9 Platinum-wire needle electrode.

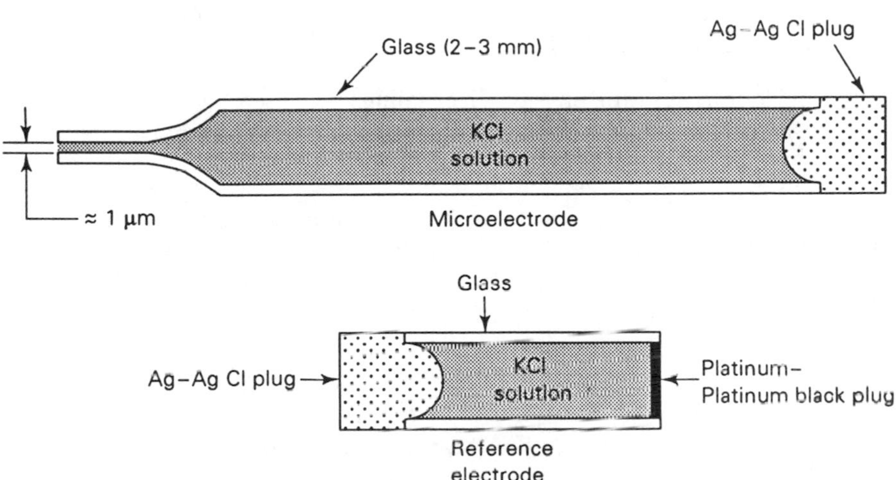

Figure 8–10 KCl-filled microelectrode and reference electrode.

the large end is capped with a Ag–AgCl plug. The small end need not be capped because the 1 μm opening is small enough to contain the fluid.

The reference electrode is likewise filled with 3 *M* KCl, but it is very much larger than the microelectrode. A platinum plug contains fluid on the interface end, while a Ag–AgCl plug caps the other end.

Figure 8–11 shows a simplified equivalent circuit for the microelectrode (disregarding the contribution of the reference electrode). Analysis of this circuit reveals the signals acquisition problem due to the *RC* components. Resistor R_1 and capacitor C_1 are due to the effects at the electrode-cell interface and are (surprisingly) frequency dependent. These values fall off to a negligible point at a rate of $1/(2\pi F)^2$ and are generally considerably lower than R_s and C_2.

Resistance R_s in Fig. 8–11 is the spreading resistance of the electrode and is a function of the tip diameter. The value of R_s in metallic microelectrodes without the glass coating is approximated by the expression

$$R_s = \frac{P}{4\pi r} \tag{8–8}$$

where R_s is the resistance in ohms

P is the resistivity of the infinite solution outside the electrode (e.g.,

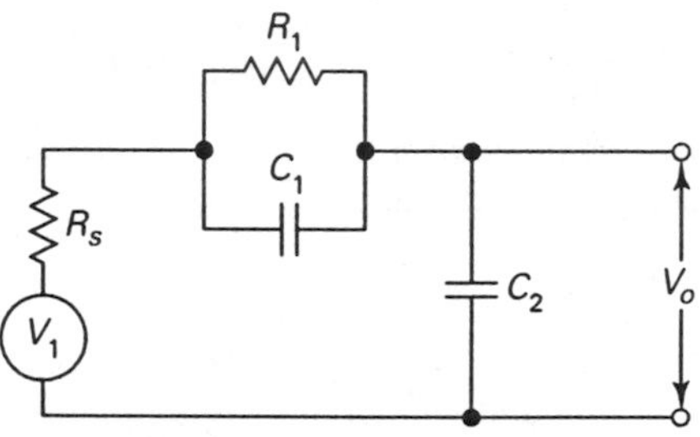

Figure 8–11 Equivalent circuit for microelectrode.

70 ohm · cm for physiological saline)
r is the tip radius (typically 0.5 μm for a 1 μm electrode)

Example 8–1

Assuming the typical values given in Eq. (8–8), calculate the tip spreading resistance of a 1 μm microelectrode.

Solution

$$R_s = \frac{P}{4\pi r}$$

$$R_s = \frac{70 \text{ ohm} \cdot \text{cm}}{4\pi\ [0.5\ \mu\text{m}\ (10^{-4}\ \text{cm}/1\ \mu\text{m})]}$$

$$R_s = 111{,}465 \text{ ohms}$$

The impedance of glass-coated metallic microelectrodes is at least one or two orders of magnitude higher than this figure.

For fluid-filled KCl microelectrodes with small taper angles ($\pi/180$ rad), the series resistance is approximated by the expression

$$R_s = \frac{2P}{\pi r \alpha}$$

where R_s is the resistance in ohms
P is the resistivity (typically 3.7 ohm · cm for 3 M KCl)
r is the tip radius (typically 0.1 μm)
α is the taper angle (typically $\pi/180$)

Example 8–2

Find the series impedance of a KCl microelectrode using the values given in Eq. (8–9).

Solution

$$R_s = \frac{2P}{\pi r \alpha}$$

$$R_s = \frac{(2)\ (3.7 \text{ ohm} \cdot \text{cm})}{3.14\ [0.1\ \mu\text{m}\ (10^{-4}\ \text{cm}/1\ \mu\text{m})]\ (3.14/180)}$$

$$R_s = 13.5\text{M ohms}$$

The capacitance of the microelectrode is given by

$$C_2 = \frac{0.55e}{\ln(R/r)} \text{ pF/cm} \tag{8-9}$$

where e is the dielectric constant of glass (typically 4)
R is the outside tip radius
r is the inside tip radius (r and R in same units)

Example 8–3

Find the capacitance of a microelectrode if the pipette radius is 0.2 μm and the inside tip radius is 0.15 μm.

Solution

$$C_2 = \frac{0.55e}{\ln(R/r)} \text{ pF/cm}$$

$$C_2 = \frac{(0.55)\ (4)}{\ln(0.2\ \mu\text{m}/0.15\ \mu\text{m})} \text{ pF/cm}$$

$$C_2 = \frac{(0.55)\ (4)}{1.33} = 1.65 \text{ pF/cm}$$

How do these values affect the performance of the microelectrode? Resistance R_s and capacitor C_2 operate together as an RC low-pass filter. For example, a KCl microelectrode immersed in 3 cm of physiological saline has a capacitance of approximately 23 pF. Suppose it is connected to the amplifier input (15 pF) through 3 ft of small-diameter coaxial cable (27 pF/ft, or 81 pF). The total capacitance is (23 + 15 + 81)pF = 119 pF. Given a 13.5M ohm resistance, the frequency response (at the −3 dB point) is

$$F = \frac{1}{2\pi RC} \tag{8-10}$$

where F is the −3 dB point in hertz
R is the resistance in ohms
C is the capacitance in farads

Evaluating Eq. (8–11) with the values given shows a 100 −3 dB point in the frequency response. A 100 Hz frequency response, with a −6 dB/octave characteristic above 100 Hz, results in severe rounding of the fast-rise-time action potentials. A strategy must be devised in the instrument design to overcome the effects of capacitance in high-impedance electrodes.

Neutralizing Microelectrode Capacitance

Figure 8–12 shows the standard method for neutralizing the capacitance of the microelectrode and associated circuitry. A neutralization capacitance C_n is in the positive feedback path along with a potentiometer voltage divider.

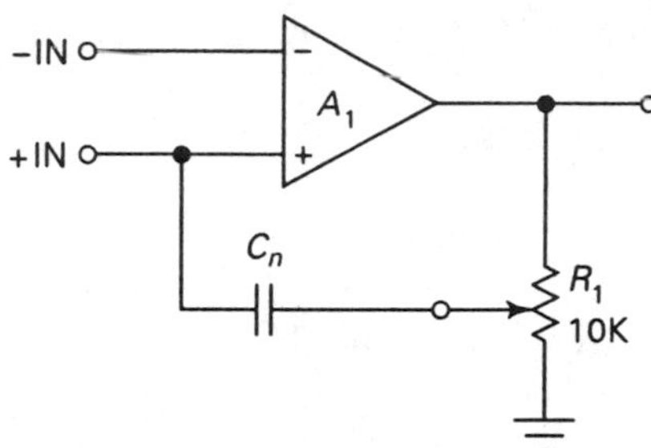

Figure 8–12 Neutralization capacitor for reducing effects of microelectrode capacitance. (See Fig. 8–11.)

The value of this capacitance is given by

$$C_n = \frac{C}{A - 1} \tag{8–11}$$

where C_n is the neutralization capacitance
C is the total input capacitance
A is the gain of the amplifier

Example 8–4

A microelectrode and its cabling exhibit a total capacitance of 100 pF. Find the value of neutralization capacitance (Fig. 8–12) required for a gain-of-10 amplifier.

Solution

$$C_n = \frac{C}{A - 1}$$

$$C_n = \frac{100 \text{ pF}}{10 - 1}$$

$$C_n = \frac{100 \text{ pF}}{9} = 11 \text{ pF}$$

Proximity and Presence Sensors

9

Proximity and *presence sensors* are used both for intruder alarms and in industrial processes to detect when an object or person (or animal) is within a certain field of regard. Several different phenomena are used for presence/proximity sensing, including capacitance, inductance, light, ultrasonics, and microwaves. In this chapter we examine all these classes.

CAPACITANCE PROXIMITY SENSORS

Capacitance is an electrical property that exists between any two conductors that are separated by an insulator (see Chapter 3). The classic textbook capacitor is a pair of parallel metallic plates separated by a dielectric consisting of a tiny air gap or other insulating material. The capacitance is a function of the area of the conductive plates, the spacing between them, and the nature of the insulating material. Figure 9–1(a) gives the expressions for several different geometries of capacitor plates, including those most useful in proximity detectors. Because most of our proximity detector sensors will be single wires to ground, we take a closer look at that expression:

$$C \simeq \frac{7.354}{\log 4(h/d)} \frac{\text{pF}}{\text{in.}} \qquad (9\text{–}1)$$

where C is in picofarads per inch.

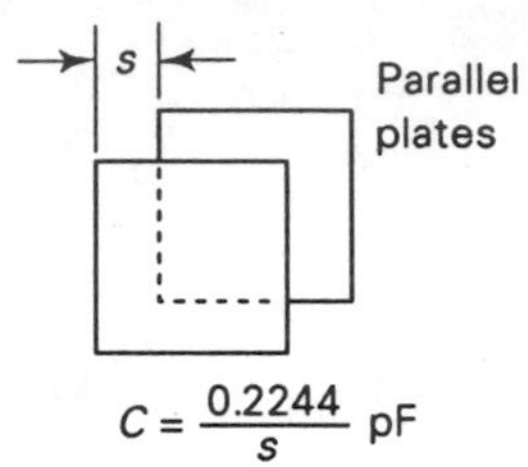

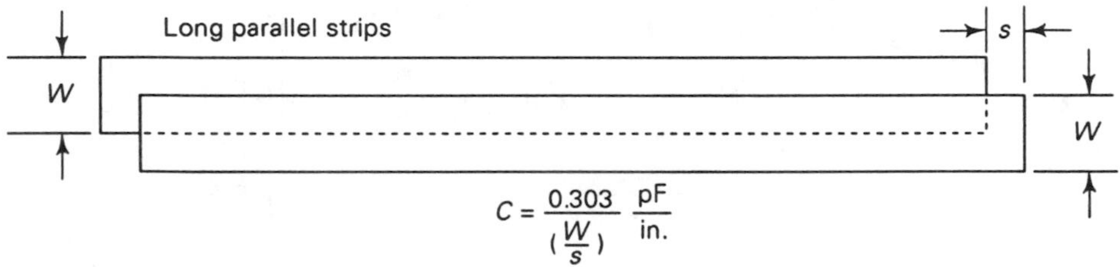

Single wire to ground

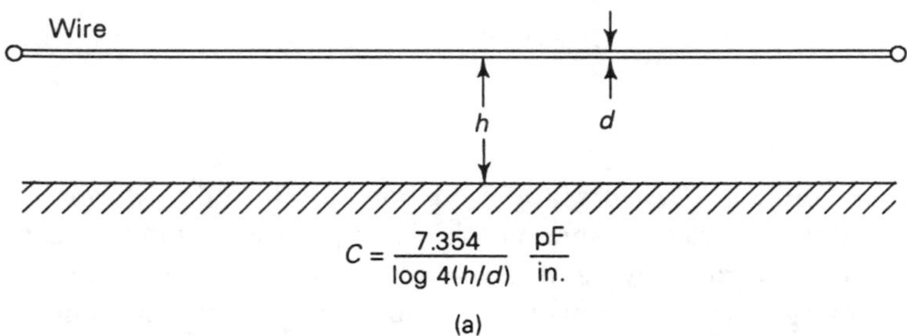

(a)

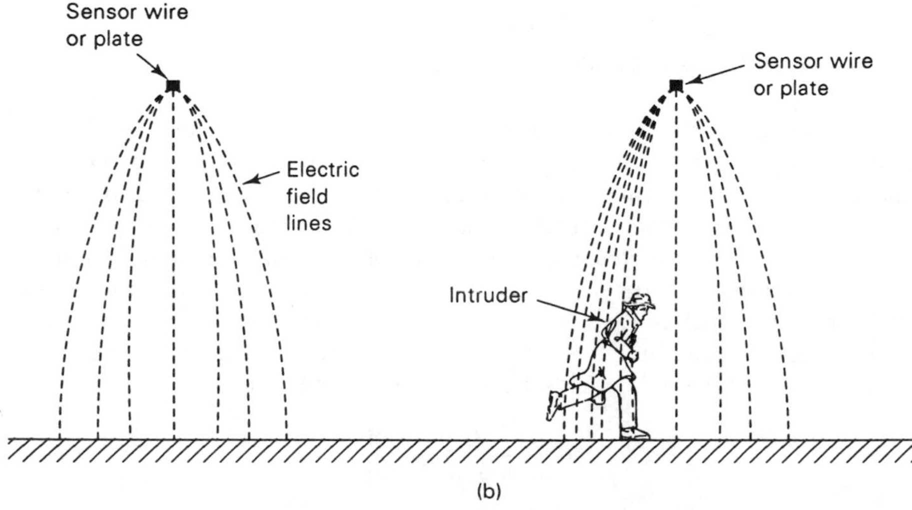

(b)

Figure 9–1 (a) Capacitance formed by parallel plates, parallel strips, and wire with respect to ground. (b) Dielectric constant of intruder is higher than that of air, causing an increase in voltage field flux.

Example 9–1

A 0.25 in. diameter capacitive sensor wire is placed at a constant height of 2 ft above the ground. What is the capacitance of a 40 ft run of this sensor wire.

Solution

$$C \simeq \frac{7.354}{\log 4(h/d)} \frac{\text{pF}}{\text{in.}} \times L$$

$$C \simeq \frac{7.354}{\log 4\left[\dfrac{2 \text{ ft} \times (12 \text{ in.}/1 \text{ ft})}{0.25 \text{ in.}}\right]} \frac{\text{pF}}{\text{in.}} \times 40 \text{ ft} \times \frac{12 \text{ in.}}{1 \text{ ft}}$$

$$C \simeq \frac{7.354}{\log(400)} \frac{\text{pF}}{\text{in.}} \times 480 \text{ in.}$$

$$C \simeq \frac{7.354}{2.6} \frac{\text{pF}}{\text{in.}} \times 480 \text{ in.} = 1358 \text{ pF}$$

The two factors that are most important (and controllable by the designer) are the diameter of the wire used and its height above the ground. Also affecting the capacitance (but not accounted for in the simplified equation) are the humidity and temperature of the air dielectric between the wire and ground.

HOW DO CAPACITIVE PROXIMITY DETECTORS WORK?

All sensors rely on a transducible property, that is, some physical parameter that can be converted into a representative voltage or current. Capacitive proximity detectors rely on the fact that the capacitance is dependent on the dielectric of the capacitor. Dry air has a dielectric constant close to 1.006, while the human body has a dielectric constant of about 80. Thus, when a human enters the field of a capacitor [Fig. 9–1(b)], the capacitance will increase by up to 80 times the former value. The electrical flux between the sensor wire and ground thus increases dramatically.

Several different types of instruments and circuits utilize the capacitance phenomenon, for example, *LC* oscillators, electrometers, and bridge circuits.

LC OSCILLATOR CIRCUITS

An *LC oscillator* (Fig. 9–2) produces an output frequency that is determined by the inductance and capacitance of the resonant tank circuit. All *LC* oscillators operate on a frequency that is determined by the resonant frequency of an inductor-capacitor (*LC*) circuit (see Chapter 6 for a discussion of the nature of these circuits). Varying either the inductance or the capacitance of the circuit will cause the operating frequency produced by the oscillator to vary. The particular circuit shown in Fig. 9–2 is called a *Colpitts*

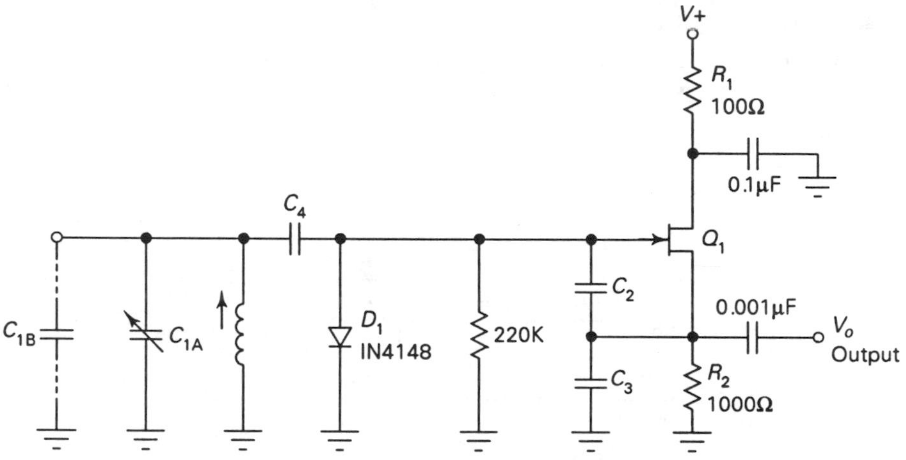

Figure 9–2 RF oscillator using wire capacitance.

oscillator, after the inventor. Other forms of *LC* oscillators are also used, but the Colpitts is considered particularly adaptable to capacitive sensors.

The output frequency of an *LC* oscillator is usually in the range above human hearing, the range sometimes called *radio frequencies*. In Fig. 9–2 the inductance is L_1, while the capacitance comprises variable capacitor C_{1A} and the sensor capacitance C_{1B}.

There will also be stray capacitance and inductance in the circuit, but they are not shown in Fig. 9–2 as discrete components. The circuit shown here will oscillate at frequencies from about 20 kHz to 10 MHz, depending on the values of the *LC* tank circuit components and the particular values of feedback capacitors C_2/C_3 that are selected.

Figure 9–3 shows a method by which the oscillator can be used in a proximity detector circuit. Two sense wires are used in this circuit [Fig. 9–3(a)] because changes in temperature and humidity will easily change the oscillation frequency as they change the capacitance between the wire and ground. The use of two wires allows environmental effects to be nullified in the circuitry that follows. Each sense wire is a length of #12 or #14 insulated wire mounted on insulated supports such as fence posts.

Figure 9–3(b) shows a scheme for using the capacitances in a heterodyned "beat-frequency" circuit. Two oscillators are selected so that the third harmonic of oscillator 1 will be close to the fourth harmonic of oscillator 2. For example, a common scheme sets oscillator 1 to a frequency of 100 kHz, while oscillator 2 is set to 74.875 kHz. The third harmonic of oscillator 1 is 300 kHz, while that of oscillator 2 is 299.5 kHz. Therefore, the beat note at the output of the mixer stage will be the difference frequency, or 500 Hz. This seemingly complicated method results in a very large percentage change of beat note for relatively small changes in operating frequency. For example,

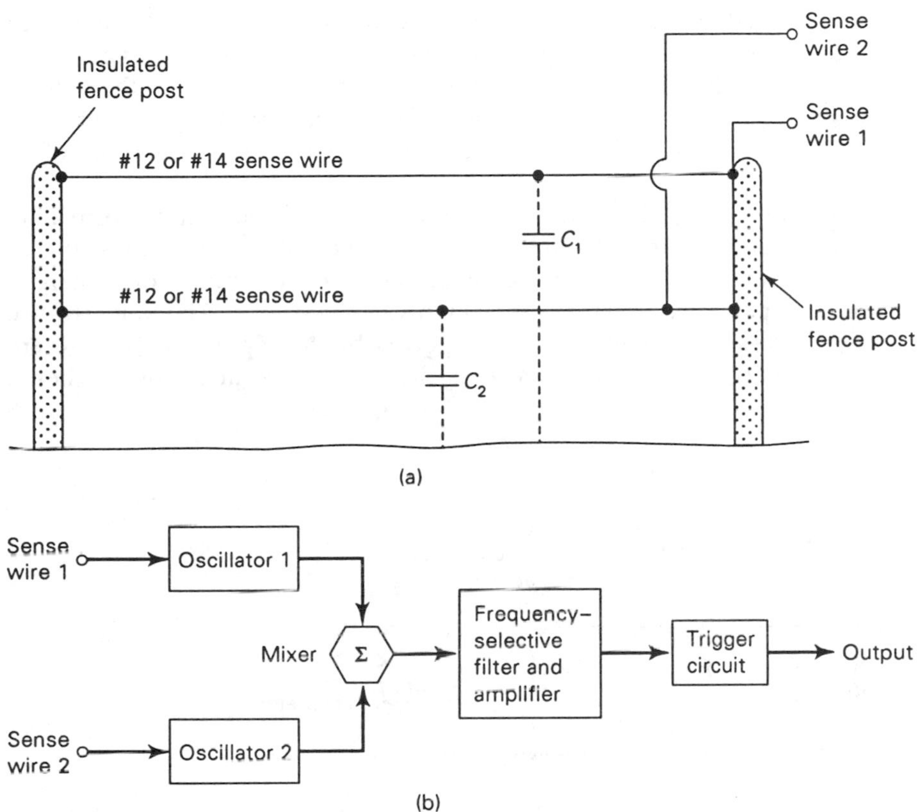

Figure 9–3 (a) Parallel-wire proximity sensor; (b) RF implementation of detector circuitry.

causing oscillator 1 to shift to 100.010 kHz causes the beat note frequency to shift to 530 Hz. The difference between the "same frequency" method and the harmonic method is that a 10 Hz change represents a 0.01 percent change in operating frequency and a 6 percent change in harmonic beat note.

The filter circuit in Fig. 9–3(b) is a sharply tuned op-amp bandpass filter centered on the normal beat note (500 Hz in our example). As long as the output of the filter sees a signal, the trigger circuit will not be alarmed. But if the operating frequencies of the oscillators change, the beat note is no longer within the filter's pass band, so the trigger circuit sounds the alarm.

ELECTROMETER CIRCUITS

An electrometer is an amplifier with an extremely high input impedance. Thus, when an electrometer is shunted across a capacitance, its input impedance will not try to discharge the capacitor. Traditionally, electrometers have been used to measure the outputs of devices such as very high impedance transducers and capacitor charges. The high input impedance does not bleed the charge off the capacitor as it measures its charge voltage, so the voltage across the sensor capacitor is a function of the parameter being measured or, in the case of a proximity or presence sensor, the presence or absence of an object or intruder.

Figure 9–4(a) shows the basic circuit for an electrometer proximity detector. The sensor wire produces a capacitance (shown in Fig. 9–14(a) as C_1) at the input of the electrometer amplifier (A_1). The capacitor is charged from a dc power supply through a high-value resistor. The value of the voltage across the charged capacitor is given by the expression Q/C, where Q is the charge in coulombs and C is the capacitance in farads. When a

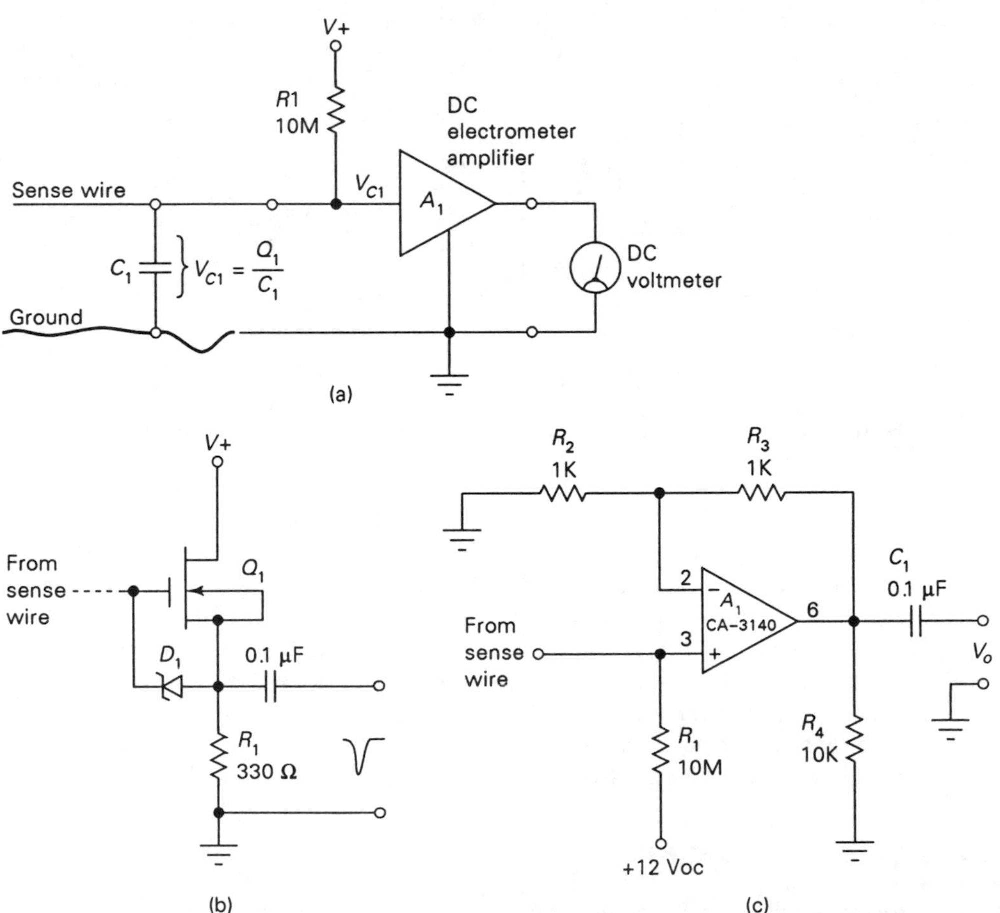

Figure 9–4 (a) Electrometer sensor circuit; (b) electrometer element using MOS-FET transistor; (c) op-amp implementation.

human enters the field of the protected area, the capacitance increases tremendously, so the voltage across the capacitor takes a momentary dip. That change of voltage is amplified by the electrometer and used to indicate intrusion.

Examples of electrometer circuits are shown in Figs. 9–4(b) and 9–4(c). The circuit shown in Fig. 9–4(b) uses a single MOSFET transistor as the electrometer amplifier. These transistors have input impedances in the range of 10^9 to 10^{12} ohms. In most cases a CMOS inverter or one of the transistors in the CMOS 4017 complementary transistor array device can also be used. The protection diode shown in Fig. 9–4(b) is inherent in B-series CMOS devices and gate-protected discrete MOSFETs, but for other discrete MOSFETs it must be provided. The purpose of the diode is to shunt harmful high voltage from electrostatic potentials harmlessly around the delicate gate structure.

Another variant electrometer is shown in Fig. 9–4(c). This circuit is based on a special class of op-amps that have MOSFET transistors in their input circuits. The device shown here is part of the BiMOS line of op-amp devices. The CA-3140 device is an ordinary op-amp with industry standard pinouts and an input impedance of 1.5×10^{12} ohms. The dc power connections are not shown for the sake of simplicity.

BRIDGE METHODS

A basic ac bridge circuit is shown in Fig. 9–5. Similar to a Wheatstone bridge, this circuit is basically two voltage dividers in parallel: Resistors R_1/R_2 form one voltage divider, while capacitors C_1/C_2 form the other. An oscillator operating in the 100 kHz region drives the bridge and is its voltage source. As long as the ratio of the resistors R_2/R_1 is equal to the ratio X_{C1}/X_{C2}, the output voltage V_o across terminals A and B is zero. But if either

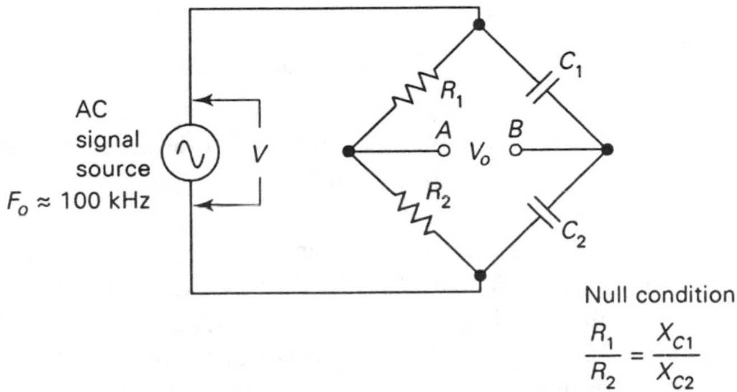

Figure 9–5 Resistance-capacitance bridge circuit.

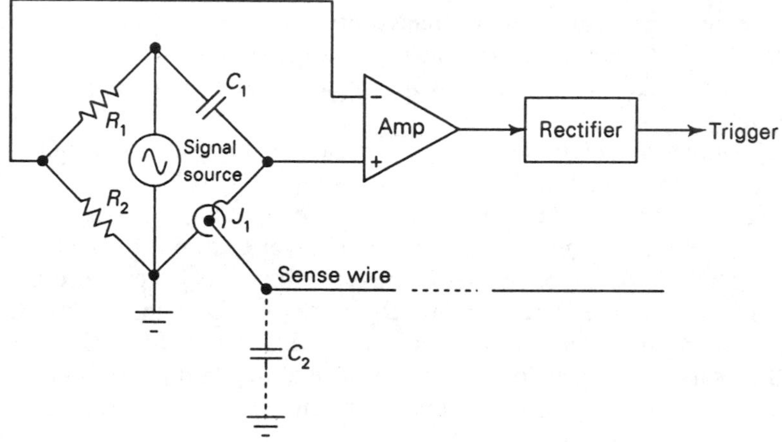

Figure 9–6 Circuit for single sensor wire.

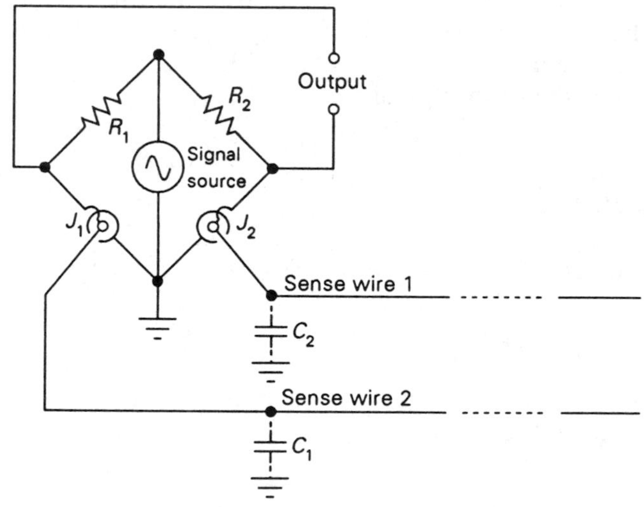

Figure 9–7 Bridge circuit for parallel sensor wires.

ratio changes, then V_o is nonzero. In a proximity detector the capacitances are sense wires, and the capacitance changes when a person or object comes close to them.

Figure 9–6 shows a method for using a single sensor wire in a bridge circuit. Resistors R_1 and R_2 are fixed (although in some circuits balance adjustment is possible), as is capacitor C_1. A sensor wire is plugged into jack J_1 to complete the bridge. The output voltage is applied to a differential rectifier circuit and then to a trigger circuit. Alternatively, the output of the bridge can be applied to a differential tuned amplifier before being applied to the rectifier. The purpose of this is to selectively accept signals. A long sensor wire will act as an antenna, so it will pick up signals that can pass through the untuned rectifier with ease. A tuned amplifier will pass only those from the bridge circuit.

The variant circuit shown in Fig. 9–7 uses two sense wires in the same manner as the system shown earlier, and for exactly the same reasons (temperature and humidity). In this case, however, the unbalanced bridge circuit has one end of the output grounded. The output terminal can be a coaxial connector that takes the signal to the electronics package.

SINGLE-OBJECT PROTECTION

The protection of specific objects in a space is a problem that proximity detectors can solve. Figure 9–8 shows a method that is similar to the lamp my clowning friend at the beginning of this chapter used to spoof us. It can also be used to thwart the criminal who would try to break into the protected cabinet or safe.

In the system of Fig. 9–8 if the protected object is conductive (i.e., metallic), then it can form one plate of the capacitor in lieu of the sensor wire. The ground forms the other plate in some systems, in which case the circuit is the same as other proximity detectors. But in many cases, especially indoors, the floor is nonconductive, so it will not suffice as a ground. But a metal-backed mat can be placed on the floor, or a large area of either metal screen or foil can be inserted underneath a carpet, dust mat, or masonite walkboard. The idea is to place a large metallic surface underneath and insulated from the protected object. The capacitance can be used in any of the circuits discussed earlier.

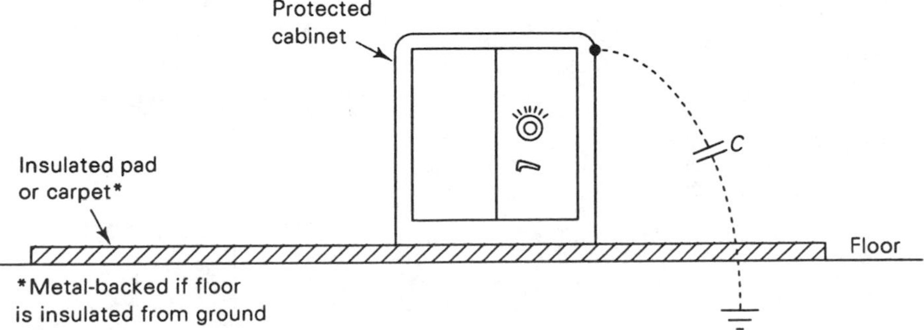

Figure 9–8 Single-object protection.

CAPACITANCE PRESENCE SENSOR

Figure 9–9(a) shows another form of capacitive presence sensor, while a typical circuit is shown in Fig. 9–9(b). This type of sensor is often used on assembly lines and in controlling industrial processes. The capacitor in Fig. 9–9(a) is formed by two plates (*A* and *B*) embedded in an insulating dielectric material. A signal applied to plate *A* is coupled to plate *B*, where it causes a current to flow in an external circuit [Fig. 9–9(b)]. When no object is present, the coupling between *A* and *B* is minimal, so only a small current flows in the output circuit connected to *B*. The voltage across the load resistor *R* is small, but not zero, under the no-object condition. This is called the *baseline response*. When an object is in the field of the capacitor formed by *A* and *B*, the capacitance between the plates changes, which causes a change in the amount of coupled signal.

What happens when an object approaches depends on the nature of the object. A grounded object dissipates the flux lines between the plates, causing a decrease in the amount of signal coupled to the output. An ungrounded metal object sharply increases the flux concentration, causing an increase in capacitance and therefore an increase in the current coupled between the two plates. As the object shades more and more of the plates, a shielding effect is seen that reduces the signal, and as the object moves out of the field, the signal returns to baseline. A dielectric object causes a very large increase in capacitance and, surprisingly, operates in a manner similar to the ungrounded metal object.

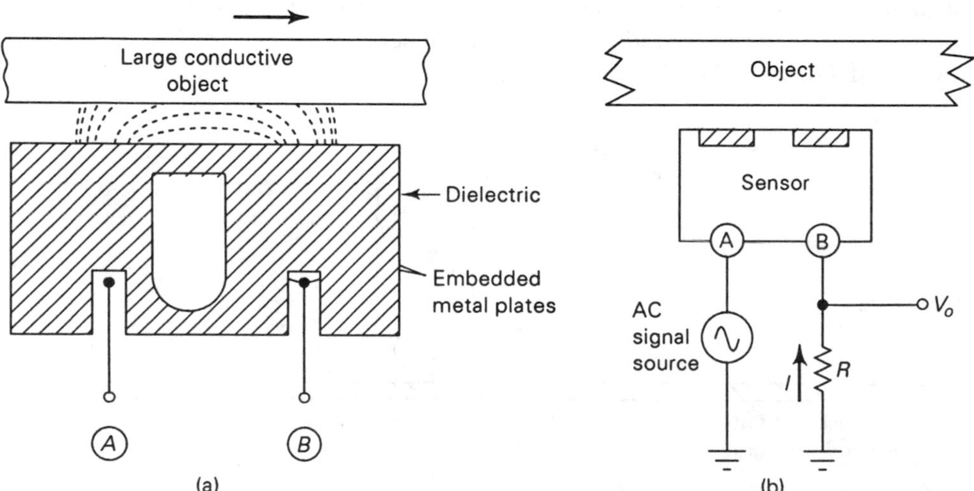

Figure 9–9 Large-object capacitive sensing: (a) mechanism; (b) sensor circuit.

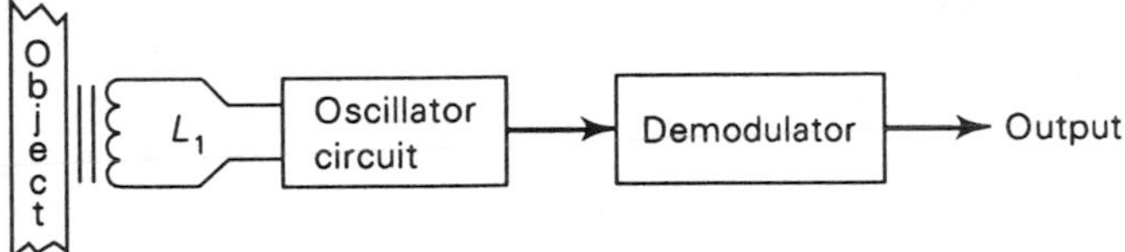

Figure 9–10 Oscillator circuit implementation.

INDUCTIVE PROXIMITY SENSORS

Inductive proximity sensors depend on changing a magnetic field surrounding a sensor element by inducing eddy currents in a ferrous or nonferrous target brought into close proximity with the sensor element. This type of sensor is also sometimes called a *proximity switch*. It offers fast, accurate, and highly repeatable (3 percent of target distance, typically ±0.0005 in.) operation at relatively low cost. Eddy current proximity sensors will operate at temperatures up to about 250°F and at frequencies to 10 kHz for some designs, and 2 MHz for others. They can differentiate metallic object thickness variations on the order of 0.03 in.

The eddy current inductive sensor circuit (Fig. 9–10) consists of an ac oscillator circuit, an inductor coil L_1, and an output voltage detector (or demodulator circuit) that senses the amplitude of the output waveform produced by the oscillator. When nothing is near the inductor sensor head, the oscillator amplitude will stabilize to some nominal value. But when a metallic target is in the field of the coil, the magnetic field of the coil will induce eddy currents in the target. These eddy currents create a magnetic field around the target, and this field will change the resistance component of the impedance of the sensor coil. This interaction will cause a sudden drop in the output amplitude of the oscillator, and this drop can be used to indicate the presence of the object.

The reduction of amplitude is usually quite sharp with respect to position. At some minimum *deadband* gap (usually 5 to 20 mils) the amplitude will drop to zero.[1] Below this distance, sensing will not occur.

Other inductive proximity sensors operate by allowing the metallic target to create a change in the oscillator frequency. If a frequency-selective circuit is monitoring the output, then its output level is a function of target position.

In both amplitude and frequency versions, the inductor sensor coil is calibrated using a standard target. In many cases a 30 mm × 30 mm × 1 mm mild steel disk at a specified sensing range is used as the standard.

MAGNETIC PROXIMITY SWITCHES

Reed switches are a class of switches used in certain areas of electronics. These devices consist of a set of magnetized switch contacts inside an evacuated glass envelope (Fig. 9–11). When a magnet is brought near the glass

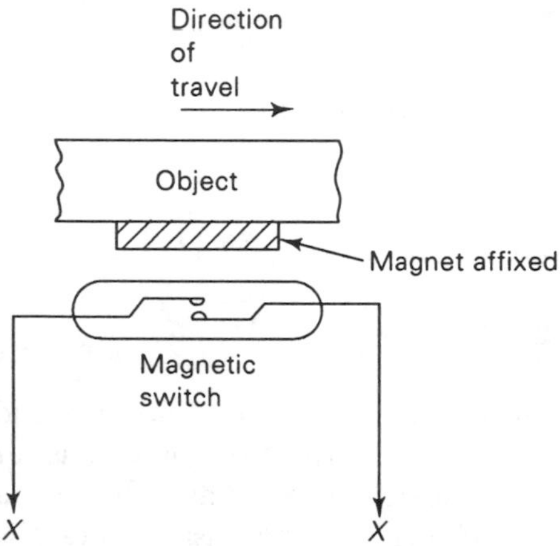

Figure 9–11 Magnetic reed switch proximity sensor.

envelope, the switch contacts either move apart or come together, depending on which pole of the magnet is nearest. Some reed switches are *normally closed* (N.C.), so they open when the magnet is present. Others are *normally open* (N.O.), so they close when the magnet is close. Still others are single-pole, double-throw (SPDT) types, so they provide both N.C. and N.O. functions.

Some reed switches are used in explosive atmospheres (natural gas, gasoline vapors, etc.) where a regular electrical switch would spark and cause detonation or fire. These explosion-proof reed switches have dc electromagnet coils surrounding them. When current is applied to the coil, a magnetic field builds up and actuates the reed switch.

Another application of magnet-operated reed switches is in intruder alarm systems for windows and doors. A close look at some such intruder alarms will reveal two pieces placed in close proximity, one on the moving part of the window and the other on the window frame, yet neither piece touching the other. The component on the moving part of the window is a permanent magnet, whereas the component on the fixed portion is a reed switch. When the window is closed, the magnet actuates a switch and closes the internal contacts. But when the window is raised, the magnet moves away, and the switch contacts open—causing the alarm to sound if everything is in good working order.

In the magnetic proximity sensor shown in Fig. 9–11, the object being detected is fitted with a permanent magnet. When the magnet comes close to the reed switch, it causes the switch to actuate and send a signal to the electronic circuits monitoring the process. Figure 9–12 shows two examples of such circuits. In Fig. 9–12(a), the reed switch is connected in series with a light-emitting diode (LED) and a current-limiting resistor R_1. When the switch closes, the dc power supply V_1 is connected to the LED/R_1 circuit,

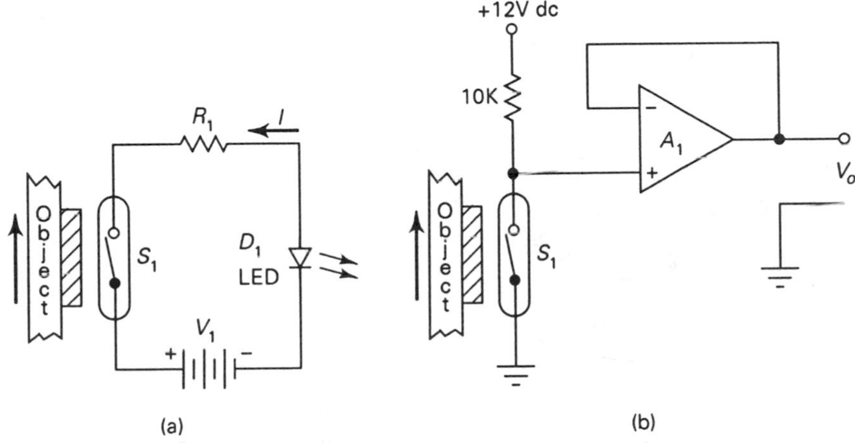

Figure 9–12 Reed switch circuits: (a) simplest form; (b) operational electronic form.

causing the LED to light up. The lighted LED indicates that the object being detected is present and in position.

A second version is shown in Fig. 9–12(b). If a N.O. reed switch is used, then the noninverting input (+) of the op-amp (see Chapters 19 and 20) will be placed at a logical HIGH level, that is, about +12 V with the configuration as shown. This voltage level is reflected as a HIGH at the output V_o. But when the reed switch closes, the noninverting input is grounded, so the op-amp sees an input voltage of zero. This LOW condition is reflected as a LOW output. Thus, the output terminal of the op-amp will switch between HIGH and LOW as the object approaches, and back to HIGH when the object passes. The polarity can be reversed by using a N.C. switch instead of a N.O., as shown.

LIGHT-OPERATED PROXIMITY/PRESENCE SENSORS

A number of different light sensors are available on the market (see Chapter 15). Devices such as photoconductive cells, photovoltaic cells, photodiodes, and phototransistors are used in various applications. Wavelength sensitivities can run from the far infrared to the far ultraviolet or near–X-ray regions of the electromagnetic spectrum. These sensors provide an output voltage, output current, or change of resistance in response to an impinging light. We will not dwell here on the nature of these electro-optical (E-O) sensors but will discuss how they are used for proximity sensing.

The simplest form of light-based sensor system is shown in Fig. 9–13. In this system the sensor (a phototransistor) is positioned opposite a light

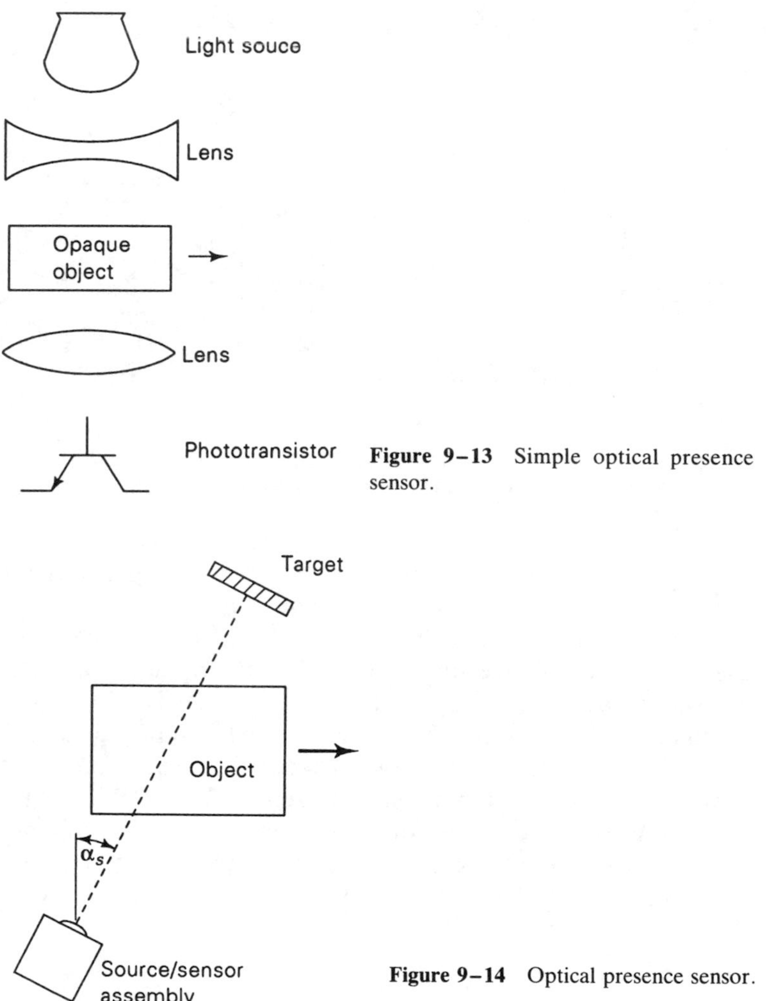

Figure 9–13 Simple optical presence sensor.

Figure 9–14 Optical presence sensor.

source. A condensing lens system forms parallel light beams that are directed toward the focusing lens. It, in turn, directs the light to the sensor. When an opaque object passes between the sensor and the light source, the light is cut off, and that changes the sensor state.

A problem with the type of sensor system shown in Fig. 9–13 is that it requires the light source and sensor to be on opposite sides of the object's path of travel. A solution to the problem is the retroflective system shown in Fig. 9–14. In this system the sensor and source are collocated in the same package, while a reflective target is positioned on the opposite side of the object pathway.

When shiny objects are being sensed, the photosensor may be overloaded by the direct reflection at close ranges. The solution to this type of problem is often very simple: The optical path is canted at a small angle α_s with respect to the pathway of the objects being sensed.

An increasingly popular version of the reflective proximity sensor is shown in Fig. 9–15. In this sensor system *fiber optics* are used as light conduits. Fiber optics are to light what a waveguide is to microwaves. The fiber efficiently transfers light from its input end to the output end. Thus, light introduced into fiber *A* will be transmitted with only a small loss to the output end, where it is aimed at the object's pathway. When an object is close to the ends of the fiber optic cords, light is reflected from its surface into the end of fiber *B*. This light is transferred through fiber *B*, where it impinges on a photosensor device.

Background light is a continual problem in some photosensor-based proximity detector systems. If the ambient light level is high, then the background light level will keep the sensor biased in a high excitation state and cause it to be either desensitized or not work at all. Fortunately, there are ways to rid the system of light.

Of course, the obvious solution is to reduce the background light level to a point where the sensor is dominated by the light source. But, except in a few instances, the simple solution is elegant, appealing, low cost, and *wrong*. Why wrong? Because it is not usually practical (or safe) to reduce ambient light levels. In addition, the variance in "ambient" levels with time

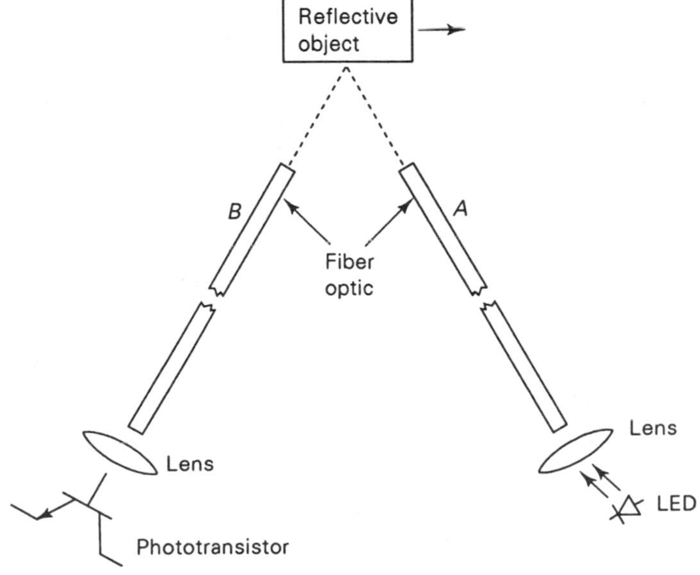

Figure 9–15 Fiber-optic presence sensor.

would change the operation of the sensor circuit. Alternative methods are shown in Fig. 9–16. Either method may be used alone or in combination with the other.

The system in Fig. 9–16(a) uses a modulated light beam in place of the steady beam used in previous systems. Although a number of methods will result in a chopped or modulated light, the method proposed in Fig. 9–16(a) is a pulse waveform applied to D_1, an LED. The LED flashes on and off as the pulse excursions pass. Although the light beam flashes, if the flash rate is above the *flicker frequency* for human eyes (about 8 to 10 flashes per second), it appears to be constant.

If the output circuit of the light sensor is equipped with a capacitor-coupling network (C_1/R_2), then only *variations* in light level are passed. The circuits following the sensor can be filtered or tuned to admit only those

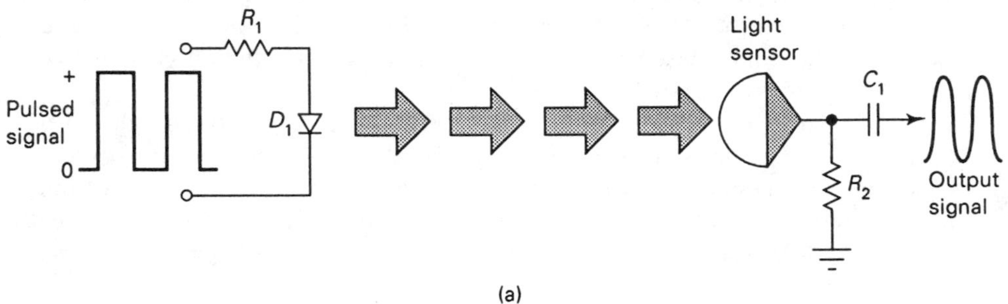

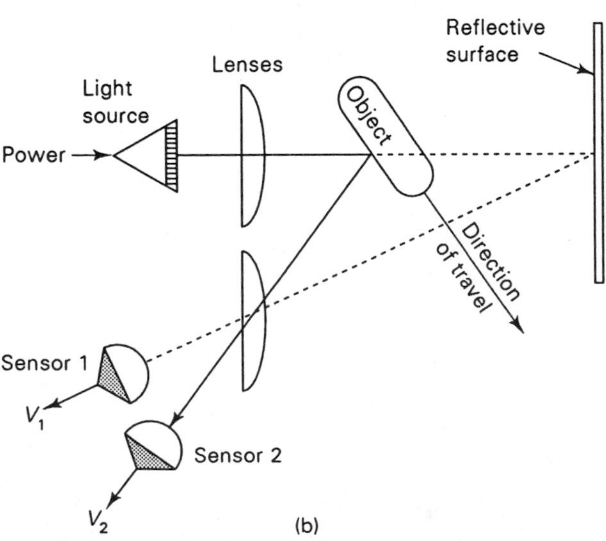

Figure 9–16 (a) Light-pulse operation of sensor; (b) reflective presence optical sensor.

signals that exactly match the source. That will prevent ambient light modulated by such sources as ceiling fans or tree branches outside the window blowing in the wind from also exciting the alarm. It is also possible to encode the light beam so that multiple sources can be used in the same system. If a proper decoder is provided, then the alarm circuit will recognize only the correct code.

The system shown in Fig. 9–16(b) is based on the principle that light will impinge on two different sensors under different conditions. When there is no object present, the light beam will reflect from the reflective surface on the far side of the object path, through the second lens, to sensor 1. But when an object is in the path, the light is reflected through a much shorter path to sensor 2. Ambient light is suppressed because the transducible event is the *relative* light levels of sensors 1 and 2. As long as the light level detected by sensor 2 is much higher than the light level detected by sensor 1, there is an object present in the optical path.

MICROWAVE PROXIMITY/PRESENCE SENSORS

Microwaves are electromagnetic radio waves that lie in the portion of the electromagnetic spectrum from approximately 1 GHz (1 GHz = 1000 MHz) to about 300 GHz (after which point the waves are labeled *infrared*). Microwaves are also called *centimetric waves* in some countries because their short wavelengths are conveniently measured in centimeters, rather than meters. In some contexts, microwaves about 30 GHz are called *millimeter microwaves*, or just *millimeter waves*, because the wavelengths are conveniently measured in millimeters (also, they seem to have some properties not found at lower frequency, for example, they become more optical in their behavior). Two basic forms of microwave proximity sensors are *simple radar* and *Doppler radar*.

INTRODUCTION TO RADAR DEVICES

Although by common usage now regarded as a noun, the word *radar* was originally an acronym for *ra*dio *d*etecting *a*nd *r*anging. Thus, the radar set is an electronic device that can detect distant objects and yield information about the distance to that object. With appropriate design, radar can also give either relative or compass bearing to the object and, in the case of aircraft, height of the object. In radar jargon the object detected is called a *target*.

Modern radar sets can be land-based, ship-based, aircraft-based, or spacecraft-based. They are used for weather tracking, air traffic control, marine and aeronautical navigation, ground mapping, scientific studies, velocity measurement, military applications, and remote sensing from outer space. Readers with an unfortunate love of driving fast cars will also recognize at least one velocity measurement application: police speed radar guns. Proximity-sensor radar units are miniature editions of the same type of system (although rarely as complex).

Scientists and engineers were working on radar in the 1920s and 1930s, but World War II caused the sudden surge of development effort that led to modern radar as we know it today. One of the earliest experiments in radar was accidental. In 1922 scientists at the Naval Research Laboratory (NRL) in Washington, D.C., were experimenting with 60 MHz communications equipment. A transmitter was set up at Anacostia Naval Air Station, and a receiver was installed at Haines Point on the Potomac River (a distance of about 0.5 mi). Signal fluctuations were noted when a wooden vessel, the *USS Dorchester*, passed up the river. Further experiments at 300 MHz showed that *reflections* of the radio waves by passing ships were the cause of those signal fluctuations. Distances of 3 mi were achieved in those trials. The first practical radar units were fielded by a British team under Sir Robert Watson-Watt in the 1930s, just in time for World War II. The British Chain Home radar system is credited with giving the British a much-needed edge in the infamous Battle of Britain.

The basic principle behind radar is simple: A short burst of RF energy is transmitted, and then a receiver is turned on to listen for the returning echo of that pulse reflected from the target [Fig. 9–17(a)]. The transmitted pulse [(Fig. 9–17(b)] is of short duration, of width T, with a relatively long receiver time between pulses. It is during the receiver time that the radar listens for echoes from the target. The number of pulses per second is called the *pulse repetition rate* (PRR) or *pulse repetition frequency* (PRF), whereas the time between the onset of successive pulses is called the pulse repetition time (PRT). Note that

$$\mathrm{PRT} = \frac{1}{\mathrm{PRF}} \tag{9–2}$$

The pulse propagates at the speed of light c, which is a constant, so the range can be measured indirectly from the time required for the echo to return:

$$R = \frac{ct}{2} \tag{9–3}$$

where R is the range to the target in meters
c is the speed of light (3×10^8 m/s)
t is the time in seconds between the original pulse transmitted and arrival of the echo from the target

The factor $\frac{1}{2}$ in Eq. (9–3) reflects the fact that the echo must travel to and from the target, so the apparent value of t is twice the "correct" value.

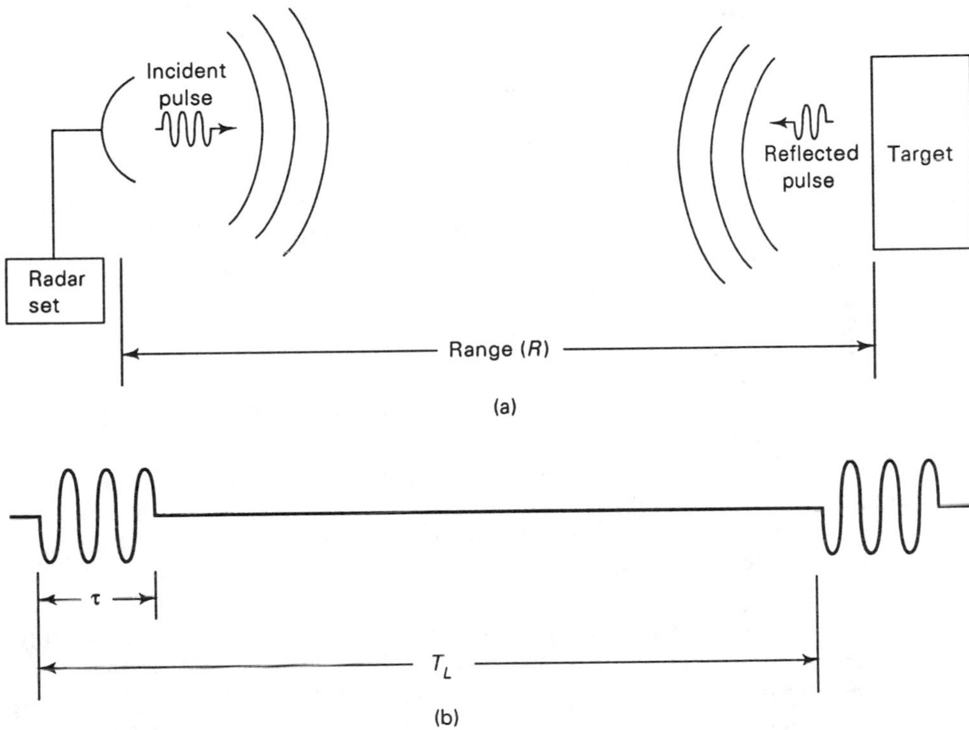

Figure 9–17 (a) Principle of radar. (b) Interpulse time determines maximum range or distance to target.

A popular radar rule of thumb derives from the fact that a radio wave travels 1 nautical mile (naut mi), or 6000 ft, in 6.18 μs. Because of the round-trip travel, radar range can be measured at 12.36 μs/naut mi. In equation form:

$$R = \frac{T}{12.36} \text{ naut mi} \tag{9–4}$$

Unfortunately, real radar waves do not behave ideally, for at least two reasons. First, the pulse may rereflect and thus present a second-time-around error echo for the same target. Second, the maximum unambiguous range is limited by the interval between pulses. Echoes arriving late fall into the receive interval of the next transmitted pulse, so the second (erroneous) "target" appears much closer than it really is. If T_i is the interpulse interval (or receiver period as we called it earlier), then the maximum unambiguous range (R_{max}) is given by

$$R_{max} = \frac{cT_i}{2} \tag{9–5}$$

where R_{max} is the maximum unambiguous range in meters
c is the speed of light (3×10^8 m/s)
T_i is the interpulse interval in seconds

The pulse width T represents the time the transmitter is on and emitting. During this period the receiver is shut off, so it cannot hear any echoes. As a result, there is a dead zone in front of the antenna in which no targets are detectable. The dead zone is a function of transmitter pulse width and the transmit-receive (T-R) switching time. If T-R time is negligible, then

$$R_{dead} = \frac{cT}{2} \tag{9-6}$$

Obviously, Eq. (9–6) requires very short pulse widths for nearby targets to be detectable.

Another radar property that is dependent on pulse width is range resolution ΔR. Simply defined, ΔR is the minimum distance along the same azimuth line by which two targets must be separated before the radar can differentiate between them. At distances less than the minimum, two different targets appear as one.

For example, suppose each pulse has a width T, and because it propagates through space with velocity c, it also has *length*. Let's assume a 1 μs pulse on a frequency of 1.2 GHz. The wavelength of this signal is 0.25 m. The number of cycles in the pulse is the product of pulse width T and operating frequency. Thus, in 1 μs the transmitter sends out a pulse containing (1.2×10^9 cps) $\times$ (1×10^9 s), or 1200 cycles, each of which is 0.25 m long. Thus, the pulse length is (1200×0.25 m), or 300 m. Two targets must be 300 m apart to not fall inside the same pulse length at the same time. Because of the round-trip phenomenon, which means that the result is divided by two, we can express range resolution as

$$\Delta R = \frac{cT}{2} \tag{9-7}$$

where ΔR is the range resolution in meters
c is the speed of light (3×10^8 m/s)
T is the pulse width in seconds

PULSE PROPERTIES

The radar transmits pulses of relatively short duration T and then pauses to listen for the echo. Important characteristics of radar pulses include the peak power P_p, the average power P_a, and the *duty cycle D*.

The output power of the transmitter is usually measured in terms of peak power P_p. The average power, which affects transmitter design, is found from either of the following equations:

$$P_a = \frac{P_p T}{\text{PRT}} \tag{9–8}$$

or

$$P_a = P_p T\,(\text{PRF}) \tag{9–9}$$

where P_a is the average power in watts
P_p is the peak pulse power in watts
T is the pulse width in seconds
PRF is the pulse repetition frequency in hertz
PRT is the pulse repetition time in seconds

Note that the average power is very much smaller than the peak power. Thus, radar transmitters pack a large peak power into a small space. In other words, the duty cycle is low on radar transmitters. The duty cycle is defined as

$$D = \frac{P_a}{P_p} \tag{9–10}$$

or, as a percentage:

$$D = \frac{P_a \times 100\%}{P_p} \tag{9–11}$$

The duty cycle can also be expressed in terms of the average power equation:

$$D = \frac{T}{\text{PRT}} \tag{9–12}$$

or,

$$D = T \times \text{PRF} \tag{9–13}$$

THE RADAR RANGE EQUATION

Basic to understanding radar is understanding the radar range equation. We study it by examining each of its components: transmitted energy, reflected energy, system noise, and system losses. The maximum RF energy on target is essentially the effective radiated power (ERP) of the radar reduced by the inverse square law $(1/R)^2$. The ERP of a pulsed transmitter is the product of pulse width T, peak power P_p, and antenna gain G: P_pTG. The power density produced by this energy over an area can be found from the equation

$$P = \frac{P_pTG}{4\pi R^2} \tag{9–14}$$

The reflected energy is only a fraction of the incident energy p, so it can be calculated from Eq. (9–14) and a factor called the radar cross section (RCS) of the target. In essence, RCS (σ) is a measure of how well radio waves reflect from the target. RCS determination is very difficult because it depends on the properties of the target. A midsize ocean liner has an RCS of about 50,000 m^2, whereas a commercial airliner may have an RCS of 20 to 25 m^2. The reflected energy reaching the receiver antenna is also reduced by the inverse square law. Thus, we can rewrite Eq. (9–14) in the form

$$S = \frac{P_pTGA_e\sigma}{16\pi^2R^4} \tag{9–15}$$

where S is the signal energy in watt · seconds
P_p is the peak power in watts
T is the pulse width in seconds
G is the antenna gain
A_e is effective aperture of the receiver antenna in square meters
σ is the radar cross section (RCS) of the target in square meters
R is the range to the target in meters

For an ideal radar system Eq. (9–15) suffices, although in the practical world losses L and noise become important. Examples of miscellaneous losses include weather losses as signal is attenuated in moist air and transmission lines. Adding these factors into Eq. (9–15) yields the classical radar range equation:

$$\frac{S}{N} = \frac{P_pTGA_e\sigma}{16\pi^2R^4KTL} \tag{9–16}$$

where S/N is the signal-to-noise ratio
K is Boltzmann's constant (1.38×10^{-23} J/K)
T is the noise temperature in Kelvins
L is the sum of all losses in the system
All other terms are as previously defined

RADAR AND MICROWAVE SYSTEMS

Several different configurations of microwave or radar systems are used to provide proximity detection. Figure 9–18 shows the simplest form. A single radar antenna serves both the microwave source or transmitter and the re-

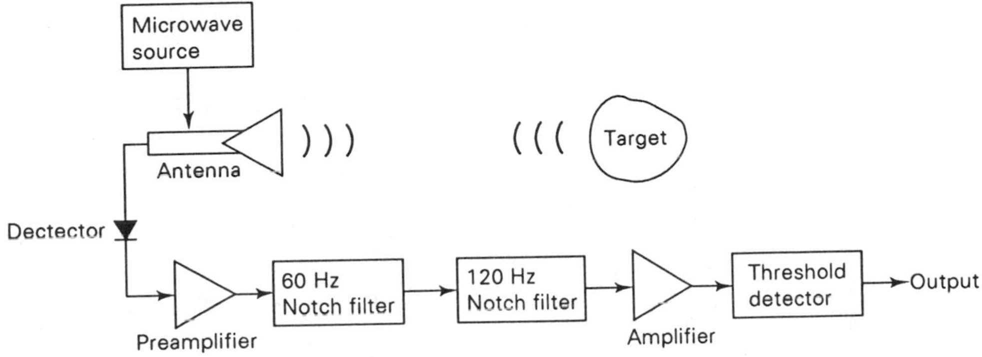

Figure 9–18 Simple radar presence detector.

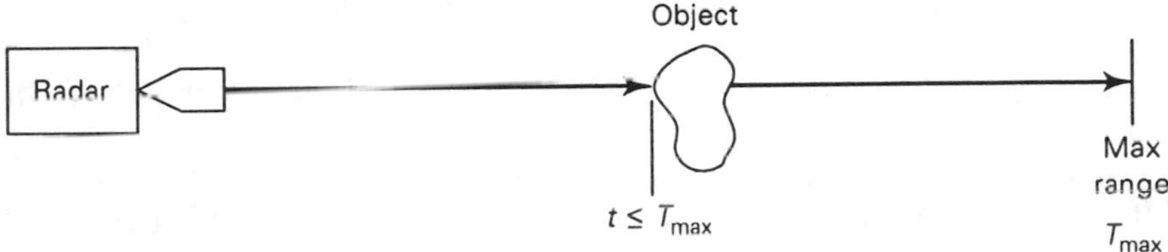

Figure 9–19 Time of pulse return determines the range to object.

ceiver. The signal is transmitted into space. If no object is present, then the signal continues to propagate, and no reflection is received. But should a target come into range, the energy is reflected and this backscatter energy is propagated back to the receiver. A detector, which can be a simple microwave diode device or a more complex circuit, demodulates (rectifies) the received signal. This signal is then amplified and filtered before being applied to a threshold detector. If the signal is above the set threshold, then the system declares a detection and passes a signal to the output.

Two filters are used in this circuit. The 60 Hz notch filter is designed to remove interfering signals picked up from the 60 Hz power wiring in the vicinity. If full-wave-rectified dc power supplies are used in the detection equipment, then it may be necessary also to include a 120 Hz filter to remove spurious modulation of the radar signal caused by the residual ripple on the power supply output.

Some microwave radar proximity detectors are range gated. The maximum unambiguous range is a function of the pulse repetition rate or, more specifically, the interpulse interval. That sets the maximum range that an emitted pulse can reach before a second pulse is emitted from the source. Because the signal travels at a fixed rate, the maximum range can be translated into a maximum time T_{max} (Fig. 9–19). The distance to the object is found

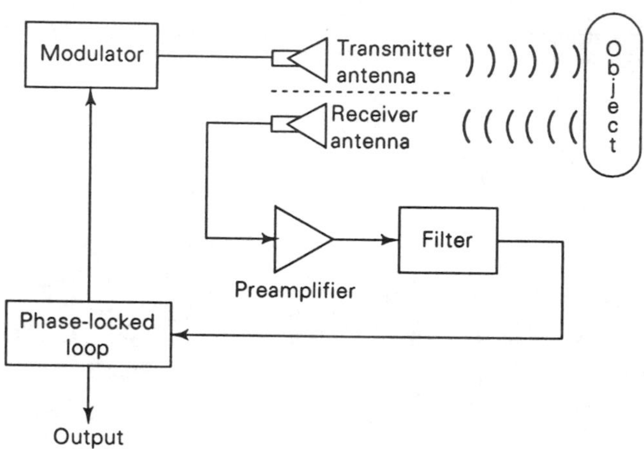

Figure 9–20 Bistatic microwave presence sensor.

by measuring the time t between the emitted pulse and the return pulse. In all cases $t \leq T_{max}$.

A very common form of radar proximity detector is shown in Fig. 9–20. This device is a *presence* sensor in that it can detect both moving and nonmoving targets. In this type of system there are separate antennas for the receiver and transmitter (i.e., the system is *bistatic*). These antennas may or may not be collocated, but there must be a high degree of isolation between them. The goal is to prevent the receiver from picking up any energy at all unless an object is present in the field of regard.

The microwave source is a phase-locked loop (PLL) oscillator, which produces a microwave output frequency that is modulated to distinguish it from other signals. When an object is present in the field of regard, it reflects signal back into the receiver antenna. This signal is amplified and then filtered. If the signal passes through the filter, then it can be used to phase-lock the PLL—an event that is detectable.

THE DOPPLER EFFECT AND VELOCITY MEASUREMENT

As many speeders ruefully admit, radar can be used to measure velocity. The phenomenon utilized in this instrument is the *Doppler effect*. The apparent frequency of waves changes if there is a nonzero relative velocity between the wave source and the observer. The classic example of the Doppler effect is the train whistle. As the train approaches, the track-side observer hears the pitch of the whistle rise an amount proportional to the train's speed. As it passes, however, the pitch abruptly shifts to a point lower than its zero-velocity value.

Thus, the direction of the relative velocity determines whether frequency increases or decreases, whereas velocity determines how great a change takes place. The Doppler frequency F_d is the amount of frequency shift:

$$F_d = \frac{2v}{\lambda} \tag{9–17}$$

$$F_d = \frac{2vF}{c} \tag{9–18}$$

$$F_d = \frac{89.4v}{\lambda} \tag{9–19}$$

$$F_d = \frac{2vF}{c} \cos \theta \tag{9–20}$$

where F_d is the Doppler shift in hertz
v is the relative velocity in meters per second
λ is the radar wavelength in meters
c is the speed of light (3×10^8 m/s)
θ is the angle of arrival.

Figure 9–21 shows a simple Doppler velocity measurement scheme. The motion of the target relative to the radar antenna will produce a Doppler shift of the original radar signal. If a sample of the original signal F_o is mixed with the Doppler-shifted reflected signal F_d, then the difference signal $F = F_o - F_d$ is output to the filter circuit. This circuit limits the signal bandwidth to the expected Doppler-shifted ΔF component. A counter or discriminator circuit will yield the velocity data.

Note that the accuracy of the Doppler measurement of velocity is a function of the relative angle θ between the antenna and target. If this angle becomes too large, then a serious error will emerge.

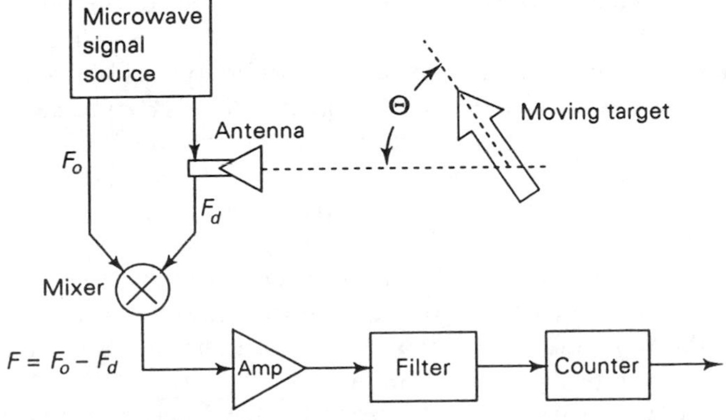

Figure 9–21 Moving-target/velocity sensor.

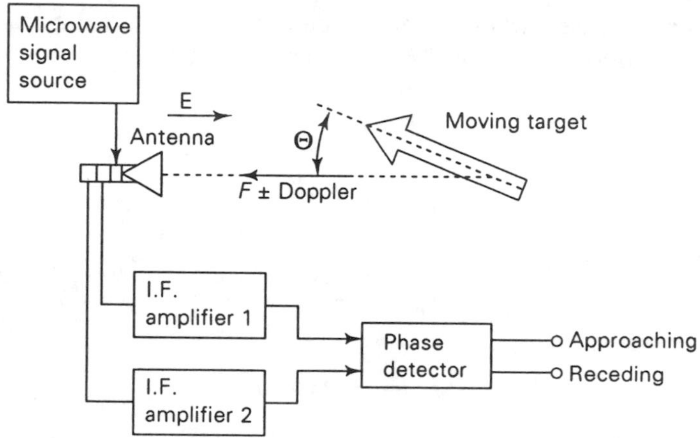

Figure 9–22 Adaptation showing approaching or receding target.

A direction-of-travel version of Doppler radar is shown in Fig. 9–22. The receiver signal is split into two channels, I.F. amplifiers 1 and 2. The output of those two channels are compared in a phase detector. The phase difference is a measure of the direction of travel of the target.

ULTRASONIC PROXIMITY DETECTORS

Ultrasonic waves are acoustical vibrations in the range above human hearing. Although often thought of as "radio" frequencies, these same frequencies can be generated as sonic waves. The difference between a radio wave (such as used in microwave detectors) and ultrasonic waves is that the radio wave is electromagnetic, whereas the ultrasonic wave is acoustical. That is, the ultrasonic wave consists of variations of the air pressure (the same as sound). Thus, a 40 kHz electromagnetic wave is a radio wave, whereas a 40 kHz acoustical wave is an ultrasonic wave. For the purposes of proximity detection, acoustical frequencies between 20 kHz and 60 kHz are used.

The ultrasonic proximity detector uses an ultrasonic transducer to create acoustical waves that flood the protected space [Fig. 9–23(a)]. A receiving transducer (which may be the same type as the transmitter, for they are bidirectional devices) picks up the ambient ultrasonic signal. The receiver may be only a few inches from the transmitter, or across the room.

When ultrasound waves propagate throughout the room, some travel directly to the receiver, while others follow an indirect route after reflecting from floor, ceiling, walls, and stationary objects in the room. When all the direct and reflected waves reach the receiver, they interfere with one another, creating a pattern of *standing waves* (Fig. 9–23(b)]. The wavelength of the

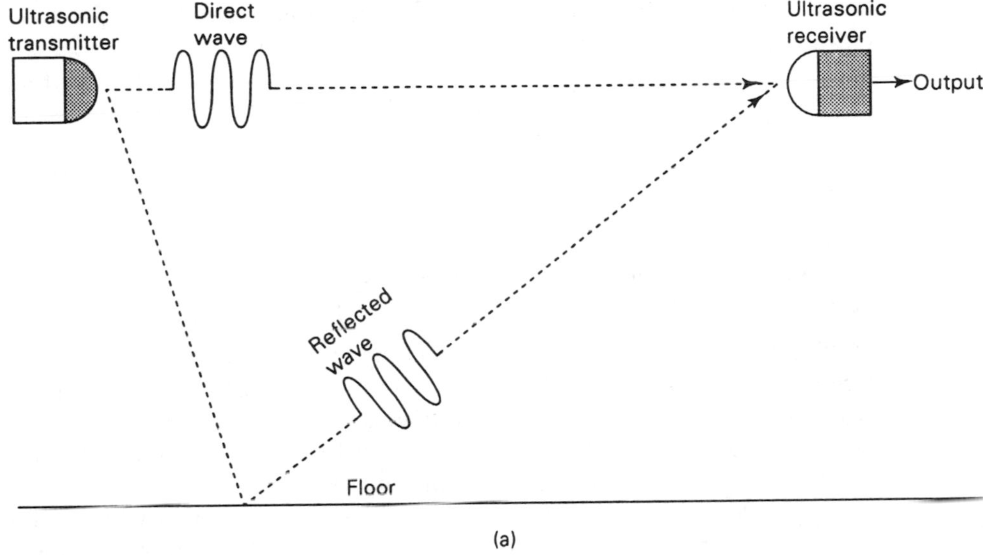

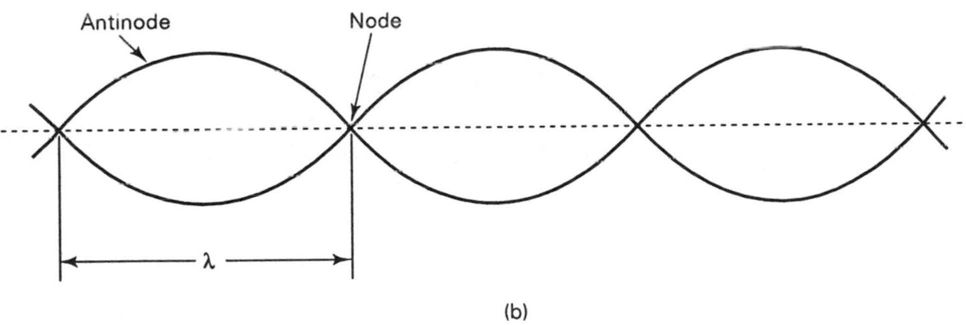

Figure 9–23 (a) Ultrasonic presence detector; (b) standing wave pattern.

standing wave is a function of the ultrasonic frequency and the velocity of travel in air (≈12,480 in./s):

$$\lambda = \frac{12{,}480 \text{ in./s}}{2F_s} \tag{9–21}$$

The factor 2 in the denominator appears because the standing wave is created by the interaction of direct and reflected waves, rather than by the same wave.

The simple ultrasonic proximity detector measures the amplitude of the standing wave. As long as no one enters the room, the standing wave remains constant. But if an intruder enters, a new reflection path is created, which alters the standing wave at the receiver. The amplitude change thus produced alerts the system of an intruder.

The Doppler effect can also be used in ultrasonic proximity detectors. The Doppler frequency produced is given by

$$F_d = \frac{2V_i}{\lambda_s} \tag{9–22}$$

where F_d is the Doppler frequency
V_i is the velocity of the intruder object
λ_s is the ultrasonic transmitting wavelength

When no intruder is present, the Doppler frequency is zero, because all reflected and direct waves are of the same frequency. But when an intruder comes into the space, the Doppler component is produced. If the receiver is equipped with a filter circuit that will pass only these frequencies, then the system will discriminate moving targets.

A problem that is frequently seen in these systems is that vibrations of walls and windows, moving curtains, and even small rodents can create enough motion to set off a misadjusted alarm. (The usual fault is that the sensitivity is too high and the filter frequency is too low).

REFERENCE

1. Carr, William W., SenTech Corporation. "Eddy-Current Proximity Sensors." *Sensors* (November 1987): 23ff.

Flow and Flow-Rate Sensors 10

Flow is defined as the motion of a fluid. The material being measured might be blood in an artery; respiratory gases into and out of a ventilator tube; water passing through a pipe, a sludge, or slurry in an open channel; gasoline or petroleum in a pipeline; or any of a number of other types of fluids. Flow can be either turbulent or smooth. The accurate measurement of turbulent flow is dauntingly difficult to model mathematically, and some of those difficulties translate into measurement problems. Fortunately, most mechanical flow sensors have inertia, so they tend to integrate out small variations due to turbulence (inertia of the sensor tends to act like a low-pass filter, or time-averager). Ultrasonic sensors, however (covered later in this chapter), do not possess the inertia of the mechanical sensors, so they sometimes tend to be more than a bit sensitive to turbulence in the system.

LAMINAR AND TURBULENT FLOW

The flow of liquids and gases can be either laminar or turbulent. *Laminar* flow [Fig. 10–1(a)] is smooth, orderly, and regular. Each volume cell in a laminar system flows parallel to the vessel wall and to other cells. Small discontinuities, which could give rise to turbulence, are damped out rapidly only a short distance after the point where they are created. A subset of laminar flow, *uniform flow*, exists "when all particles of the fluid across a section are flowing at the same velocity."[1] Uniform flow is idealized, and is only realized for short distances.

Turbulent flow [Fig. 10–1(b)], in contrast, is anything but regular, with

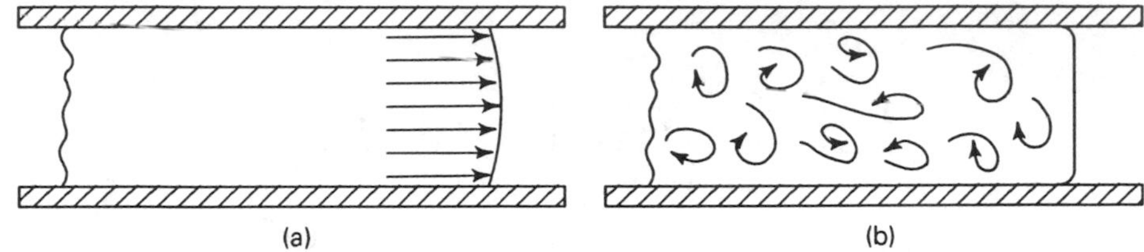

Figure 10–1 (a) Laminar flow; (b) turbulent flow.

whorls, eddies, and vortexes arising and disappearing in a seemingly random manner. In the past, turbulent flow was modeled as a random phenomenon, but more recent research has demonstrated that it is actually *chaotic* rather than random. Although chaotic phenomena appear to be random, they are in fact deterministic, but with a strange attractor at the center.[2]

Flow in a capillary is described by the Hagen-Poiseuille law (early nineteenth century) and can be found from the equation

$$W = \frac{\pi \Delta P D^4 K'}{128 \mu L} \tag{10–1}$$

where W is the flow in cubic centimeters per second
ΔP is the pressure difference between the ends of the capillary
D is the internal diameter of the capillary in centimeters
μ is the viscosity of the fluid in dyne · seconds per square centimeter
K' is a proportionality constant
L is the capillary length in centimeters

Turbulent and laminar flow can be discriminated in the system by examination of the *Reynolds number* (Re) of the system, which is:

$$\text{Re} = \frac{\zeta \overline{V} D}{\mu} \tag{10–2}$$

where Re is the Reynolds number
ζ is the fluid density
$\overline{V}$ is the mean fluid velocity

In general, Reynolds numbers above 4000 indicate turbulent flow, whereas numbers below 2000 indicate laminar flow.

FLOW, TOTAL FLOW (FLOW VOLUME), FLOW RATE

Many flow sensors actually measure *flow rate*, of which there are two general categories: *mass flow rate* and *volumetric flow rate*.[3] Mass flow rate is expressed in units of *mass transfered per unit of time* (e.g., kilograms per second or pound-mass per second). Volumetric flow rate is expressed in terms of

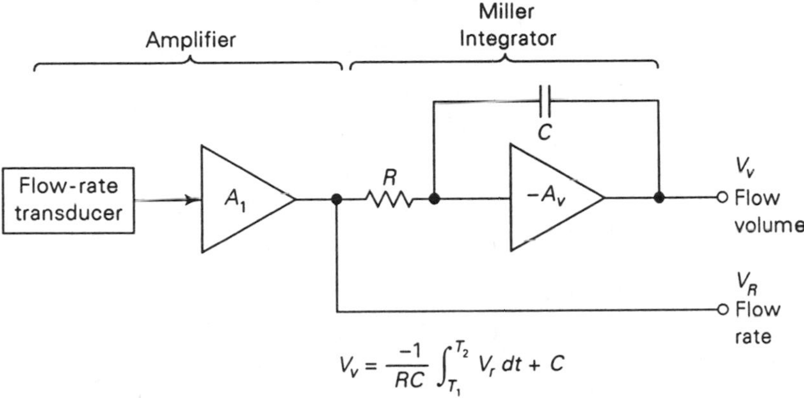

Figure 10–2 Integration of flow signal produces flow volume.

volume of material transfered per unit of time (e.g., cubic meters per second or gallons per minute). In gas systems the mass and volume flow rates are both expressed in terms of volume flow, but mass flow is referenced to standard temperature and pressure (STP) and converted to the equivalent volume flow. Examples of practical flow measurement instruments include a medical respiratory flowmeter that measures the patient's inspiration or expiration in liters per minute (l/min) and a certain medical fluid sensor that measures fluid flow in cubic centimeters per second (cc/s).

The *total flow* or *flow-volume* data can be derived from flow-rate data by integrating the flow signal, as in Fig. 10–2. The output of the flow sensor is amplified (and possibly further processed) in amplifier A_1. The output voltage of A_1 is proportional to the flow rate. This same signal is applied to an electronic integrator circuit that produces the output:

$$V_v = \frac{-1}{RC} \int_{t_1}^{t_2} V_r \, dt + c \tag{10–3}$$

Important considerations are the time constant RC of the integrator, which should be long compared with variations in the flow rates, and the ratio $-1/RC$, because it can lead to extremely high integrator gain for commonly available values of R and C.

FLOW DETECTORS

A *flow detector* is a circuit that informs of the *presence* of flow but does not quantify either the flow rate or the flow volume. Two such systems are shown in Fig. 10–3. In Fig. 10–3(a) is a thermistor-bridge version. Two thermistors, RT_1 and RT_2, are placed inside the flow container (in this case a tee plumbing connector). The purpose of this circuit is to detect whether

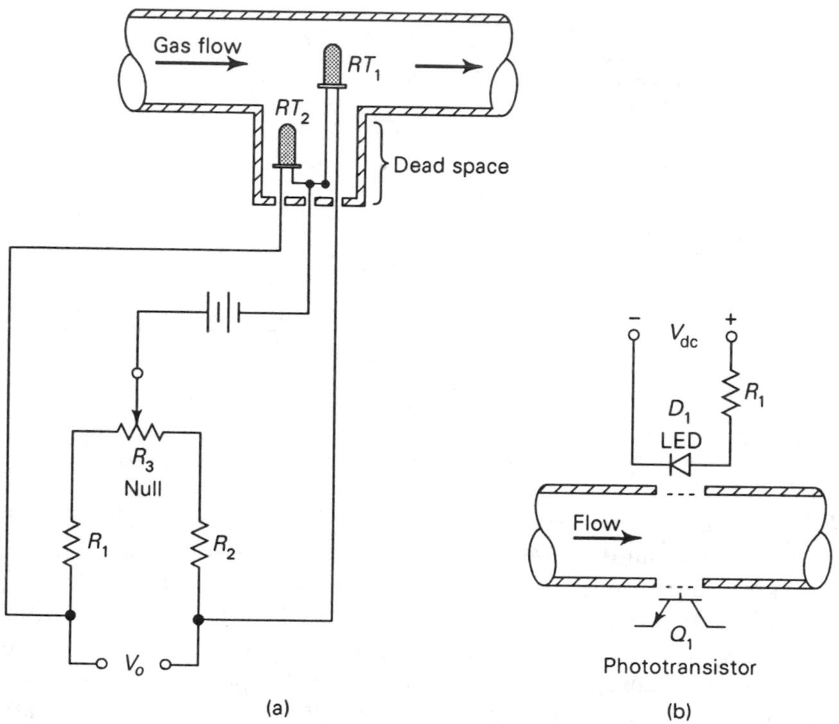

Figure 10–3 (a) Thermistor flow sensor; (b) electro-optical flow sensor.

or not air is getting through the line—"how much" is less important because this sensor is a simplified alarm. Thermistor RT_1 is placed in the airstream, while RT_2 is in a dead space off the main flow (it makes the ambient measurement).

The two thermistors in Fig. 10–3(a) are connected in an ordinary Wheatstone bridge circuit and are excited by a dc source to the point of self-heating. When $R_1/RT_2 = R_2/RT_1$, the output voltage V_o is zero. The voltage source V biases the thermistors only to the point of self-heating, but not more. At this point the thermistor resistance is the most sensitive to changes in airflow, which cools the surface of the device. When airflow changes the device resistance, the output voltage changes also.

The output voltage is proportional to the temperature change caused by the flowing air. Although it is possible to calibrate this temperature change over a reasonable range of flow rates, it is not necessary in this application. The flow-existence sensor is used in instruments such as respirator alarms, where it is not necessary to know either the flow rate or the flow volume, but where the cessation of flow indicates a potential medical emergency.

Figure 10–3(b) shows an electro-optical device that is used to detect the flow of opaque fluids. Here a light source (an LED in this case) shines across the flow path to a detector (a phototransistor in this case). The liquid or gas must be translucent or at least must pass certain wavelengths of light. For example, ordinary visible red LEDs (or other light sources) can be used for an opaque liquid. Gaseous carbon dioxide (CO_2), on the other hand absorbs infrared energy. Thus, the presence of CO_2 can be detected by making the light source and the detector IR-sensitive. Other gases and liquids may absorb other wavelengths, so each system must be designed with those properties in mind. The existence of inclusions in the liquid or gas makes it easier for this sensor to work.

POTENTIOMETER SYSTEMS

Figure 10–4(a) shows a system that uses a potentiometer displacement sensor to measure flow volume (not flow rate). This system is used on certain syringe pumps in medical devices. An ordinary glass or plastic syringe is placed in a saddle with a worm-gear pump and motor system. As the worm gear advances, it pushes the syringe plunger, thereby expressing fluid out of the tubing. A rectilinear potentiometer is ganged to the pressure plate on the end of the worm gear. Thus, the displacement of this plate is a measure of volume expressed. For example, a 50 cc syringe is connected to a potentiometer that is, in turn, connected to a $+V_{ref}$ of +5.00 V. Figure 10–4(b) shows the output function of this system. The output voltage is 5 V at 50 cc, so the scaling factor is 5 V/50 cc, or 100 mV/cc.

A mechanical flap or vane system is shown in Fig. 10–5. In this flow measurement system, the vane inserted into the flow path is deflected an amount proportional to the flow rate. In purely mechanical versions of this system, the vane is connected to a pointer or indicator that records the amount of vane deflection, hence the flow rate. In electrical output systems, the vane is ganged to a potentiometer. One side of the potentiometer is connected to a reference potential V_{ref}, and the other side is grounded. When the vane is deflected through any angle θ_d, the potentiometer wiper resistance setting changes ΔR, so the output voltage also changes by a proportional amount. The output voltage V_o appearing at the potentiometer wiper is proportional to the deflection of the vane such that $0 \leq V_o \leq V_{ref}$, and $V_o \propto \theta_d$.

OTHER MECHANICAL SENSOR SYSTEMS

Another gas flow system is shown in Fig. 10–6. Two versions are presented. Fig. 10–6(a) is a magnetic system, while Fig. 10–6(b) is an optical one. In Fig. 10–6(a) a small magnet is introduced into the flow stream. This form of sensor is used for both liquids and gases and is also usable in closed systems where it is difficult to introduce other forms of sensors. Coils L_1 and L_2 are

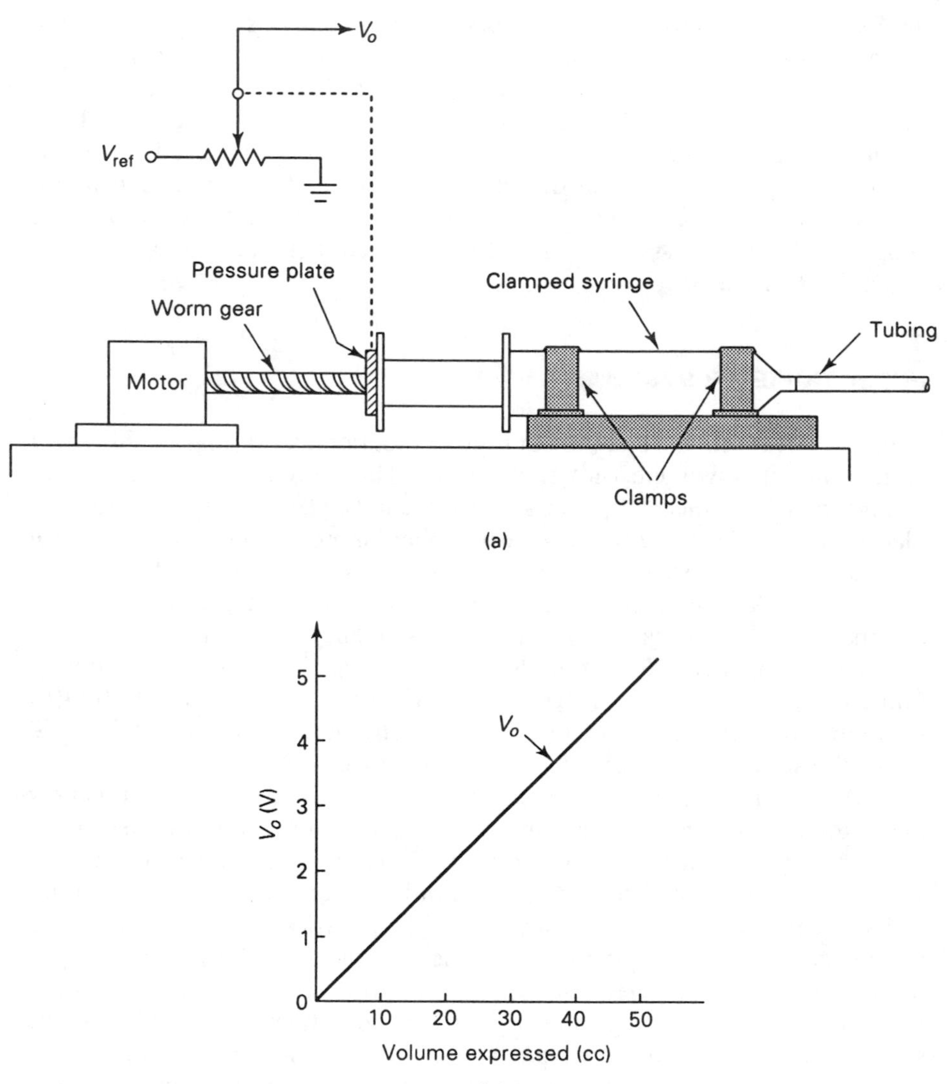

Figure 10–4 (a) Potentiometric flow sensor for syringe pump; (b) output characteristic.

placed at right angles to the flow path, at the point where the magnetic rotor is placed. When a moving magnetic field cuts across the turns of a coil, a current is introduced into the coil, causing a voltage to appear. Thus, a voltage V_o is found at the output of the series-aiding connected coils. The amplitude of the voltage is proportional to the magnetic field, whereas its

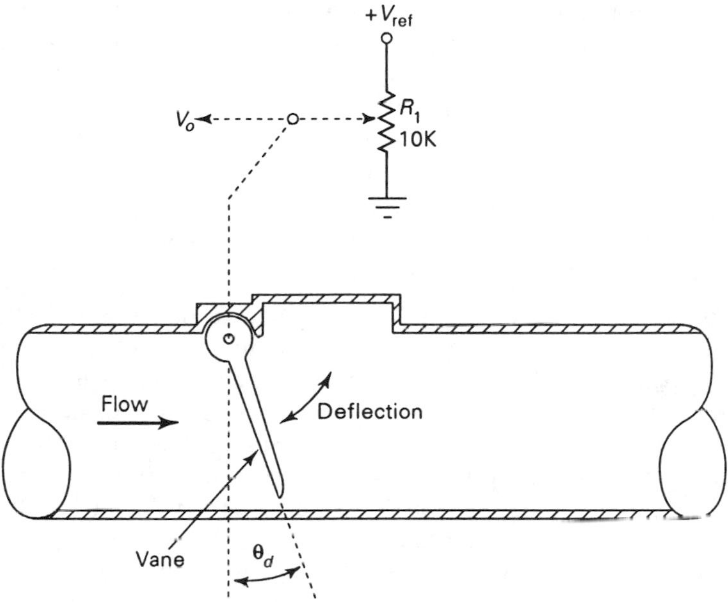

Figure 10–5 Vane flow meter.

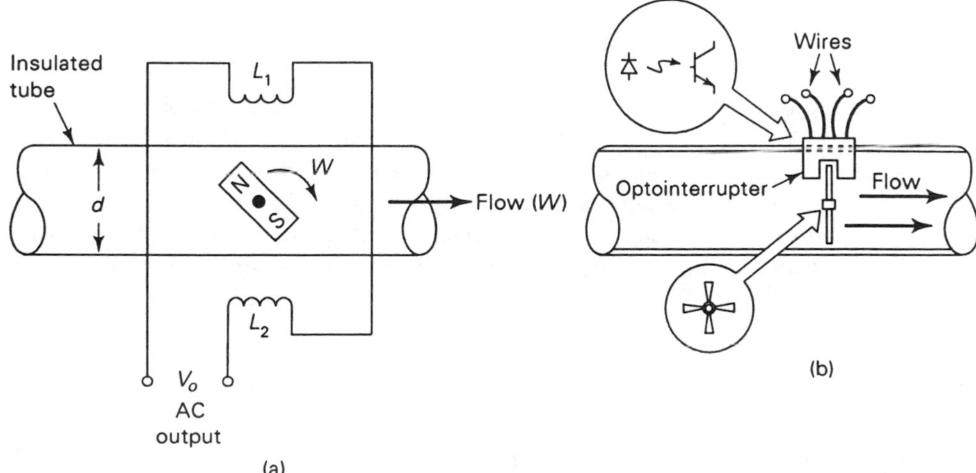

Figure 10–6 (a) Magnetic flow sensor; (b) rotor wheel flow sensor.

frequency is proportional to the rotational frequency of the magnet. The rotational frequency of the magnet is related to the flow rate by the equation

$$V_o = \frac{1.27 \times 10^{-4} HW}{d} \tag{10–4}$$

where V_o is the output potential in microvolts
H is the magnetic field strength in gauss
W is the flow rate in cubic centimeters per second
d is the diameter of the flow path

Sensors of this type are used in a wide variety of instrumentation applications.

Figure 10–6(b) shows the optical version. An optointerrupter is a device that places an LED and a phototransistor across an open path. When the path is blinded, the phototransistor is darkened; when the path is not blinded, the phototransistor is illuminated. Similar interrupters are used in applications such as the PAPER OUT sensor in computer printers and tape sensors in recorders. The light path in the sensor of Fig. 10–6(b) is interrupted by a multibladed fan placed in the flow path. Again, the frequency of the output signal is proportional to the flow rate.

With the right circuitry a rotating flow-rate sensor can also produce a flow-volume signal. An ac signal can be used to trigger a one-shot multivibrator to produce a train of pulses that are of constant duration and constant amplitude. The only variation is the pulse repetition rate (which is equal to the ac signal frequency from the sensor). Integration of these pulses produces a dc voltage that is proportional to the total area under all the pulses per unit of time, in other words the flow volume.

PRESSURE-DROP SYSTEMS

If a fluid or gas flowing in a closed pipe or vessel passes through a narrowing segment, or constriction (*vena contracta*), the pressure will drop between the proximal (upstream) and distal (downstream) sides of the constriction.[4] Figure 10–7 shows an example of a common form of flow-rate sensor in which a wire-mesh obstruction is placed across the flow path. The total pressure P_t is the sum of the static pressure head P_s and the dynamic velocity pressure P_d:[5]

$$P_t = P_s + P_d \tag{10–5}$$

According to Bernoulli's law, the dynamic velocity pressure is given by

$$P_d = \frac{\phi V^2}{2} \tag{10–6}$$

where ϕ is the fluid mass density (e.g., $\text{lb} \cdot \text{sec}^2/\text{ft}^4$)
V is the fluid velocity in feet per second

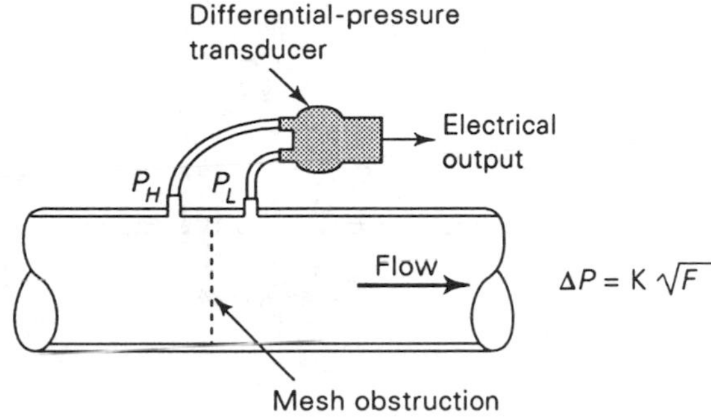

Figure 10–7 Pressure-drop flow measurement using wire mesh.

Combining Eqs. (10–5) and (10–6) yields

$$P_t = P_s + \frac{\phi V^2}{2} \tag{10–7}$$

Solving for V, we obtain

$$V^2 = \frac{2(P_t - P_s)}{\phi} \tag{10–8}$$

$$V = \sqrt{\frac{2(P_t - P_s)}{\phi}} \tag{10–9}$$

Thus, an obstruction placed in thc gas path will create a pressure drop P_d that is proportional to the square root of the flow rate, or

$$P_d = K\sqrt{F} \tag{10–10}$$

where P_d is the pressure drop
K is a sensitivity constant
F is the flow rate

The obstruction shown in Fig. 10–7 is a mesh of wire or plastic cloth stretched tautly across a constant-diameter pathway. The usual mesh density for medical respiratory measurements is 400 grid/in. across a 50 mm diameter, resulting in a ΔP of approximately 0.09 cm H_2O for a flow of respiratory gases (N_2, O_2, and water vapor) of 10 l/min.[6] For gases other than air, and other flow rates, the mesh density may be different. The pressure drop ΔP between the high-pressure side P_H and the low-pressure side P_L is measured by a *differential-pressure sensor*. Such sensors have two ports, one on each side of the diaphragm. Most other pressure sensors measure gauge pressure, so they are open on one side to the atmosphere and are therefore not suitable for this type of measurement.

A *cannula pressure-drop sensor*, also called a *Pitot-static tube*, is shown in Fig. 10–8. The cannula is a double-lumen tube in which the distal openings

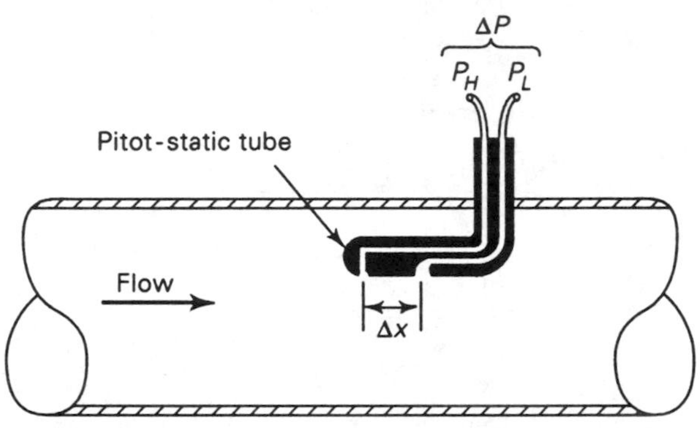

Figure 10–8 Cannula Pitot tube for pressure-drop flow sensing.

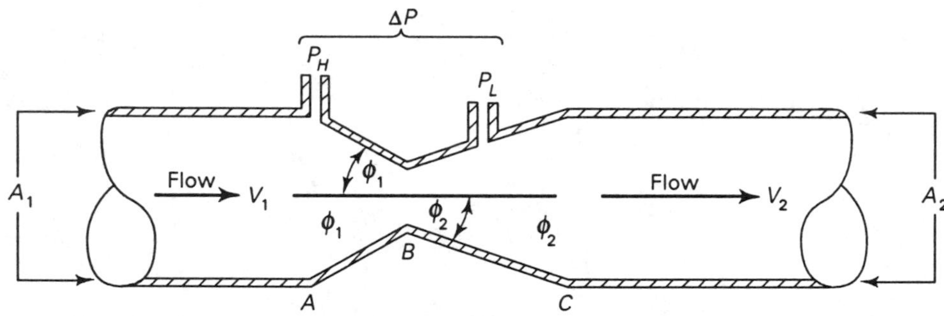

Figure 10–9 Diameter-change pressure-drop flowmeter.

of the lumen channels are spaced Δx apart. These openings are able to measure the pressures at two points, so a ΔP related to distance Δx can be determined. The proximal ends of the lumen channels are connected across a differential-pressure sensor. The expression for the pressure drop is[7]

$$\frac{-\Delta P}{\Delta x} = \left(\frac{1.1\xi}{\pi g a^2}\frac{dF}{dt}\right) + \left(\frac{12.8\mu F}{\pi g a^4}\right) \qquad (10\text{–}11)$$

where $\Delta P/\Delta x$ is the change of pressure over distance Δx in centimeters of water per centimeter.
ζ is fluid density in grams per cubic centimeter
a is the inner diameter of the vessel in centimeters
F is the fluid flow rate in cubic centimeters per second
μ is the fluid viscosity
g is 980 cm/sec^2

A *venturi tube* flow sensor is shown in Fig. 10–9. In this type of path-constriction sensor, a pair of back-to-back conical segments (AB and BC), with sides angled ϕ_1 and ϕ_2, form a region in which the density of the flowing

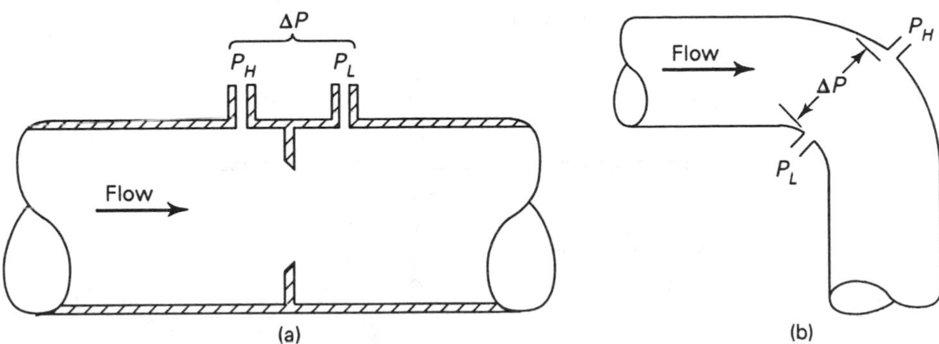

Figure 10–10 (a) Obstruction pressure-drop flowmeter; (b) direction-change pressure-drop flow sensor.

gas changes by virtue of the change in path diameter (ϕ_1 and ϕ_2). The velocity also changes from V_1 to V_2 in this region, but the relation $A_1V_1 = A_2V_2$ remains constant. The pressures at the two ends of the venturi region (P_H and P_L) are related by the expression

$$P_L + \frac{\phi_2 V_2^2}{2} = P_H + \frac{\phi_1 V_1^2}{2} \tag{10–12}$$

Thus,

$$V_2^2 - V_1^2 = \frac{2(P_H - P_L)}{\phi} \tag{10–13}$$

Because $A_1V_1 = A_2V_2$:

$$V_2 = \sqrt{\frac{2(P_H - P_L)}{\phi(1 - A_2^2/A_1^2)}} \tag{10–14}$$

Again, the volumetric flow rate is proportional to the square root of the differential pressure.

Two other pressure-drop sensor configurations are shown in Fig. 10–10. Figure 10–10(a) shows the *orifice plate* obstruction, in which a solid plate with a hole in it that is smaller than the flow vessel is used as the obstruction. An *elbow section* (or *centrifugal section*) is shown in Fig. 10–10(b). In this type of sensor the pressure gradient ΔP is caused by the centrifugal force generated when the flow changes direction abruptly.

THERMAL SYSTEMS

There are several different approaches to measuring flow using thermal sensors. As early as 1911, C. C. Thomas described a thermal flowmeter in which the temperature of the flowing material was measured both proximally and

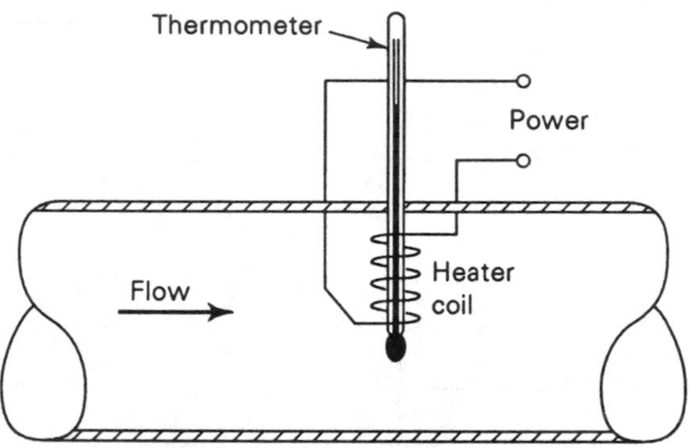

Figure 10–11 Self-heated thermometer flow sensor.

distally to a heater element.[8] In the 1920s the *self-heating anemometer* was invented. These devices are similar to the flow detector of Fig. 10–3(a) in that they depend on the cooling of a thermistor that is normally kept at a point just short of self-heating (i.e., the point where thermistor body temperature increases over ambient). In the 1930s the Willson Safety Products Company marketed a device similar to that shown in Fig. 10–11. In this sensor a mercury thermometer was wrapped with a resistance heater coil and placed in the gas flow stream. The applied heater voltage was maintained at a constant level. The change in temperature was compared with a nomograph supplied by the company, which had generated the data empirically in wind tunnel tests.[9]

In 1944, tests were done at Rutgers University, under the auspices of the U.S. Office of Science Research and Development, to see if the newly developed AT&T thermistor (see Chapter 4) could be used as a flowmeter.

One class of thermal anemometer flowmeter depends on King's law, which states that[10]

$$\frac{P}{\Delta T} = \alpha + \beta(\rho V)^n \tag{10–15}$$

where P is the power dissipated by the probe to maintain probe temperature at a constant level

ΔT is the difference between the probe temperature and the fluid temperature

ρ is the fluid density

V is the fluid velocity

n is a constant (0.5 is a common value) that depends on the probe characteristics and fluid viscosity

α and β are constants

Figure 10–12 shows a simple form of resistance anemometer in which

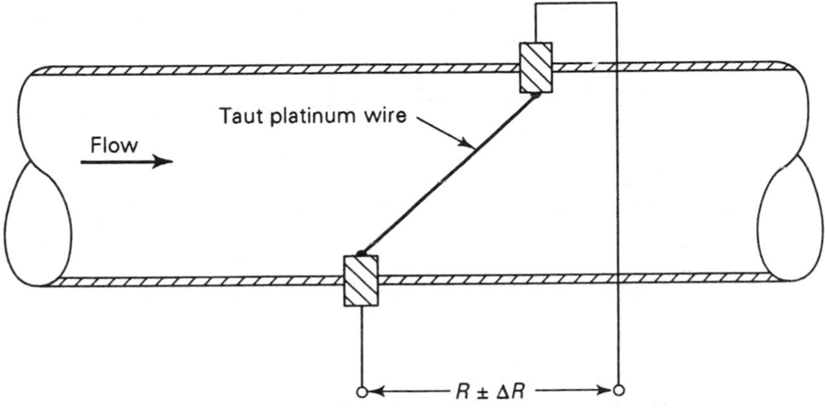

Figure 10–12 Taut platinum wire resistive flow sensor.

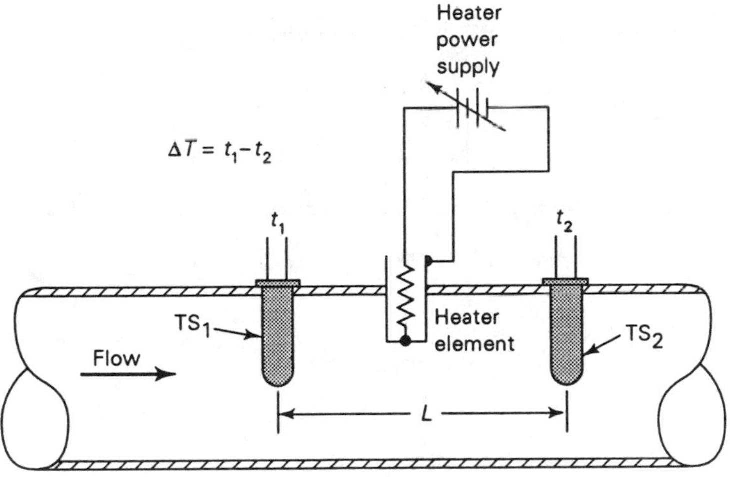

Figure 10–13 Temperature-drop flow sensor.

a platinum resistance element is stretched taut across the flow path. The resistance R of the element is proportional to both the length of the wire and its cross-sectional area, both of which change when the wire is heated or cooled. Thus, when electrical power is applied to the wire to bring it just below the point of self-heating, gases flowing across its surface will cause the sensor temperature to vary. The variations in resistance—reflecting changes in flow—can be used to indicate flow.

Figure 10–13 shows the basic Thomas thermal flow sensor. In this sensor configuration a resistive heating element is immersed in the fluid or gas flowing through the vessel. The proximal temperature sensor TS_1 measures the temperature of the incoming fluid prior to heating t_1, while the distal

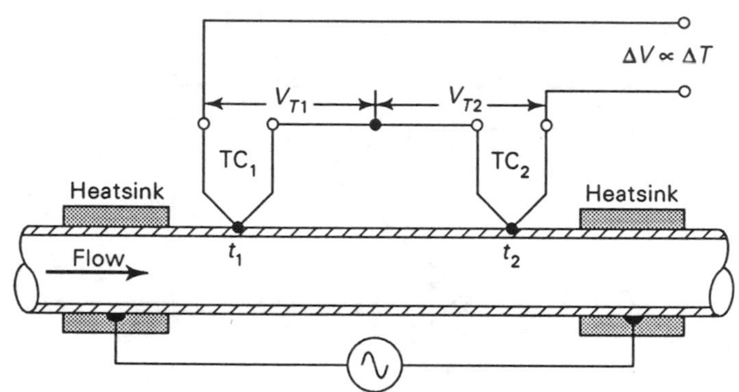

Figure 10–14 Thermocouple heat-drop flow sensor.

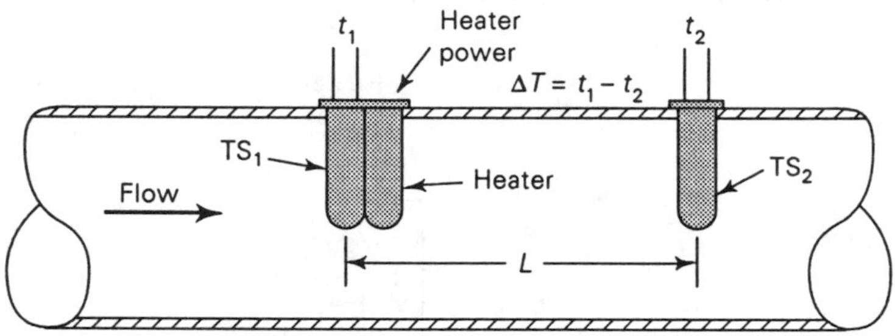

Figure 10–15 Heater-element flow sensor.

sensor TS_2 measures the temperature after heating t_2, over path length L. The heat flows according to the equation[11]

$$\frac{q}{\Delta T} = \overline{C}_p W \tag{10–16}$$

where q is the heat input over length L
ΔT is the temperature rise over L, that is, $t_2 - t_1$
$\overline{C}_p$ is the specific heat of the fluid flowing in the vessel
W is the mass flow rate

A variation on this system is the *heated-conduit flowmeter* of Fig. 10–14, which can be used for small-diameter flow vessels. In this configuration a segment of thermally conductive conduit is mounted with a pair of heatsinks that, when connected to power, cause the conduit temperature to rise. A pair of thermocouples (TC_1 and TC_2) are mounted at points along the heated portion of the conduit. The output voltage of each thermocouple (V_{t1} and V_{t2}) is proportional to the temperature at each measurement point (t_1 and t_2). The heat rise ($\Delta T = t_2 - t_1$) is proportional to the flow rate.

Still another variant is shown in Fig. 10–15. In this configuration the

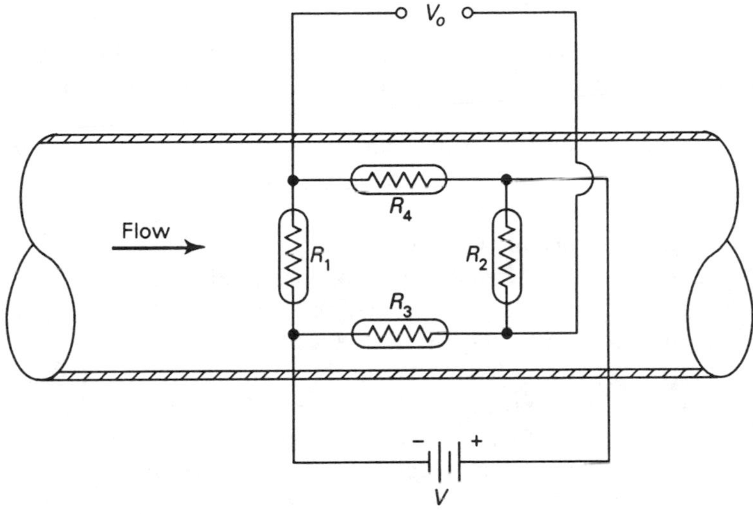

Figure 10–16 Resistive bridge sensor.

constant-wattage heater element is mounted in intimate contact with the first temperature sensor TS_1, while a second sensor TS_2 is positioned downstream. This second sensor is usually given a little extra thermal mass so that it matches the sensor/heater pair upstream.[12] This mass flowmeter depends on the relationship

$$\rho V = k\left(\frac{q}{\Delta T}\right)^{1.67} \qquad (10\text{–}17)$$

where ρ is the fluid density
V is the fluid velocity
K is a constant (empirically derived)
q is the heat flow rate
ΔT is the temperature rise over length L

Newer sensors and newer sensor configurations are being developed with silicon integrated circuit technology. For example, a bridge thermal flow sensor (Fig. 10–16) has been developed that uses silicon thermistors.[13] Other silicon sensors are also being developed.[14]

ULTRASONIC FLOW MEASUREMENT SYSTEMS

Ultrasonic waves are *acoustic* waves (i.e., mechanical vibrations in a medium such as liquid or gas) that have a frequency above the range of human hearing (>20 kHz). In measurement applications, *ultrasonic* can mean anything from

20 kHz to about 20 MHz. It is important to remember that ultrasonics are acoustic waves, not electromagnetic (i.e., radio) waves. Several types of ultrasonic sensors are available, but most of them are of either the *dynamic* or *piezoelectric* category.

A dynamic ultrasonic transducer is similar to a dynamic microphone: a thin, low-mass, metal diaphragm tautly stretched over a passive electromagnet. These sensors are used for relatively low frequencies (<100 kHz) and are often found in ultrasonic intruder alarms.

Piezoelectricity is a natural phenomenon, observed in some natural (quartz) and synthetic (ceramic) crystals, in which an electrical potential is generated when the crystal is mechanically deformed. Figure 10–17(a) shows a piezoelectric element mounted with electrodes and wires to measure the voltage appearing across the opposing faces. When the crystal is at rest [Fig. 10–17(b)], the output voltage is zero. But when the crystal slab is mechanically deformed to position $+\Delta x$ [Fig. 10–17(c)], the output voltage becomes nonzero, and the polarity is such that A is negative with respect to B. When the crystal is deformed to a position $-\Delta x$ [Fig. 10–17(d)], the voltage is still nonzero, but the polarity is reversed: A is positive with respect to B.

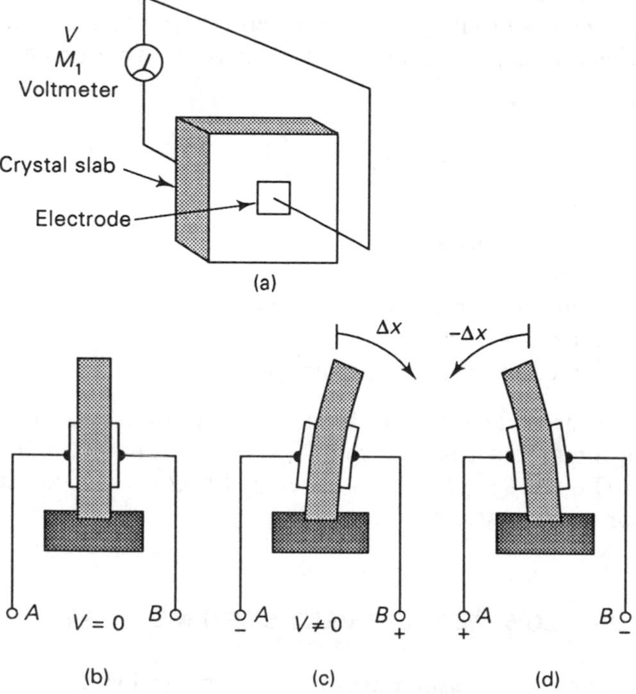

Figure 10–17 (a) Piezoelectric crystal sensor element; (b) at rest; (c) right deflection; (d) left deflection.

If a piezoelectric element is momentarily struck, it will vibrate at a resonant frequency (like a bell) that is determined mostly by the crystal slab's mechanical dimensions. The voltage appearing across the electrodes attached to the slab will be a sinusoidal waveform at the resonant frequency of the slab.

When a piezoelectric element is electrically stimulated, it will deform—a behavior opposite that shown in Fig. 10–17. Thus, when an electrical potential is applied to the crystal slab electrodes, the slab deforms by $-\Delta x$ or $+\Delta x$ according to the applied electrical polarity. In a class of circuits called *oscillators*, the crystal is periodically stimulated to keep the slab mechanically vibrating, while the output voltage is tapped for use as an ac signal. Radio transmitters, digital clocks, and a host of other devices are frequency- or time-controlled with piezoelectric crystal elements.

The piezoelectric crystals used as ultrasonic transducers are similar to crystal microphones. A piezoresistive crystal will vibrate when an ac electrical signal of its resonant frequency is applied. This vibrational energy is applied either to a diaphragm or directly to the medium being measured. Similarly, the same crystal will produce an ac signal at its resonant frequency when vibrated by an acoustic wave of that frequency. Thus, the piezoelectric crystal can be used for both transmit and receive functions (this dual usage is why the term *transducer* is preferred over the term *sensor* here). In ultrasonic flow meters, lead zirconate titanate (PZT) ceramic crystals are commonly used as transducers.

TYPES OF ULTRASONIC FLOWMETERS

Ultrasonic flowmeters can be classified into three categories: *Doppler*, *transit time*, and *echo sounding*. The Doppler form can be further classified as *flow detector* and *flow measurement* types.

Ultrasonic Doppler Flowmeters

The Doppler effect is discussed in several places in this book, because it is so widely used in instrumentation and measurement applications. As defined earlier, the Doppler effect is a change of frequency of an incident wave train due to relative motion between the source and the observer. The change of frequency:

$$\Delta f = \frac{2fV \cos \theta}{c_s} \qquad (10\text{–}18)$$

where f is the incident ultrasonic frequency in hertz
Δf is the change of frequency in hertz

V is the velocity of an object in the ultrasonic field*
θ is the angle between the ultrasonic beam and the velocity vector of the object causing the frequency change
c_s is the velocity of sound in the medium being measured*

The "object" in the ultrasonic field is easy enough to see in proximity detector systems, but it is perhaps a bit ambiguous in flowmeters and detectors. For a Doppler flowmeter system to work, it is necessary to have a so-called acoustic density interface (ADI), that is, an abrupt change in the acoustic density of the medium flowing, instead of an actual object. The object can be entrained air, solids, particulate matter, or even hydraulic density changes due to the medium's chemistry.[15] When particulate matter is the object that causes the Doppler shift, the particles need be only a few microns in diameter and have populations on the order of $\approx$100 parts per million (ppm). Thus, Doppler flow detection works with slurries, sludges, blood, or any other liquid that contains inclusions or air bubbles, but it does not work with pure liquids such as deionized distilled water.

Figure 10–18 shows a Doppler blood flow detector (not flow *meter*) based on ultrasonic piezoelectric crystals. Blood contains inclusions—red and white blood cells and other objects—that present an ADI to the ultrasonic wave. Two crystals are used in the Doppler sensor, one each for transmit and receive. The frequency of the incident signal f is changed by the Doppler effects as blood flows beneath the sensor. The reflected energy will contain frequency components of $f \pm \Delta f$, where Δf is the Doppler shift.

In blood flow monitors based on this technique, it is nearly impossible to calibrate this sensor for blood flow rate. The system is used to check for vessel patency, that is, to see whether or not blood is flowing, but not to measure blood flow rate. In other systems calibration is possible, however, because the Doppler frequency shift is proportional to the fluid velocity. By filtering the Doppler information it is possible to calibrate the system. The problem in simple medical systems is that blood vessels are distensible, that is, they flex as blood flows, and blood flow is pulsatile. In addition, blood vessels are located at varying distances beneath the surface.

For most blood flow detectors, frequencies in the 2 to 12 MHz region are used, although for special purposes narrower ranges are selected. For example, to examine only vessels close to the surface, ultrasonic frequencies in the 8 to 10 MHz range are preferred because they are rapidly attenuated in tissue. Thus, there is less interference from more deeply embedded blood vessels. For those deeper vessels, frequencies in the 5 to 6 MHz range are preferred. At those frequencies, with normal blood flow rates (<20 cm/s), the expected Δf components are in the audio range (<1500 Hz).

Physicians use the Doppler flowmeter to detect the presence of blood flow in underlying arteries in arms and legs, especially after surgery or when

*V and C_s are in the same units, for example, m/s

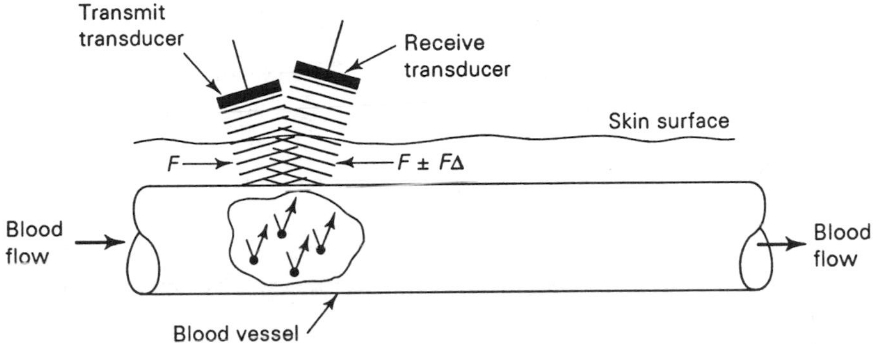

Figure 10–18 Ultrasonic Doppler flowmeter.

treating a crushing injury. A "sloshing" audio sound created when the Doppler return signal is heterodyned with the incident signal (producing an audio frequency difference signal) tells them that the vessel is patent.

A Doppler flowmeter system is shown in Fig. 10–19. This system uses the frequency change of a wave scattered from particulate matter flowing in the fluid path and is particularly useful in blood flowmeters, as well as in many other applications. It has been shown that the change in frequency Δf is given by

$$\Delta f = \frac{Vf_s(\cos \phi_1 + \cos \phi_2)}{C_s} \tag{10–19}$$

where Δf is the Doppler-shift frequency in hertz
f_s is the source frequency in hertz
V is the fluid velocity*
C_s is the velocity of sound in the medium*
ϕ_1 is the angle of the receive crystal to the flow axis
ϕ_2 is the angle of the transmit crystal to the flow axis

Ultrasonic Transit-Time Flowmeters

Transit-time flowmeters were patented by Rutten in Germany in 1928, but it wasn't until the early 1950s that the electronic circuits needed to process the signals became available for practical use. An example of a transit-time flow sensor is shown in Fig. 10–20. This system depends on the difference in the upstream and downstream transit times of acoustic pulses. A pair of piezoelectric crystal sensors are aimed at each other in an oblique path across the flow path. The angle between the crystal path and the flow path is θ. Both crystals are used for both transmit and receive functions. The system first fires a pulse downstream from A to B and measures its transit time; then

*V and C_s are in the same units, for example, cm/s

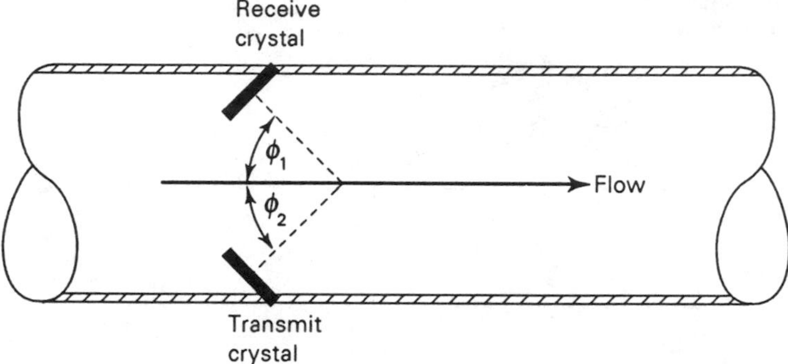

Figure 10–19 Ultrasonic flowmeter.

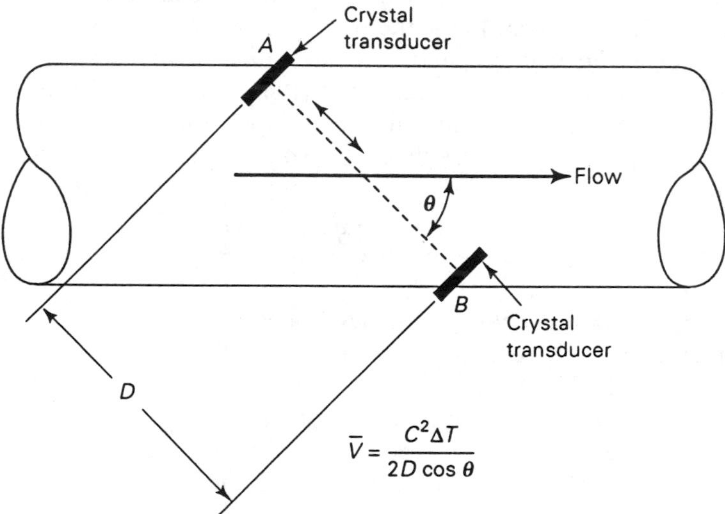

Figure 10–20 Ultrasonic transit-time flowmeter.

it fires an upstream pulse from B to A and measures its transit time. The average flow velocity is proportional to the difference between the upstream and downstream transit times ΔT:

$$\overline{V} = \frac{C_s^2(t_u - t_d)}{2D \cos \theta} \tag{10–20}$$

$$\overline{V} = \frac{C^2 \Delta T}{2D \cos \theta} \tag{10–21}$$

where $\overline{V}$ is the average flow velocity
C_s is the speed of the signal in the medium
t_u is the upstream velocity
t_d is the downstream velocity
ΔT is the difference between downstream and upstream transit times
D and θ are as defined in Fig. 10–20

Transit-time flowmeters complement Doppler flowmeters because they work on pure, inclusion-free liquids. In fact, the same air bubbles that make Doppler flowmeters work well can attenuate the signal in transit-time flowmeters so much that they become useless. In any event, inclusions add an error component to transit-time measurements.

Echo-Sounding Flowmeters

An open-channel system is one in which the flow channel is open and unrestricted in at least one dimension—for example, a channel in which the top is open to the atmosphere (Fig. 10–21). If the velocity and cross-sectional area of the fluid are known, then the flow rate is given by

$$Q = kAV \tag{10–22}$$

where Q is the flow volume
A is the cross-sectional area ($w \times h$)
V is the flow velocity
k is a proportionality constant related to the type of fluid in the channel

The common way to measure the flow rate is (as shown in Fig. 10–21) to pass the fluid through a restricted weir, flume, or nozzle that has a known width.[16] The height h of the fluid, or *head*, is found by using an echo-sounding method in which an ultrasonic pulse is fired into the fluid and the time of the return echo measured. This principle is similar to that used in radar and sonar systems. If the velocity of propagation of sound in the medium is

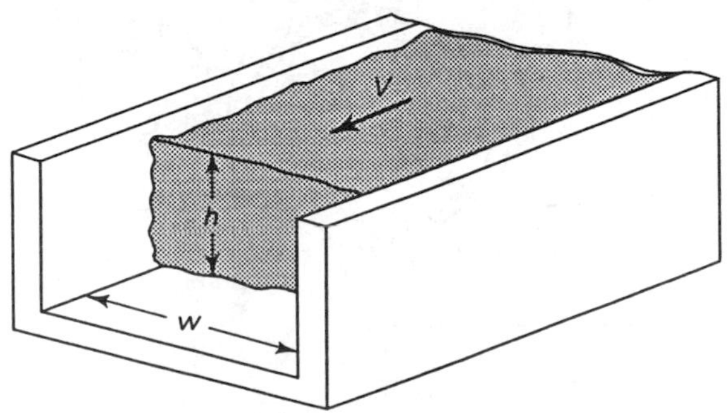

Figure 10–21 Open-channel flow sensing.

known, then the round-trip time represents the depth. Common depth finders used by boaters and anglers are examples of simple versions of this type of echo sounder.

This method of flow measurement requires two pieces of data: the head h, provided by the echo sounder, and the velocity V, provided by a Doppler flowmeter.

ELECTROMAGNETIC FLOWMETERS

An electromagnetic flow transducer uses as the transducible property the electromotive force that a conductor moving in a magnetic field creates across its length. Figure 10–22 shows how this property is used to measure the flow of conductive fluids. A vessel of radius a carries the fluid, with the flow coming out of the page, as shown by the vector symbol inside the vessel. A magnetic field is set up by a horseshoe electromagnet so that the magnetic field cuts across the fluid path. A pair of electrode contacts are placed in contact with the fluid orthogonal to the magnetic field. The EMF appearing across these electrodes is given by

$$V = \frac{FB}{50\pi a} \tag{10–23}$$

where V is the EMF appearing across the electrodes in microvolts
F is the fluid volumetric flow in cubic centimeters per second

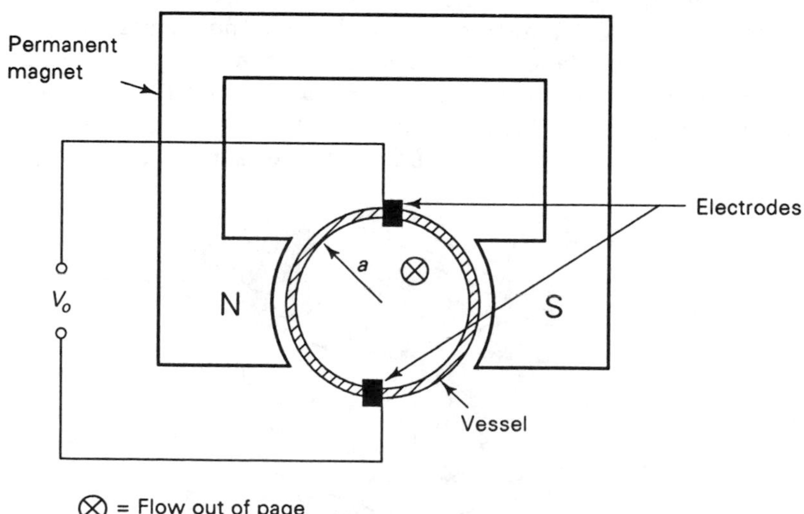

Figure 10–22 Magnetic flow sensing.

B is the magnetic field in gauss
a is the radius of the vessel in centimeters

Example 10–1

Calculate the potential appearing across the electrodes on a 1.75 cm radius vessel, when a magnetic field of 450 gauss is present and the flow rate is 240 cm³/s.

Solution

$$V = \frac{FB}{50\pi a}$$

$$V = \frac{(240\ \text{cm}^3/\text{s})\ (450\ \text{gauss})}{(50)\ (3.14)\ (1.75\ \text{cm})}$$

$$V = \frac{108{,}000}{274.75} = 393\ \mu V$$

The potentials associated with the electromagnetic flow sensor are easily handled in electronic circuitry, making this method quite feasible for a broad range of applications.

Capacitive Flowmeters

Figure 10–23 shows a method for measuring the flow of a thin liquid film. In this case the sensor is a coaxial, cylindrical capacitor. The nonconducting film flowing across the surface changes the dielectric constant of the capacitor, hence its capacitance. The actual capacitance can be measured in any of several ways, for example, with an electrometer or an FM oscillator.

Injectate Dilution Methods

An injectate dilution method is shown in Fig. 10–24. In life sciences and medicine this method is known as *Fick's method* of flow measurement. It consists of adding a fixed amount of tracer or dye to or subtracting it from a system upstream and then measuring the change in concentration as it flows down-

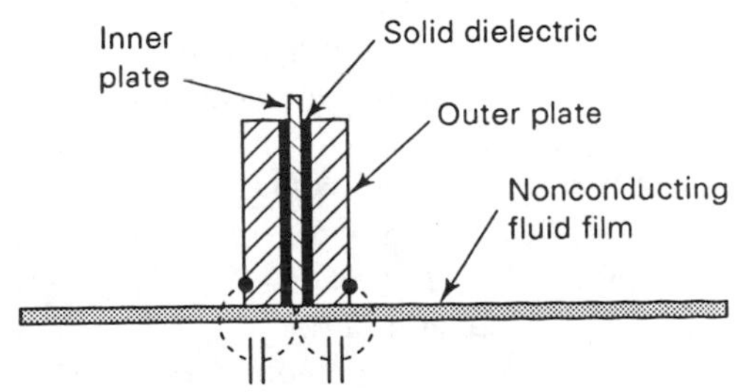

Figure 10–23 Capacitance-film flow sensing.

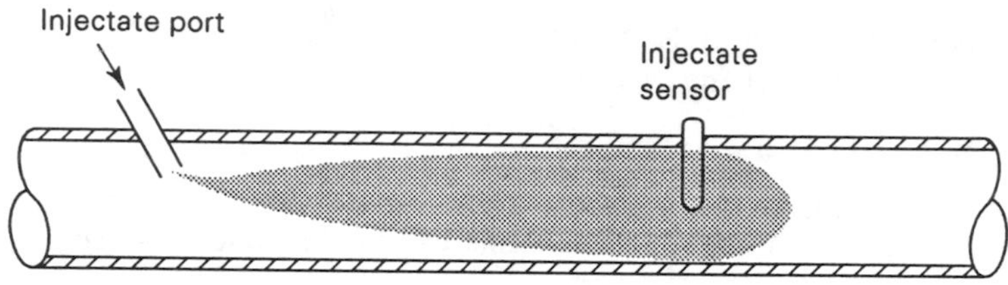

Figure 10–24 Dye injection/dilution flow measurement.

stream. According to Fick's principle, the volumetric flow rate is given by[17]

$$Q = \frac{M_i/\text{min.}}{\Delta C_i} \tag{10–24}$$

where M_i/min is the amount of dye or indicator added to, or removed from, the system per minute
ΔC_i is the change of concentration of the indicator over a fixed length path

An improvement on the Fick method, which still uses the configuration of Fig. 10–24, is to inject a *bolus* (a calibrated quantity rapidly injected) into the injectate port and then measure the change of concentration at the injectate sensor. In physiological settings indigo blue (an optical dye), radioisotopes, and ordinary chilled saline solution are used as indicator elements. If the indicator is a bolus, then the amount of indicator passing the sensor point per unit of time is given by

$$k\,dI_i = QC_t \tag{10–25}$$

where I_i is the total amount of indicator material
Q is the flow rate of the host liquid
C_t is the concentration at time t
k is an empirically determined constant

If we integrate both sides of Eq. (10–25) and solve for flow rate Q, we obtain the expression

$$Q = \frac{kI_i}{\int_0^\infty c_t\,dt} \tag{10–26}$$

A problem sometimes seen in closed-loop flow systems results when the material that is flowing passes the same point again after a period of time. If that period is short compared with the integration time, then there will exist a *recirculation artifact* (Fig. 10–25). This artifact tends to distort the integration and leads to error. A solution is to break the integration into two pieces: T_1-T_2 and T_2-T_4. The segment T_1-T_2 is integrated by ordinary means—the technique depending on whether analog or digital solutions are selected. But the segment T_2-T_4 requires the technique of *geometric inte-*

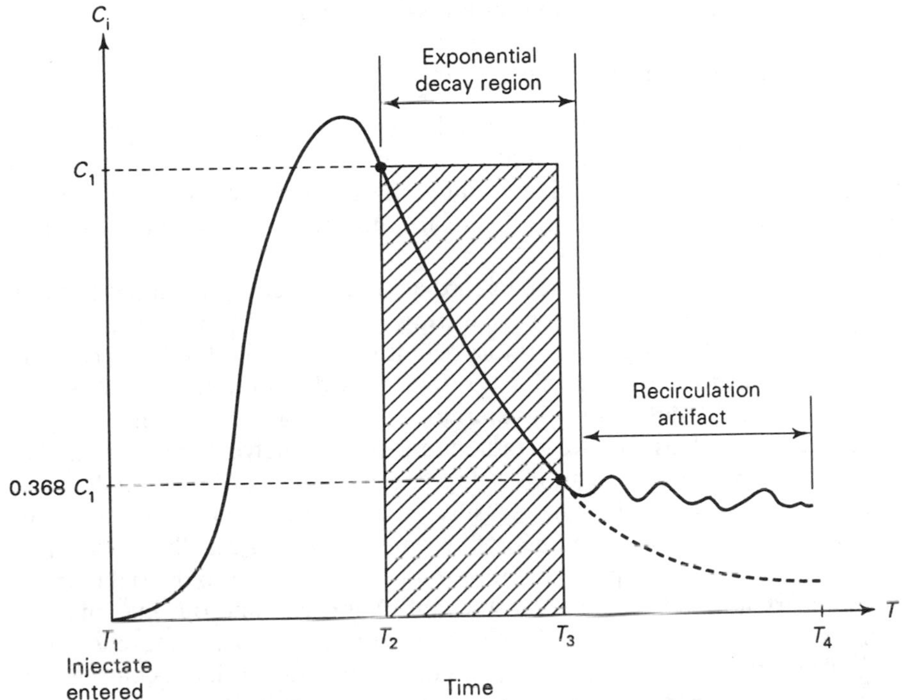

Figure 10–25 Geometric integration of the exponential decay portion of the flow waveform in dye-injection measurement.

gration. This segment is assumed to be an exponential decay, so a highly accurate estimate of the true integral can be found by erecting a rectangle from the peak and one time constant (0.37) points on the curve. The area of this rectangle $[C_1(t_3 - t_2)]$, shown as a shaded region in Fig. 10–25, is approximately equal to the area under the spoiled portion of the curve.

An example of a simple measurement system using injectate dilution is the measurement of cardiac output in human and animal subjects using chilled or room temperature saline solution as the injectate. This method is called *thermodilution*. Cardiac output (C.O.) is defined as the rate of blood volume pumped by the heart. The C.O. measurement determines how much blood is being pumped per unit of time, so it measures volumetric flow rate. Cardiac output is measured in units of liters of blood per minute of time (l/min). In healthy adults C.O. typically ranges between 3 and 5 l/min.

A quantitative measure of cardiac output is the product of the stroke volume and the heart rate. The stroke volume is merely the volume of blood expelled from the heart ventricle (lower chamber) during a single contraction of the heart. Cardiac output is calculated from the equation

$$\text{C.O.} = VR \tag{10–27}$$

where C.O. is the cardiac output in liters per minute
V is the stroke volume in liters per beat
R is the heart rate in beats per minute

It is difficult, and usually impossible (except on animals in laboratory settings), to directly measure cardiac output using any technique based on Eq. (10–27). The main problem is obtaining good stroke volume data without excessive risk to the patient. There are, however, several related indirect methods that yield C.O. data.

The thermodilution method of cardiac output measurement has become the standard indirect method for measuring cardiac output in clinical settings and is also popular among laboratory scientists. Thermodilution technique forms the basis for most clinical and research cardiac output computers now on the market. One reason why thermodilution is preferred is that no toxic injectates are used (as they are in radiopaque or optical dye dilution methods), just ordinary medical IV solutions such as normal saline or 5 percent dextrose in water (D_5W) are used. The measurement is made using a special hollow catheter that is inserted into one of the patient's veins, usually on the right arm (the brachial vein is popular), through the vena cava into the right atrium of the heart, through the valve to the right ventricle, and out into the pulmonary artery, where the actual measurement of blood temperature is made.

The catheter is multilumened, and one of the lumens has its output hole several centimeters from the catheter tip. This proximal lumen is situated so that it is outside the heart (close to the input valve on the right atrium) when the tip is all the way through the heart, resting in the pulmonary artery (Fig. 10–26). Other lumens in the catheter output at the tip. They are used to measure pressures in the pulmonary artery in other procedures. A thermistor in the tip registers a resistance change with changes in blood temperature.

Most thermodilution cardiac output computers use a version of the following equation:

$$\text{C.O.} = \frac{kC_tV_i(T_b - T_i)}{\int_{t1}^{t4} T'_b \, dt} \tag{10–28}$$

where C.O. is the cardiac output in liters per minute
k represents a collection of other constants and the conversion factor from seconds to minutes
C_t is a constant that is supplied with the injectate catheter that accounts for the temperature rise in the portion of the outside of the patient's body
V_i is the injectate volume
T_b is the blood temperature in degrees Celsius
T_i is the temperature of the injectate in degrees Celsius
T'_b is the temperature of the blood as it changes due to mixing with the injectate

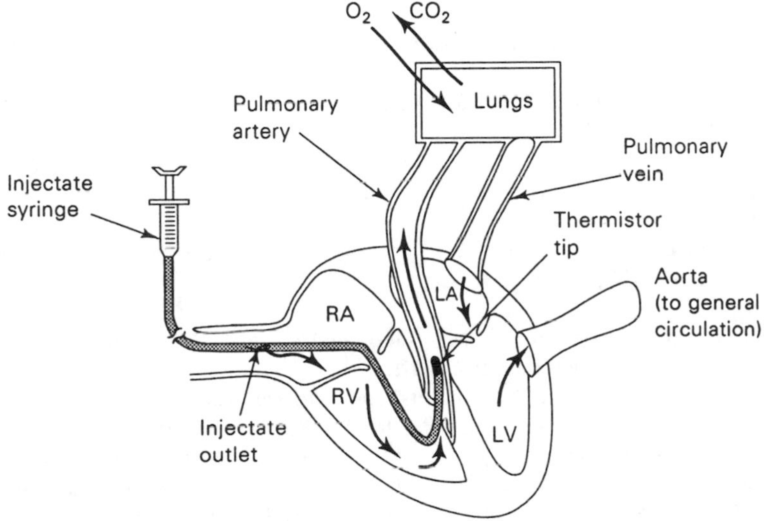

Figure 10–26 Cardiac output flow-detection system (thermodilution).

The integration of the temperature of the blood at the output side of the heart gives the time-average of the temperature.

The thermistor in the end of the catheter is usually connected in a Wheatstone bridge circuit. The dc excitation of the bridge is critical. Either the short-term stability of this voltage must be very high, or a ratiometric method must be used to cancel excitation potential drift. In addition, it is also necessary to limit the bridge excitation potential to about 200 mV for the safety of the patient (electrical leakage is especially dangerous because the thermistor is inside the heart or pulmonary artery). This low value of excitation voltage promotes both patient safety and thermistor stability through freedom from self-heating–induced drift, even though it imposes a greater burden on the amplifier design. The output of the bridge, depending upon the design, is typically 1.2 to 2.5 mV/°C, with 1.8 mV/°C being quite common. This signal is usually amplified approximately 1000 times to 1 V/°C by a preamplifier, that is isolated for patient safety. The output voltage of this circuit is integrated and used in the denominator of an equation as above.

REFERENCES AND NOTES

1. Swan, Robert H. Accura Flow Products Co., Inc. "Laminar Flow—What Is It?" *Sensors* (December 1985):37ff.
2. Gleick, James. *Chaos: The Making of a New Science* (op-cit), and Frances Moon, *Chaotic Vibrations* (op-cit).
3. Norton, Harry N. *Handbook of Transducers*. Englewood Cliffs, N.J.: Prentice Hall, Inc., 1989.

4. Ibid.
5. Oliver, Frank J. *Practical Instrumentation Transducers*. New York: Hayden Book Company, 1971.
6. Strong, Peter. *Biophysical Measurements*. Measurement Concepts Series. Beaverton, Or: Tektronix, Inc., 1973.
7. Cobbold, Richard S. C. *Transducers for Biomedical Measurements: Principles and Applications*. New York: John Wiley, 1974.
8. Norton.
9. Wiseman, Donald F. DyBec Corporation. "Variable Airflow Measurement." *Sensors* (May 1988):12ff.
10. Ibid.
11. Norton. See also Sabin, C. M. Geoscience Ltd. "Measuring Mixed Phase Flows." *Sensors* (October 1986):11ff.
12. Walsh, Terrel S. Fluid Components, Inc. *Sensors* (December 1989):27ff.
13. Henderson, H. Thurman, and Walter Hsieh. University of Cincinnati, "A Miniature Anemometer for Ultrafast Response." *Sensors* (December 1989):22ff.
14. Jerman, J. H., and J. W. Knutti. IC Sensors. "Silicon Sensors for Gas Flow and Thermal Measurements." *Sensors* (August 1987):5ff.
15. Perry, Jack A. Sam Tec, Inc. "Ultrasonic Flow Measurement." *Sensors* (December 1987):22ff.
16. Ibid.
17. Cobbold.

Liquid-Level Sensors 11

Liquid-level sensors are used in a wide variety of measurement and control applications. They come in two very general forms: *discrete* and *continuous*. The discrete forms give an indication only when the level of liquid in a vessel or tank is above or below a certain point. A tank might be fitted with several discrete sensors to let tenders know when the $\frac{3}{4}$, $\frac{1}{2}$, or $\frac{1}{4}$ full levels are reached. The continuous forms are used to constantly monitor the level and often make other measurements as well (see the discussion on Hydrostatic Tank Gauges).

DISCRETE-LEVEL SENSORS

Making discrete measurements is very simple compared with making continuous measurements. The object is to determine whether the liquid level is above or below a certain critical threshold or set point. A number of different types of discrete sensors are used: continuity elements, heat transfer, damped oscillation, optical, and ultrasonic.

Continuity Level Sensors

A *continuity level sensor* (Fig. 11–1) depends on a conductive liquid shorting together two electrodes that are placed in the tank. This condition does not exist until the liquid rises above the bottom of the electrodes, at which point the resistance between points *A* and *B* in Fig. 11–1 drops dramatically from an open circuit ("infinite" is the word used, but "extremely high" is nearer the truth) to a very low value.

Any number of sensing circuits can be used to indicate whether the level

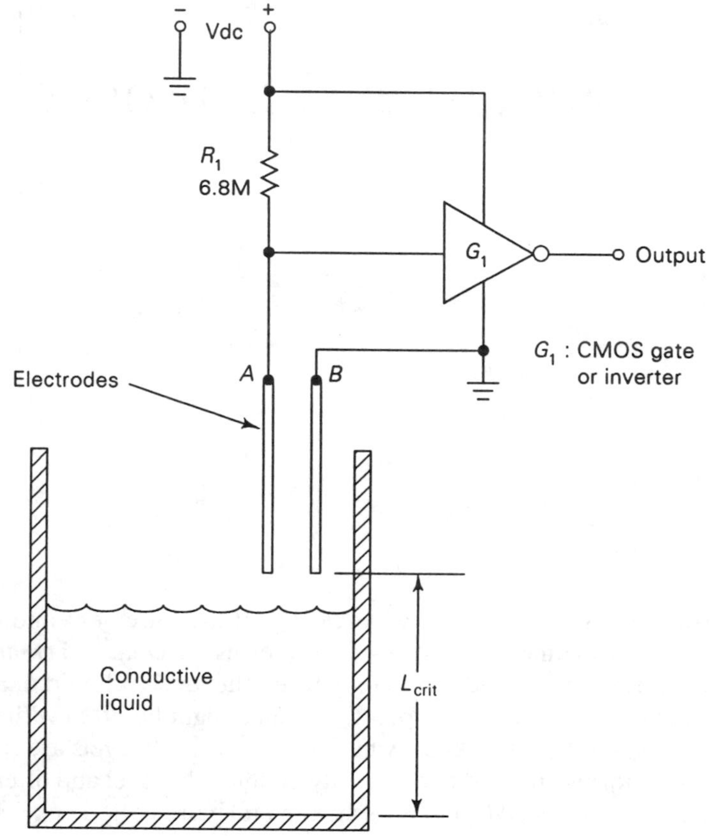

Figure 11–1 Conductive fluid level sensor uses a pair of electrodes that are shorted together when the fluid level rises to L_{crit}.

is above or below the critical-level set point L_{crit}. One such circuit is shown in Fig. 11–1. A complementary metal oxide semiconductor (CMOS) digital inverter is used to sense the level. These inverters have a simple rule of operation: The output is the opposite level of the input. Because these circuits are digital devices, they respond only to the binary levels, HIGH and LOW. A HIGH level is a positive voltage above 0.5 V dc (which is the applied power supply voltage), while LOW is a voltage closer to ground potential, that is, less than 0.5 V dc. The relationship between input and output is as follows:

Input	Output
HIGH	LOW
LOW	HIGH

When the electrodes are not shorted, the resistance across AB is very high, so the voltage applied to the input is HIGH, and the output of G_1 is LOW. Thus, a LOW at the output indicates that the liquid level is below the L_{crit} level. Alternatively, when the level is above L_{crit}, the electrodes are shorted, so the resistance across AB is very low ($R_{AB} <<< R_1$). In this condition the input of G_1 is shorted to ground, so it is LOW, and the output is therefore HIGH. Thus, the output signal indicates the liquid level:

Output Signal	Liquid Level
LOW	$<L_{crit}$
HIGH	$\geq L_{crit}$

For this circuit to work properly the capacitance between points A and B must be kept low. Otherwise, with a very high value of R_1, the RC time constant will be excessive.

A multilevel continuity level sensor system is shown in Fig. 11–2. In this system there are five level sensors, four of which are active (A through D), and one of which is the reference electrode at the bottom of the tank. (Alternatively, in a metallic tank the tank itself could be the fifth electrode.) Each electrode indicates a different critical level, so the respective outputs of circuits such as in Fig. 11–1 indicate which of the four discrete levels represents the liquid level.

The limitations of the continuity method include the complexity that more than a few discrete levels would entail and the requirement that the liquid be both an electrical conductor and noncorrosive and nonreactive with the metals or other materials used in the resistance elements. Otherwise, the liquid might become contaminated, or the sensor might deteriorate or even be destroyed.

Heat-Transfer Discrete-Level Sensors

A *heat-transfer* level sensor (Fig. 11–3) involves immersing a heat-sensitive element (see Chapter 4) in the liquid when it is at or above the critical level L_{crit}. At least two approaches can be taken. The first shown in Fig. 11–3 involves using a thermistor R_t that is heated to a point just below self-heating. The self-heating point is used because it is the most sensitive, that is, it exhibits a greater resistance change when the ambient temperature is changed. A milliammeter (M_1) in the circuit, or some other sensing mechanism such as G_1 shown previously in Fig. 11–1, can be used to indirectly indicate the resistance of R_t, hence its ambient temperature.

The goal in the resistance scheme of R_t is for the liquid in the tank to absorb a certain amount of the heat of R_t, reducing its ambient temperature and changing its resistance sufficiently to register a change in the current I indicated on M_1.

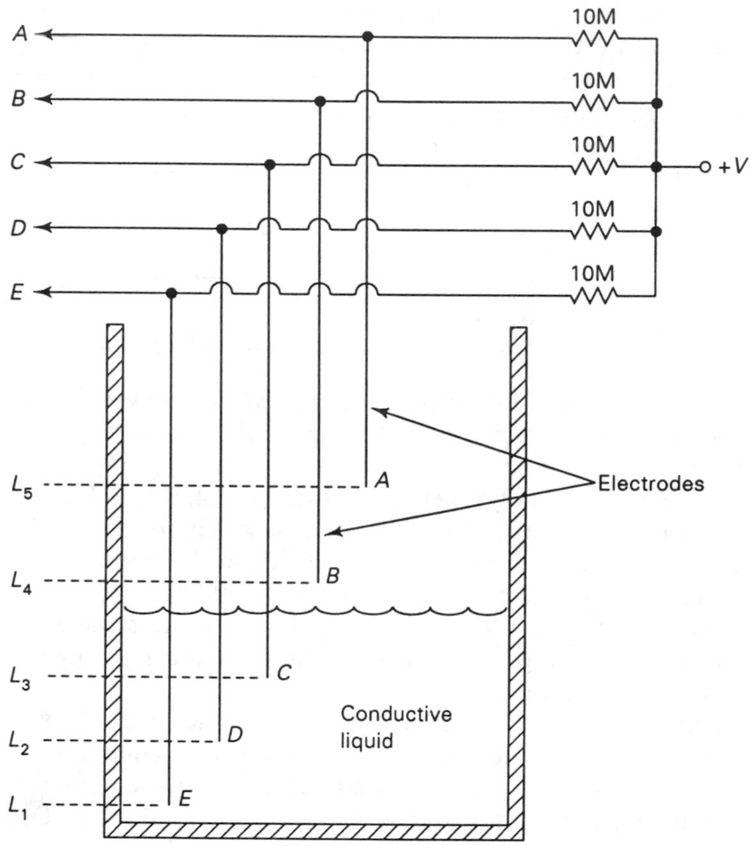

Figure 11–2 Multilevel version of the conductive level sensor uses a single common electrode and multiple-depth electrodes.

A variant on this theme, also shown in Fig. 11–3, is to use a resistance heater element R_h, maintained at a constant power by an external power source, that is intimately linked to a thermocouple element. The thermocouple output voltage V_{TC} will then indicate the body temperature of the resistive heater element R_h. When the liquid reaches the element, it will heatsink some of the resistor's body heat, causing V_{TC} to drop—thereby indicating that the liquid level has reached the critical point.

Damped-Oscillation Discrete-Level Sensors

A vibrating or oscillating element will maintain a given resonant vibration frequency in air with little excitation. But if the vibrating element is immersed in a viscous fluid, then the oscillations will either cease altogether or be severely attenuated. Figure 11–4 shows a vibrating paddle. If the

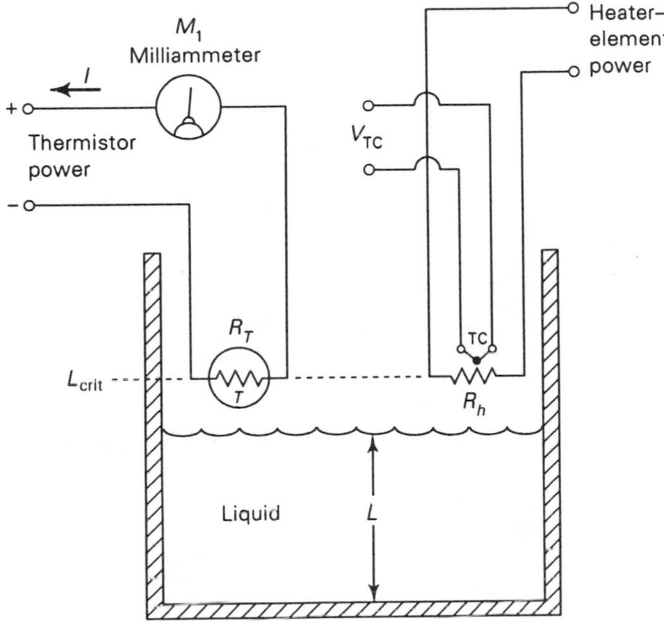

Figure 11–3 Thermal level sensing.

paddle is attached to either a piezoelectric element or a dynamic magnet-coil sensor, then it will produce an ac electrical output signal at the same frequency as the vibrations of the element. The amplitude of the ac output will be proportional to the deflection of the paddle.

Optical Discrete-Level Sensors

If the liquid in a vessel is opaque to either visible light or infrared waves, then it is possible to make a sensor system consisting of a light/IR source and a matching sensor. Figure 11–5 shows a basic system in which a sensor and source are placed opposite each other. When the liquid is below the critical level, the sensor can see the light source and will register an output. But when the opaque liquid covers the path between the source and the sensor, the sensor is blinded and produces no output signal.

A potential problem with this system is that the light sensor may be sensitive to ambient light sources and thereby give false readings. For this reason it is good practice to make the output condition protocol such that signal equals low level, so that ambient sources do not give false results quite so easily. It is also necessary to ensure that the fluid is truly opaque to the light or IR wavelengths being used. It is quite possible, for example, to find a fluid that will transmit some wavelengths in a white spectrum with little attenuation while severely attenuating others. In those cases either the sensor

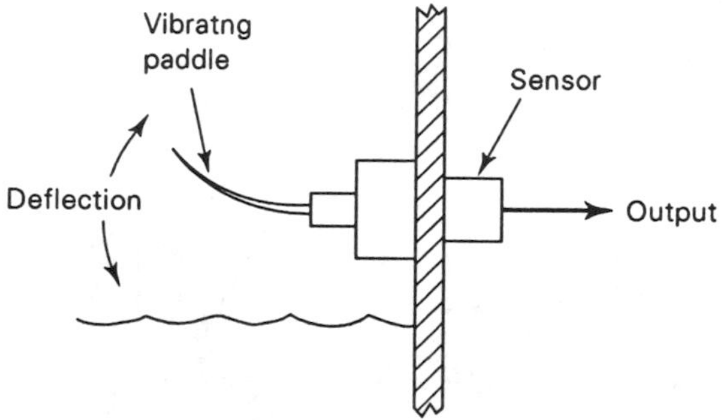

Figure 11–4 Vibrating paddle is damped when the level rises to the proper level.

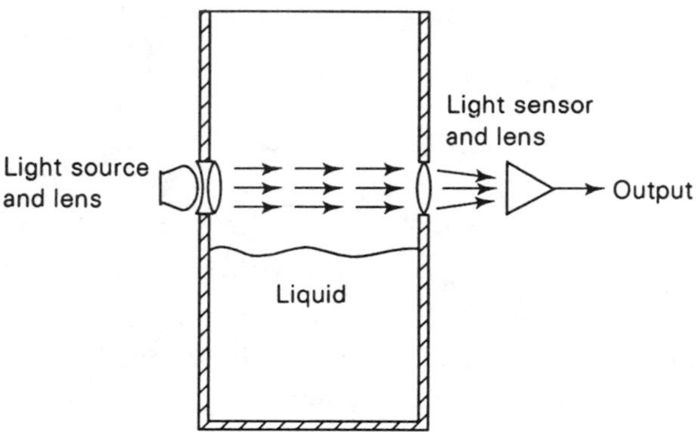

Figure 11–5 Optical level sensor.

or the source (or both) must be filtered to remove the transmittable wavelength.

The configuration shown in Fig. 11–5 requires a transmission path across the entire diameter of the vessel, but this is not always a convenient approach. A superior approach for many situations is the version shown in Fig. 11–6, in which the sensor and source are mounted in a common module that is attached to one wall of the vessel. When the liquid level rises to the critical point L_{crit}, it will blind the sensor and cause a change in the output signal.

Still another optical variant is shown in Fig. 11–7, the *optical prism method* of discrete-level sensing.[1] This sensor works on the basis of differences in the *index of optical refraction n* between the liquid, the prism, and the air or overgas in the tank.

According to Snell's law, when a light ray approaches a boundary between two media of different optical densities, that is, with different values of n, at an angle of incidence α_i, it will pass over the boundary and be refracted

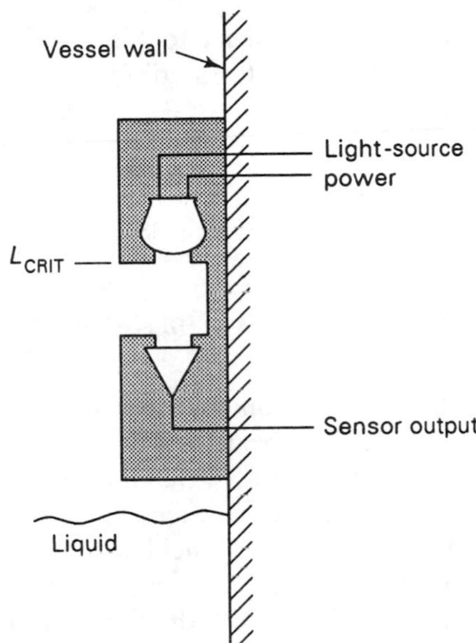

Figure 11–6 Alternative optical level sensor.

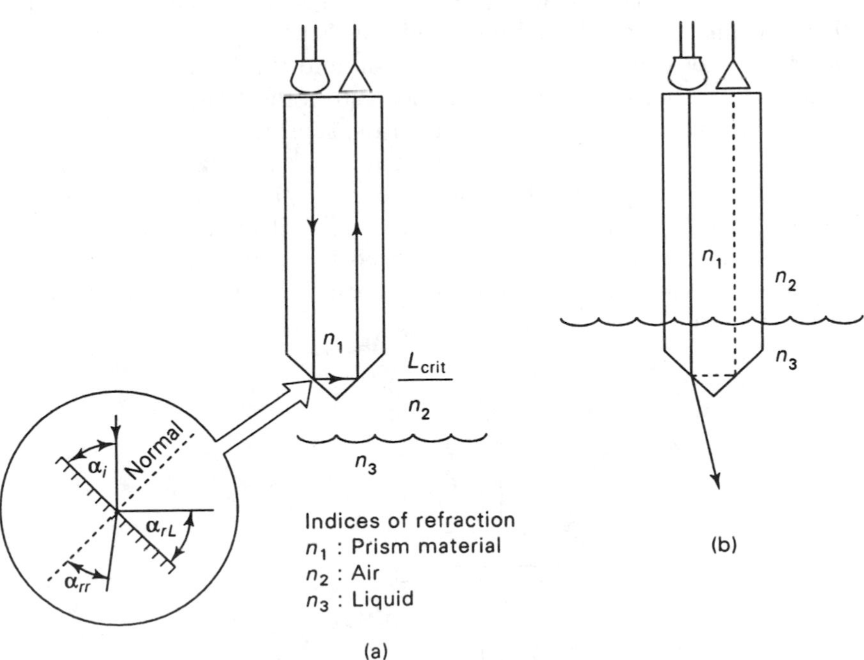

Figure 11–7 (a) Prism-type optical level sensor depends on a change in the index of refraction at the interface (b) when the liquid covers the face of the crystal.

at an angle α_{rr} [see inset to Fig. 11–7(a)]. The angles of incidence and refraction are related to the differences between the indices of refraction by the equation

$$n_1 \sin \alpha_i = n_2 \sin \alpha_{rr} \tag{11-1}$$

The angle of refraction is given by

$$\alpha_{rr} = \sin^{-1}\left(\frac{n_2 \sin \alpha_i}{n_2}\right) \tag{11-2}$$

At certain critical angles of incidence for any given ratio of indices of refraction, the light beam will be *totally internally reflected* in the originating medium [Fig. 11–7(a)]. Total internal reflection appears like, and obeys the same rules as, specular reflection. If the ratio of the indices of refraction changes, and the angle of incidence remains the same, then the light will pass through the boundary and continue propagating [Fig. 11–7(b)]. Thus, the sensor will see light when the liquid is below the critical level but not when the liquid is at or above the critical level.

In the prism method of discrete-level sensing, the ratio of the index of refraction of the prism and the air (or other gas) is such that total internal reflection takes place, as in Fig. 11–7(a). The light is totally internally reflected from the second prism face and travels back to the sensor. But the ratio of the index of refraction of the prism and the liquid does not satisfy this condition, so the light ray continues across the boundary, causing the light to travel into the liquid, rather than the sensor.

The position of the prism sensor can be vertical, as in Fig. 11–7, or horizontal as in Fig. 11–8. In either case, multiple sensors are required to indicate different liquid levels in the vessel. Still another approach (Fig. 11–9) is to couple the sensor and source ports on the prism to fiber optics and then route the fiber optics to couplings attached to the actual source and sensor elements.

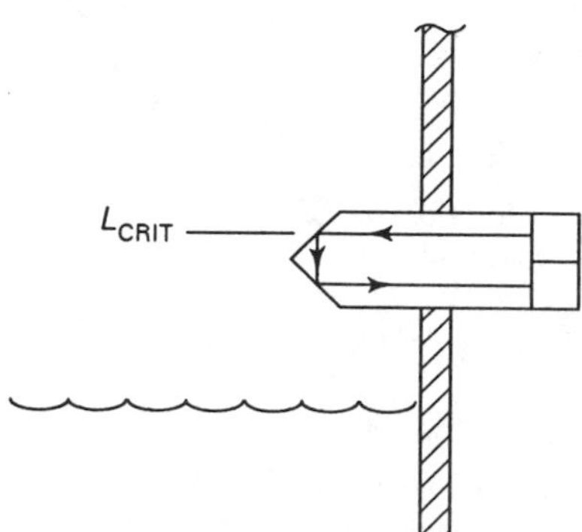

Figure 11–8 Prism placed horizontally for sensing at levels below capacity.

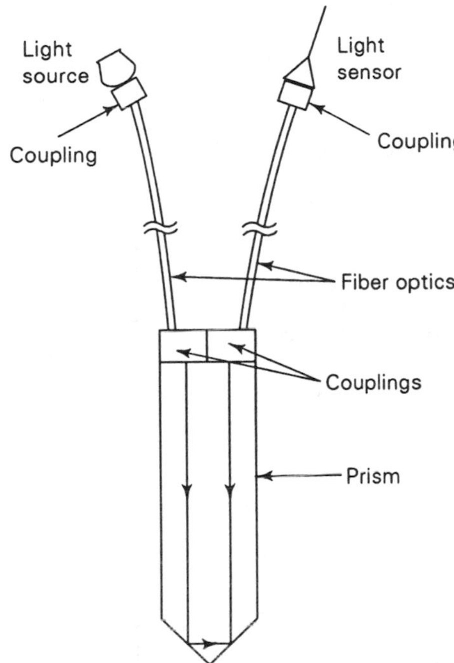

Figure 11–9 Fiber-optic version of the prism level sensor.

CONTINUOUS LEVEL SENSORS

Discrete-level sensors are perfectly well suited for many applications, but in other cases, *continuous-level sensors* are necessary. In these devices the output level is continuously read out at all points between empty and full. By examining the sensor output, it is possible to find out (to a very small resolution unit) the *exact* level, rather than simply the fact that the liquid is above or below certain critical threshold levels. The several different types of continuous (or nearly so) sensors include resistance-element sensors, float, pressure/weight, capacitive, ultrasonic, and microwave.

Resistance-Element Continuous-Level Sensors

The *resistance-element sensor* (Fig. 11–10) consists of a resistive wire element or chain of series-connected discrete resistors immersed in a *conductive* liquid, along with a low-resistance connection element.[2] The resistance of the connection element is so low compared with the resistance of the resistive element that its contribution to the total resistance R_t measured between points A and B is negligible.

If the liquid provides a conductive path between the two wire elements, then it will short out the portion of the resistance wire that is immersed in the liquid, reducing the total resistance R_t by an amount ΔR_L that is propor-

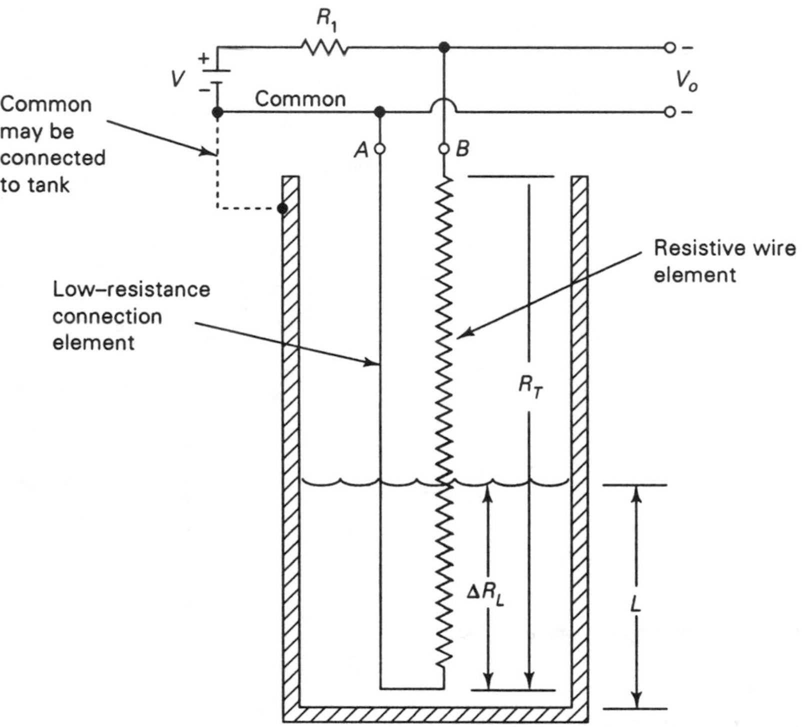

Figure 11–10 Resistance level sensor.

tional to the liquid level L. The total resistance when the tank is not empty $R_t{}'$ is given by

$$R_t{}' = R_t - \Delta R_L \tag{11–3}$$

The liquid level L is proportional to the term ΔR_L, so the term $R_t{}'$ can also be used as a measure of L by Eq. (11–3). The circuit for making this measurement can be a simple ohmmeter or the combination of the voltage source V and calibration resistor R_1 shown in Fig. 11–10. This circuit transforms the level data into an analog output voltage V_o:

$$V_o = \frac{VR_t{}'}{R_1 + R_t{}'} \tag{11–4}$$

A limitation on the method is the change in electrical resistance of any resistor as a function of temperature. The resistance is proportional to the length of the conductor and inversely proportional to its cross-sectional area. These factors change with temperature, so the electrical resistance R_t will change according to the relationship

$$R_{t''} = R_0(1 + \alpha T) \tag{11–5}$$

where R_0 is the resistance measured at 0°C in ohms
α is the temperature coefficient of resistance in ohms per degree Celsius
T is the temperature in degrees Celsius

If the range of temperature variation is small, or if the temperature coefficient α for the resistive material is small enough to be neglected for the purpose at hand, then no correction is needed. However, if the temperature change of resistance is great enough to cause concern, then the tank temperature must be measured and a correction factored into the measurement. In some cases a thermistor (see Chapter 4) placed in the tank can be connected either in series or parallel (depending on type) with R_t' to temperature-compensate the measurement.

Float-Type Continuous-Level Sensors

The *float-type sensor* (Fig. 11–11) is one of the oldest forms of level sensors. This is the type of sensor system used in most motor vehicles for the fuel gauge sender unit. Figure 11–11(a) shows the type of float sensor used in vehicle fuel tank systems. A sense arm is connected to the float and also to a pivot point. When the float rises, it causes the other end of the sense arm to rotate, moving the actuator of a position or displacement sensor. The output of this sensor is a current or voltage that is proportional to the liquid level.

A variation of this sensor is shown in Fig. 11–11(b). In this case the float rises on a pipe. Either of two mechanisms is used to indicate the level. If the system is discrete, then a permanent magnet is embedded in the float and is used to actuate a magnetic reed switch that is located at the critical level. If the system is continuous, then a position or displacement sensor is

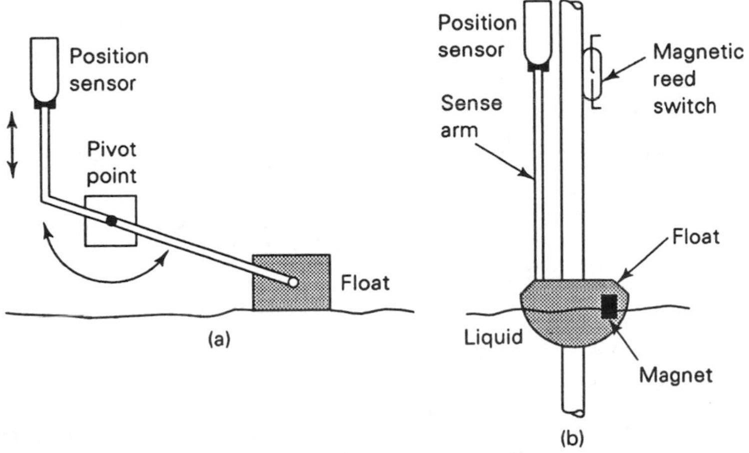

Figure 11–11 Float-type level sensor using a lever arm (a); using float ball (b).

actuated by the float and produces an output voltage or current proportional to the liquid level.

Pressure/Weight Continuous-Level Sensors

If the weight of a liquid inventory in a vessel is known, then it is possible to use a *load cell* or other electronic weight sensor to determine the level of the liquid inside the tank (Fig. 11–12), especially if the tank is open to the atmosphere and has no ullage pressure.

The load cell produces a signal that represents the total weight W, which consists of the sum of the liquid weight W_L and the empty tank weight, also known as the *tare* W_t.[3] The actual weight of the liquid is equal to the total weight less the tare:

$$W_L = W - W_t \tag{11–6}$$

The tare is known for each individual tank and can be entered into a differential amplifier's negative port as a constant. The total weight signal W from the load cell is entered into the positive port of the differential amplifier, so the output signal represents Eq. (11–6).

The pressure system (Fig. 11–13) uses a pressure sensor at the base of the vessel (P_1) and a second pressure sensor at the top to measure the ullage pressure (P_2). This system depends on knowing the *specific weight* ω of the liquid inside the tank:

$$L \propto \frac{P_1 - P_2}{\omega} \tag{11–7}$$

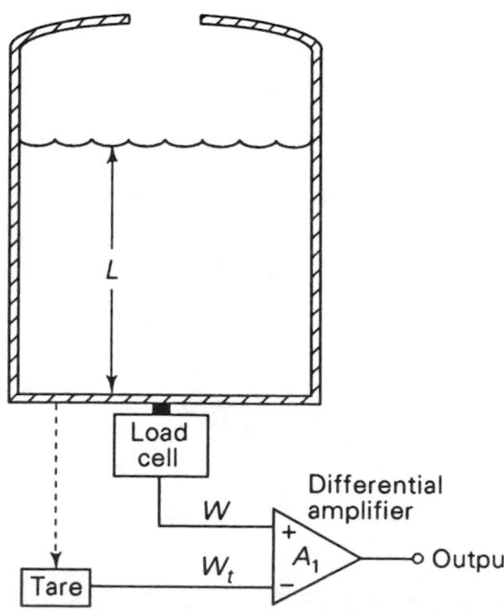

Figure 11–12 Load-cell level sensor measures the weight of the liquid and then subtracts the (tare) of the tank.

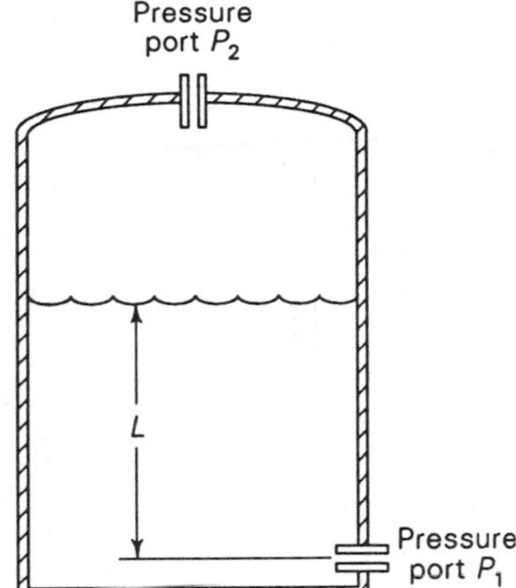

Figure 11–13 Pressure sensor is alternative means of measuring the weight of the liquid.

Capacitive Continuous-Level Sensors

The electrical phenomenon of capacitance was described in Chapter 3. The capacitor stores energy in an electrostatic field between two conductive plates. Many capacitors are based on parallel plates, but it is also possible to make a cylindrical capacitor in which two conductive cylinders are coaxial [Fig. 11–14(a)]. The capacitance of a coaxial cylindrical capacitor is found from the equation

$$C = \frac{0.2416K}{\log_{10}(r_2/r_1)} \tag{11–8}$$

where C is the capacitance per unit of length in picofarads per centimeter
K is the dielectric constant relative to a vacuum
r_1 is the radius of the inner cylinder
r_2 is the radius of the outer cylinder

If a coaxial cylindrical capacitor has an air dielectric, then the value of the K-term is around 1.006 (by convention, the value of K_{air} is often taken to be 1.000). But when a nonconductive liquid fills the space between the two cylinders [Fig. 11–14(b)], the capacitance goes up dramatically because the value of K for the liquid is very much higher than that of air. For example, common oils have dielectric constants of 2.1 to 4, while ethyl alcohol has a dielectric constant of 28.4 at 0°C. Thus, the capacitance of a coaxial capacitor with an alcohol dielectric is about 28 times higher than that of air.

The coaxial cylindrical capacitor in Fig. 11–14(b) is partially immersed in liquid, so the dielectric is partially air and partially liquid. Thus, the capacitance will be somewhere between that for an all-air and that for an all-liquid dielectric.

The method of Fig. 11–14(b) will work satisfactorily in many situations,

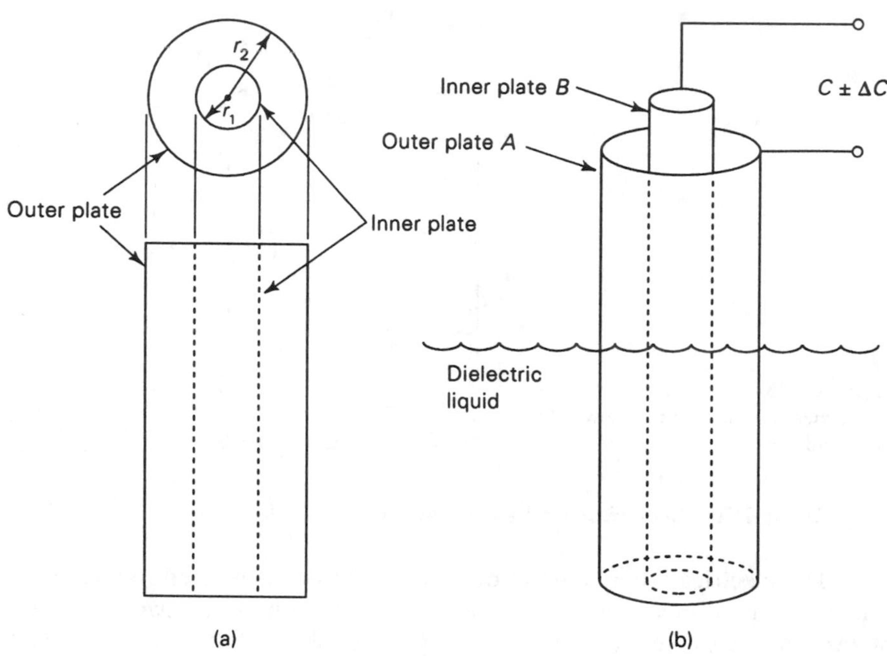

Figure 11–14 Capacitance liquid-level measurement: (a) capacitor in cylindrical form; (b) capacitor assembly.

but it fails when the dielectric parameters of the liquid change. It also fails when the space above the liquid contains varying amounts of vapor from the liquid below, which happens under temperature variations in some cases. The problem is solved by using a *reference capacitor* system such as that shown in Fig. 11–15. In this system a small coaxial cylindrical reference capacitor C_{ref} is mounted at the bottom of the tank so that it is always submerged. If the level of the liquid needed to cover the reference capacitor is L_{ref}, and C_{meas} is the capacitance of the measurement capacitor, then the following relationship obtains:

$$\frac{L}{L_{ref}} = \frac{\Delta C_{meas}}{C_{ref}} \tag{11–9}$$

The measurement and reference capacitors can be used in several different circuits: capacitive voltage dividers, capacitive Wheatstone bridges, variable frequency oscillators, and others.

A discrete-level capacitive sensor (Fig. 11–16) can also be built. In this type of measurement system the coaxial cylindrical capacitor is mounted horizontally in the tank. When the liquid rises to the critical level, the capacitance changes dramatically and causes the circuit to alarm.

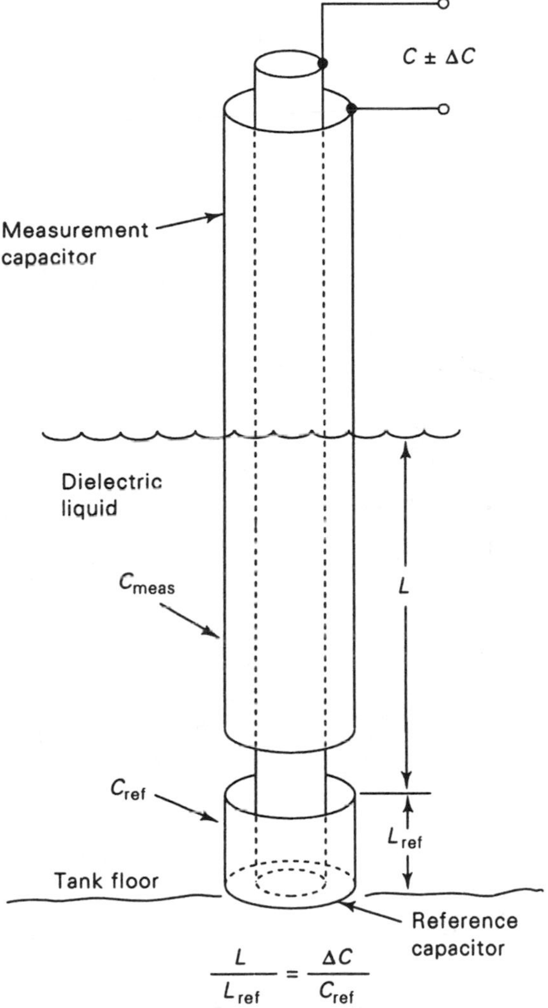

Figure 11–15 Capacitance level sensing against a reference system.

Ultrasonic Continuous-Level Sensors

Ultrasonic (and microwave radio) waves can be used to measure the level of a liquid in a vessel in both pulse and continuous methods. The *pulse method* is equivalent to radar or sonar systems in that a short-duration pulse is fired into the environment and then the reflected return pulse is timed.[4] Figure 11–17 shows the basic reflected-pulse system. The transducer can be mounted either in the dead space above the liquid (point *A*) or at the bottom of the tank (point *B*). In both cases, the incident wave will be reflected back

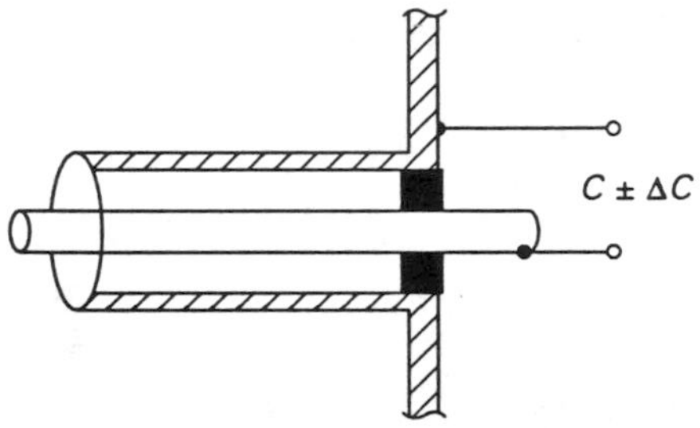

Figure 11–16 Discrete-level sensing with a capacitor sensor.

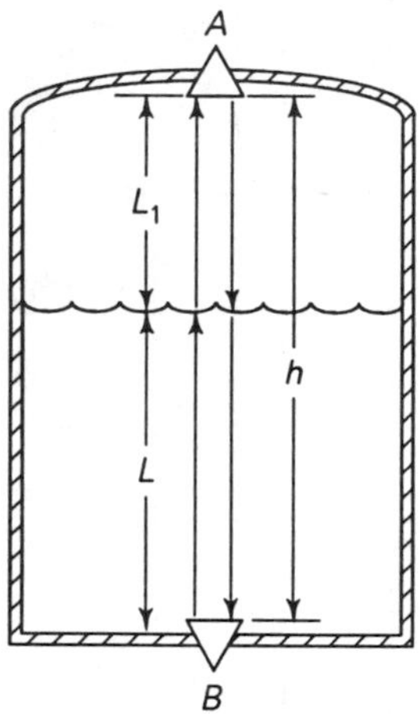

Figure 11–17 Transit-time ultrasonic level sensing.

toward the source from the liquid-gas interface. In the case of sensor A, the level of the liquid L is found by taking the difference between the sonic transmission path length L_1 and the tank height h:

$$L = h - L_1 \tag{11–10}$$

$$L = h - \frac{V_{air}T}{2} \tag{11–11}$$

where V_{air} is the velocity of the sound wave in air
T is the time required for the pulse to make the round trip
The factor 2 represents a correction for the round trip

For the case of the transducer at position B, the level L is found from:

$$L = \frac{V_{liq}T}{2} \tag{11–12}$$

Where: V_{liq} is the velocity of the sound in liquid

A continuous system is shown in Fig. 11–18. Here the transducer emits a continuous wave signal that radiates into the tank. There are two ways to

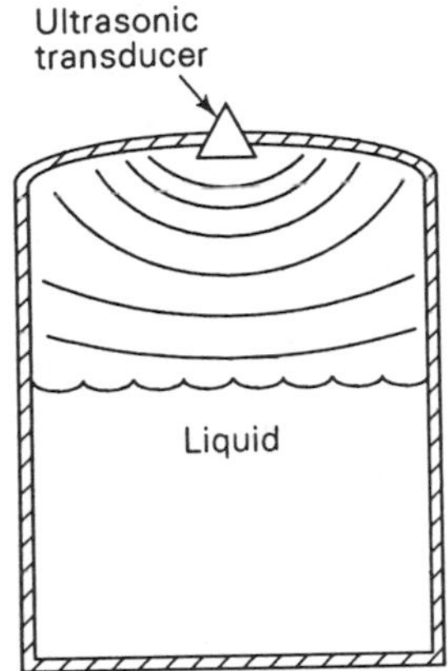

Figure 11–18 Resonant-cavity level sensing.

use the signal. One is the *standing-wave* method, which is analogous to the similar system in ultrasonic proximity detectors. When the direct and reflected waves combine, they create a standing wave that is a function of the free space in the tank above the liquid. As the liquid level changes, the relative locations of the nodes and antinodes of the standing wave also change, and these variations can be sensed.

In the *tank-resonance* method, advantage is taken of the fact that the resonance of the tank to ultrasonic waves is a function of the volume of the tank. This same phenomenon can be observed by partially filling a bottle with water and blowing across the bottle opening and creating an audio oscillation tone. If the liquid level is low (high free-space volume), then the frequency of the tone is also low. Conversely, when the liquid level is high, there is less free space, and the tone frequency is higher. In the ultrasonic case, the frequency band can be swept to note the point at which a sudden increase in the resonance occurs. This frequency corresponds to a liquid level.

Hydrostatic Tank Gauging (HTG) Systems

The task of determining the volume of liquid inventory in a storage tank has been a problem ever since liquids (probably with high alcoholic content) were first stored in bulk containers. Today, however, *hydrostatic tank gauging* (HTG) systems use a combination of pressure and temperature readings to accomplish this task. And with small XT-class personal computers (or better), the task is simplified even more.[5,6]

Figure 11–19 shows an HTG system involving a cylindrical vertical tank with an area A equal to πr^2 or $\pi D^2/4$, where r is the tank radius and D is the tank diameter. At any given time the liquid inventory will be at a level L, which varies with consumption and resupply.

The tank is equipped with four sensors: TP_1 is a temperature probe,

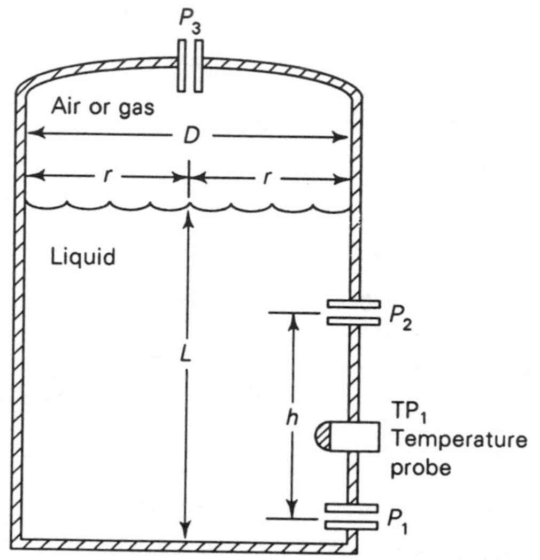

Figure 11–19 Advanced level sensing based on pressures.

while P_1, P_2, and P_3 are pressure sensors. Pressure sensors P_1 and P_2 are spaced a calibrated distance h apart, with P_1 being placed at the bottom of the tank. The *ullage* pressure, that is, the pressure of the gas or air in the space above the liquid inventory, is measured by pressure sensor P_3. The temperature sensor TP_1 is needed because volume varies somewhat with the temperature. In the HTG system, the level L and temperature T are compared with the parameters taken at a standard temperature of 60°F.

Let's now review some basic concepts that underlie the measurement relationships that make this system possible. First, the volume of a cylindrical column of liquid of cross-sectional area A and height L is

$$V = LA \tag{11–13}$$

where V is the volume in cubic meters
L is the liquid column height in meters
A is the cross-sectional area in square meters

If $A = \pi D^2/4$, then

$$V = \frac{L\pi/D^2}{4} \tag{11–14}$$

Because the diameter is fixed by the design of the tank, and $\pi/4$ is a constant under all circumstances, the volume of the liquid in the tank is a function of the level.

It is sometimes necessary to know the mass M of the liquid in the tank, and this information can be determined from the relative pressures in the tank. It is not necessary to know the density of the material or to resort to material sampling or making any temperature measurements. The measurements needed are the ullage pressure P_3 and the overall pressure at the bottom of the tank P_1. Because pressure is force per unit of area

$$P = \frac{F}{A} \tag{11–15}$$

in this situation

$$F = PA \tag{11–16}$$

The force F is basically the weight of the liquid, which is (from Newton's $F = MA$)

$$F = MG \tag{11–17}$$

where G is the gravitational constant (6.67×10^{-11} N · m/kg^2)

Substitution of Eq. (11–17) into Eq. (11–16) yields

$$MG = PA \tag{11–18}$$

The mass, therefore, is

$$M = \frac{PA}{G} \tag{11–19}$$

Because the pressure is a combination of liquid static head pressure P_1 and the ullage pressure P_3, the differential pressure ΔP_{1-3} must be used:

$$M = \frac{\Delta P_{1-3}A}{G} \tag{11–20}$$

or, combining Eq. (11–20) with the definition of area:

$$M = \frac{(\Delta P_{1-3})(\pi D^2/4)}{G} \tag{11–21}$$

The *density* δ of the liquid inventory is defined as *mass per unit of volume* and can be calculated from that information. However, it is also possible to calculate the density of the inventory material from the differential pressure ΔP_{1-2}:

$$\delta = \frac{\Delta P_{1-2}}{h} \tag{11–22}$$

The level of the material can be determined from differential pressure ΔP_{1-3} and the density:

$$L = \frac{\Delta P_{1-3}}{\delta} \tag{11–23}$$

The volume V of the liquid inventory is partially a function of temperature T, so the temperature must be measured with TP_1 and used with a set of standard look-up tables (which can be programmed into a computer) that allow determination of present volume from the known *standard volume* V_s at the standard temperature of 60°F (15.56°C). The standard volume is found by comparing the mass with the *reference density* δ_s, which is the density at the standard temperature:

$$V_s = \frac{M}{\delta_s} \tag{11–24}$$

Hydrostatic tank gauging systems represent a simple but reliable approach to measuring the mass, density, level, and volume of liquid inventory in a tank.

REFERENCES AND NOTES

1. Rakucewicz, John. Kinematics and Control Corporation. "Fiber-Optic Methods of Level Sensing." *Sensors* (December 1986):5.
2. Ehrenfried, Albert D. Metritape, Inc. "Level Sensor Key To Dispersed Plant Operation." *Sensors* (December 1987):5.
3. Norton, Harry N. *Handbook of Transducers*. Englewood Cliffs, N.J., Prentice Hall, Inc., 1989.
4. Perry, Jack A. Sam Tec, Inc. "Ultrasonic Instrumentation in Principle and Practice." *Sensors* (October 1987):15.
5. Much of the material in this section was developed from Schneider, Les. The Foxboro Company. "Hydrostatic Level Sensing." *Sensors* (January 1987):34ff. However, probably for pedagogical reasons in his tutorial, Schneider used simplified equations that ignored the gravitational constant G.
6. Piccone, Roland P. Sarasota Automation, Inc. "Combining Technologies to Computer Tank Inventory, Mass, and Volume." *Sensors* (October 1988):27.

Electro-Optical Sensors 12

Electro-optical sensors are electronic components that respond in some way or another either to light or to the other electromagnetic waves in the infrared (IR) and ultraviolet (UV) bands close to the visible light band. Photosensors that are readily available include photoemissive cells, photoresistors, photovoltaic cells, photodiodes, and phototransistors. In this chapter we examine each of these along with common forms of circuits, but first we review the theory of light.

LIGHT

Light is a form of *electromagnetic radiation* and is thus essentially the same as radio waves, infrared (heat) waves, ultraviolet, and X-rays. The principal difference among these various types of electromagnetic waves is their frequency f and wavelength λ (Fig. 12–1). The wavelength of visible light is 400 to 800 nanometers (1 nm = 10^{-9} m), which roughly corresponds to frequencies between 7.5×10^{14} and 3.75×10^{14} Hz. Infrared radiation has wavelengths longer than 800 nm, and ultraviolet has wavelengths shorter than 400 nm; X-radiation has wavelengths even shorter than ultraviolet. Frequency and wavelength in electromagnetic waves are related by the equation

$$\lambda = \frac{c}{f} \tag{12–1}$$

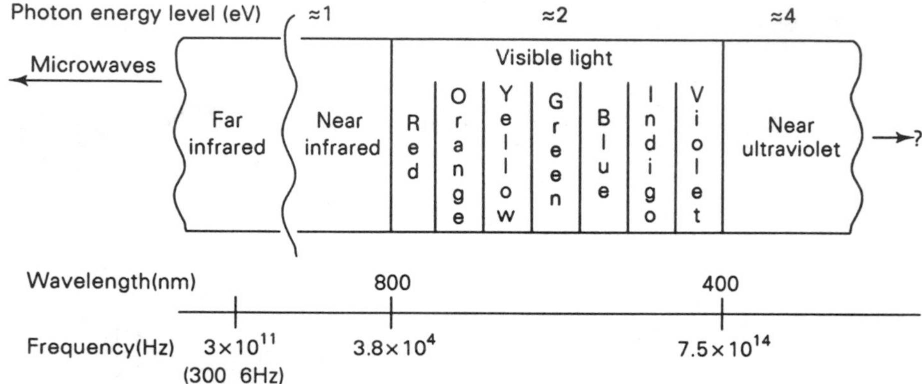

Figure 12–1 Electromagnetic spectrum showing region of interest to electro-optics: IR, visible, and UV.

where c is the velocity of light (300,000,000 m/s)
λ is wavelength in meters
f is frequency in hertz

Thus, light has a frequency on the order of 10^{14} Hz (compare with the frequencies of AM (10^6 Hz) and FM (10^8 Hz) broadcast bands in the radio portion of the spectrum).

Because IR, UV, and X-radiation are similar in both basic nature and wavelength to visible light, many of the sensors and techniques applied to visible light also work to one extent or another in those adjacent regions of the electromagnetic spectrum. Although performance varies somewhat, and some devices aren't useful at all in certain spectral regions, it is nonetheless true that people whose applications deal with those spectra may find these devices useful.

The photosensors described in this chapter depend on quantum effects for their operation. Quantum mechanics arose as a new branch of physics in December 1900, the very dawn of the twentieth century, with a now-famous paper by German physicist Max Planck. He had been working on thermodynamics problems (blackbody radiation) and had found that the experimental results reported in nineteenth-century physics laboratories could not be explained by classical Newtonian mechanics, the then-prevailing world view of physics. The solution to the problem turned out to be a simple, but revolutionary idea: Energy exists in discrete bundles, not as a continuum. In other words, energy comes in packets of specific energy levels; other energy levels are excluded. The name eventually given to these energy levels was *quanta*. The name given to energy bundles that operate in the visible light range was *photons*.

The energy level of each photon is expressed by the equation

$$E = \frac{ch}{\lambda} \tag{12–2}$$

or, alternatively,

$$E = h\nu \tag{12-3}$$

where E is the energy in electron-volts
c is the velocity of light (3×10^8 m/s)
λ is the wavelength in meters
h is Planck's constant (6.62×10^{-34} J · s)
ν is the frequency of light in hertz (Hz)

(Note: The constant *ch* is sometimes combined and expressed as 1240 eV/nm). The basis of operation of a light sensor is a device that allows at least one electron to be freed from its associated atom by one photon of light. Materials in which the electrons are too tightly bound for light photons to do this work will not work well as light sensors.

Figure 12–2 shows the effect of prisms and filters on a light beam. White light contains the entire visible light spectrum from red to violet. When the white light is passed through a prism, long wavelengths (e.g., red light) are refracted less than short wavelengths (e.g., violet). As a result, the spectrum is spread out and the individual colors become visible.

Optical filters are materials that selectively block or pass specific wavelengths. For example, if a green filter is placed in the path of a light beam, only green light will pass through the filter. This principle is applied in many electronic instruments used in science, medicine, and industry. Typical filters include colored glass, plastic, theatrical gels, and other materials. Developed color photographic film can be used to filter infrared.

The *spectral response* of a sensor is a measure of its ability to respond to electromagnetic radiation of different wavelengths. The spectral response of a sensor is generally given in the form of a graph relating relative response to wavelength. Figure 12–3 shows such a spectral output of a number of different light emitters. The vertical scale is a normalized output, meaning that full scale is arbitrarily labeled 100 percent. For reference, the monochromatic (one color) spectral outputs of the various forms of LEDs are plotted.

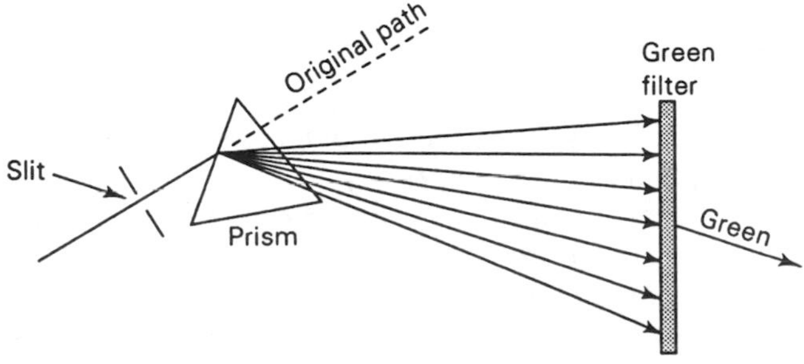

Figure 12–2 A prism breaks light into its component wavelengths (colors).

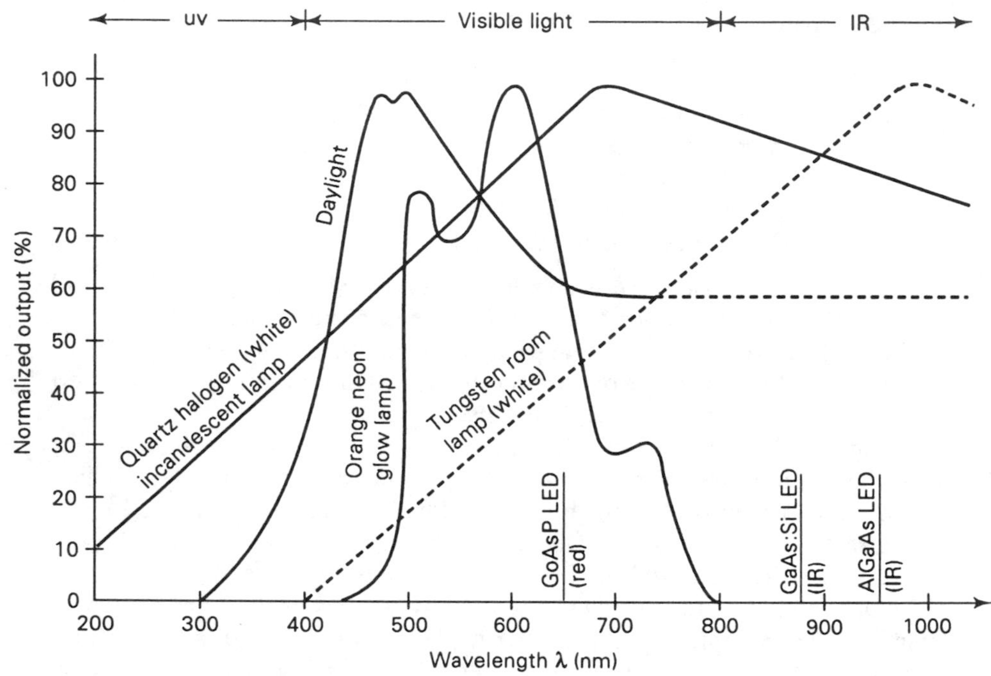

Figure 12–3 Spectra of various types of light sources.

LIGHT SENSORS

Figure 12–4 shows the approximate spectral response of some of the common solid-state light sensors used in electronics. The type of sensor selected for any given purpose is determined in part by the spectral response required for the specific application. For example, to use infrared radiation to measure carbon dioxide (CO_2), a device with a strong response within the IR region ($\lambda > 800$ nm) would be needed. Clearly, silicon solar (photovoltaic) cells and silicon phototransistors would be the choice.

PHOTOEMISSIVE SENSORS

Photoemissive sensors are specially constructed vacuum tube diodes (i.e., two-electrode devices) that output a current I_o that is proportional to the intensity of a light source that impinges on their sensitive surface. Photoemissive sensors fall into two categories: photoelectric tubes and photomultiplier tubes. These devices utilize the photoelectric effect for their operation.

The *photoelectric effect*—the photoemission of electrons—was discovered in the early to mid-nineteenth century. Scientists working on primitive

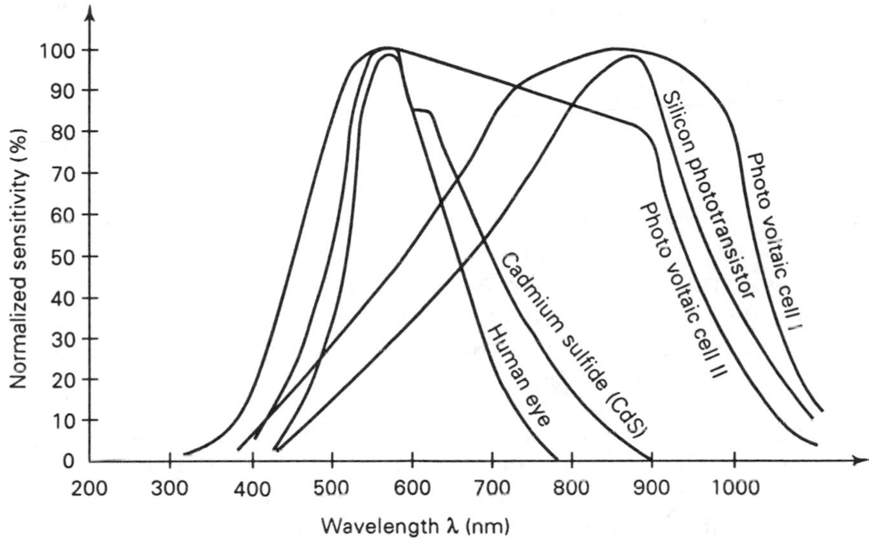

Figure 12–4 Spectral responses of several types of E-O sensors.

batteries noticed that they worked more efficiently when exposed to light. In other experiments it was found that certain metallic plates in a vacuum were able to emit electrons when exposed to visible or UV light. As early as 1887, radio research pioneer Heinrich Hertz was using spark-gap receiving apparatus in his experiments. He noted that the spark-gap receiver was more sensitive when the gap region was illuminated by UV light. William Hallwachs explained this phenomenon in 1888 by describing the photoelectric effect, also sometimes called the *Hallwachs effect*. The cause of the photoelectric effect, however, eluded scientists until early in this century.

Planck's 1900 paper that dealt with blackbody radiation did not address the photoelectric effect. It was, however, the seminal work that led Albert Einstein to the solution to the photoelectric problem. In 1905, truly a pivotal year in physics, Einstein published three major papers in *Annalen der Physik*. The subjects included the theory of relativity, an explanation of Brownian motion, and a theory of the photoelectric effect that used Planck's energy quanta. It was for his work on the photoelectric effect that Albert Einstein was awarded the Nobel Prize, not for relativity as is commonly assumed.

The photoelectric effect perplexed scientists up to 1905 because of a certain strange behavior observed in the laboratory. The *intensity* of the impinging light beam affects only the amount of current (i.e., number of electrons) emitted but not the *energy* of the emitted electrons. Oddly, however, it was noted that the *color* of the light affects the energy of the electrons. Electrons emitted from a photoemissive surface under the influence of blue light are more energetic than electrons emitted under a red light. The explanation of this behavior can be deduced from Eqs. (12–2) and (12–3). The energy expression for the photoelectric effect is

$$\frac{1}{2} mV^2 = h\nu - E_\omega \tag{12–4}$$

where m is the mass of the electron
V is the velocity of the fastest emitted electrons
h is Planck's constant (6.63×10^{-27} erg · s)
ν is the frequency of the incident light
E_ω is the energy in ergs required by an electron to escape from the photoemissive surface

The photoelectric effect is not seen at all wavelengths. For an electron to be emitted, the applied photon energy must be greater than the *work function* energy E_ω of the material. The work function is the amount of energy required to dislodge an electron from its associated atom. Therefore, for any given material or metal there is a wavelength limitation. Below a certain critical wavelength the photon energy is less than the work function energy. The maximum wavelength can be calculated from the equation

$$\lambda \leq \frac{ch}{E_\omega} \tag{12–5}$$

where λ is the wavelength in meters
c is the speed of light (3×10^8 m/s)
h is Planck's constant (6.62×10^{-34} J · s)
E_ω is the work function energy for the material illuminated

The construction of a typical photoemissive sensor tube is shown in Fig. 12–5. The device is a diode, so it has two electrodes: The photoemissive surface is called the *cathode*, while the electron-collector electrode is called the *anode*. These elements are housed inside a glass or metal envelope that is either under a high vacuum or under the partial pressure of an ionizing inert gas.

The anode is constructed of a small loop of wire or a thin rod, usually tungsten, platinum, or an alloy. The cathode is made of metal, but its surface is enhanced for improved emissivity by a special light-sensitive coating. Typical materials used include antimony, silver, cesium, and bismuth. The material is usually mixed with trace quantities of other elements.

Although photoemission does not require an external voltage source, the electrons must be collected and delivered to an external circuit before the device can be useful. To collect electrons, an electrical potential (V in Fig. 12–5) is connected across the tube so that the anode is positive with respect to the cathode. Electrons, being negatively charged, are repelled

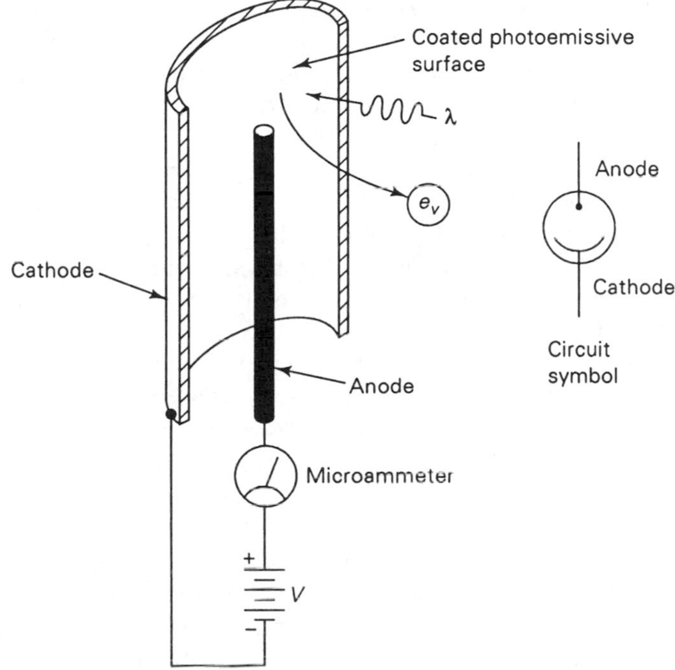

Figure 12–5 Photoemissive tube structure.

from the negative cathode and are attracted to the positive anode. For most common photoemissive sensors, $0 \leq V \leq 300$ V.

There are two general forms of tube construction (Fig. 12–6). In *side-excited* designs [Fig. 12–6(a)] the cathode is constructed so that the light must enter the side wall of the tube. The other directions may be blinded by internal blackening or silvering to prevent stray light from exciting the pho-

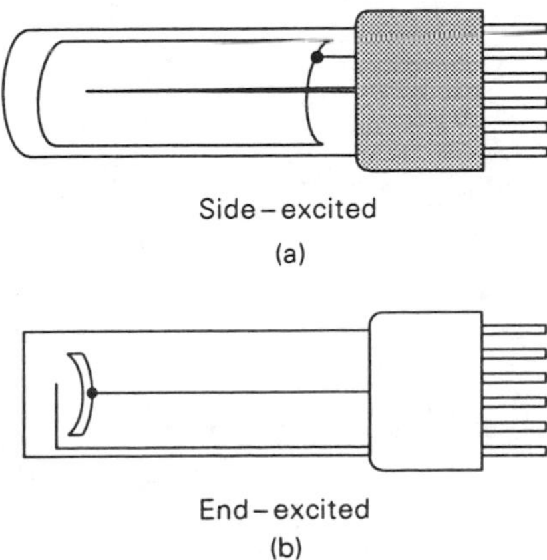

Figure 12–6 Types of photoemissive tubes: (a) side-excited; (b) end-excited.

toemissive surface. The *end-excited* design is constructed with the photoemissive surface facing the end of the housing [Fig. 12–16(b)]. Again, all surfaces other than the optical window may be blinded against stray light.

Photoemissive sensors are also categorized according to whether or not the inside is maintained under a vacuum or is gas filled. A *high-vacuum* photodetector is evacuated of all air and gases. The device outputs a small current I_o that is linearly proportional to the intensity of the impinging light.

The output current levels from the high-vacuum photodetector are generally too low to be measured directly on an ammeter, so they are typically converted to an output voltage V_o by passing the current through a load resistor (R_L in Fig. 12–7).

The output voltage V_o is the difference between the power supply potential V and the voltage drop V_1 across R_L caused by the photodetector output current I_o. Because, by Ohm's law, V_1 is equal to the product I_oR_L, the output voltage is

$$V_o = V - I_oR_L \tag{12–6}$$

A *gas-filled photoemissive sensor* is first evacuated of air and then refilled with an ionizing inert gas. Photoelectrons emitted by the cathode collide with the gas molecules and create electrons and positive ions by *secondary emission*. Thus, each of the emitted electrons creates a number of secondary electrons, so the overall current flow is 10 to nearly 1000 times greater than in a similar high-vacuum device. The gas pressure is carefully regulated to ensure that this process continues without dying out or getting out of control. The cathode-to-anode potential V_{KA} also must be kept low enough to prevent imparting sufficient kinetic energy to the electrons to cause complete ionization of the gas. If this were to happen, then the phototube would emit light like a glow lamp. The gas-filled tube produces higher output current

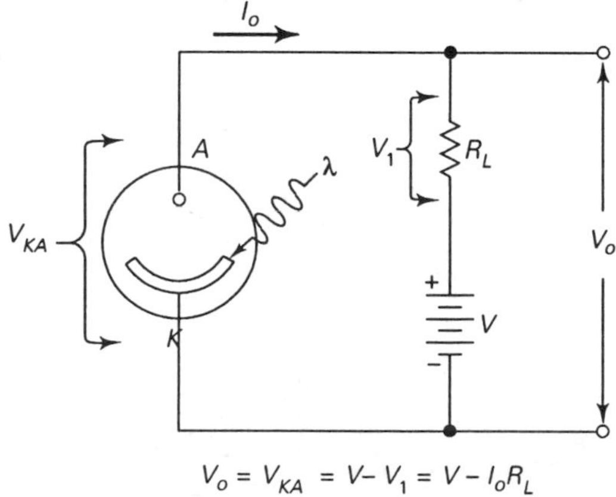

Figure 12–7 Circuit for using photoemissive tube.

than the high-vacuum types, but its light intensity versus output current is not as linear.

Photoemissive sensors have a *dark current* I_d, that is, a current that flows from cathode to anode when no light impinges on the photoemissive surface. For gas-filled tubes, I_d is on the order of 10^{-7} to 10^{-8} A, whereas for high-vacuum devices it is 10^{-8} to 10^{-9} A.

The *response time* of the photoemissive sensor is the time required for the device to respond to changes in applied light level. For high-vacuum devices, this time is on the order of 1 ns, while for gas-filled tubes it is 1 ms.

Photomultiplier Tubes

The photoemission process is not sufficiently efficient in many cases, especially under low-light-level conditions. The photoelectric tube system can be made more efficient by using a *photomultiplier* (PM) *tube* (Fig. 12–8). In this type of photosensor a number of positively charged anodes, called *dynodes*, intercept the electrons. When light impinges on the photoemissive cathode, electrons are emitted due to the photoelectric effect. They are accelerated through a positive high-voltage potential V_1 to the first dynode. The electrons acquire substantial kinetic energy during this transition, so when they strike the metal they give up their kinetic energy. Some of this kinetic energy is converted to heat, and some is converted in dislodging secondary electrons from the dynode surface. Thus, a single electron causes two or more additional electrons to be dislodged. These electrons are accelerated by high-voltage potential V_2 and reproduce the same effect at the second dynode. The process is repeated several times (each voltage step being about 75 to 100 higher than the preceding step), and each time several more electrons join the cascade for each previously accelerated electron. Finally, the electron stream is collected by the last dynode or a separate anode and can be used in an external circuit.

Spectral Response of Photoemissive Sensors

The *spectral response* of a photosensor describes its relative sensitivity to various wavelengths of light. The spectral response is typically reported in the form of a graph of either output current or relative output level versus wavelength. Figure 12–9 shows a typical family of spectral response curves. In addition to the graphical data, standardized S-numbers are also used to designate photosensors. Examples of these S-numbers are given in Table 12–1.

Most photoemissive sensors have a spectral response in the range from UV (200 nm) to visible light wavelengths, and a few also respond in the near-IR region close to 800 nm. Most of those that respond in the UV region,

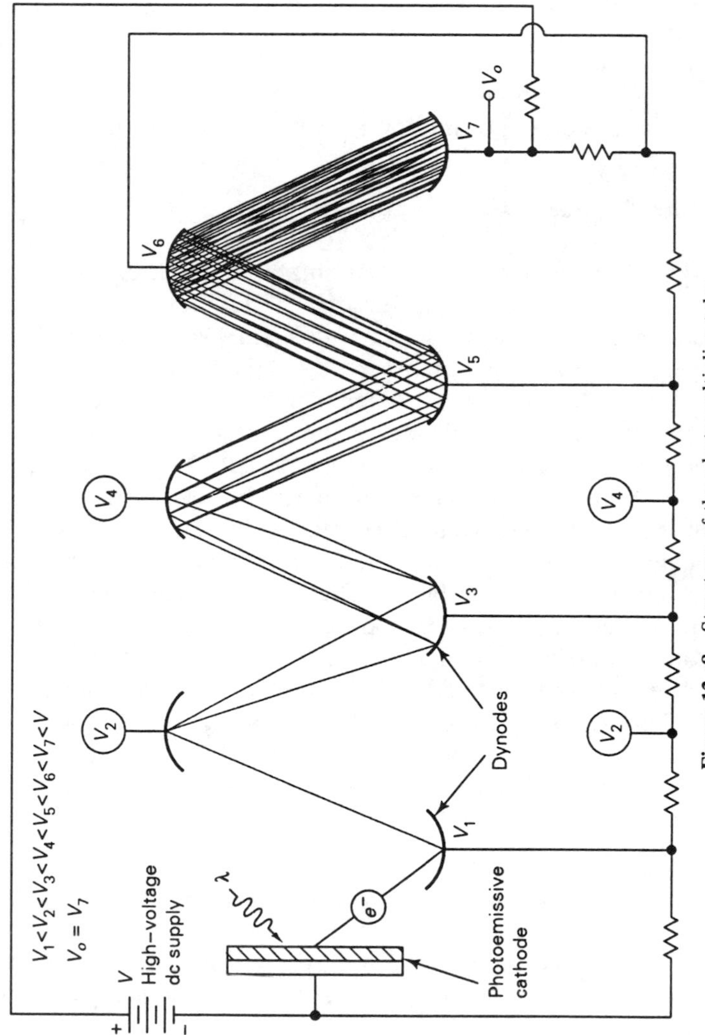

Figure 12–8 Structure of the photomultiplier tube.

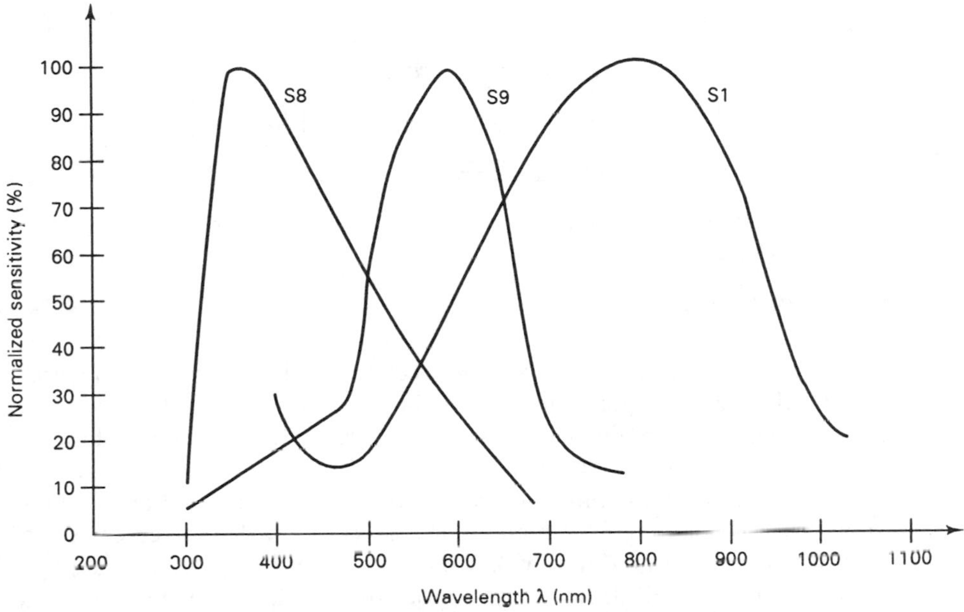

Figure 12–9 Sensitivity curves for several forms of PM tube.

however, are often limited by the lack of UV transmissivity of the glass window of the sensor tube.

Only the S14 sensor in Table 12–1 has a narrow band response. When a narrower bandpass characteristic is needed, it is necessary to install the photosensor behind a filter window that passes only the desired wavelengths of light (Fig. 12–10). The tube must be installed in a light-tight housing that is internally blacked against stray reflections.

TABLE 12–1

Spectral Designator (S-Number)	Wavelength for Peak Response (nm)	Half-Points (nm)
S1	800	620, 950
S3	420	350, 640
S4	400	320, 540
S5	340	230, 510
S8	370	320, 540
S10	450	350, 590
S11	440	350, 560
S12	500	(Narrow Band)
S13	440	260, 560
S14	1,500	760, 1,730
S20	420	325, 595
S21	450	260, 560

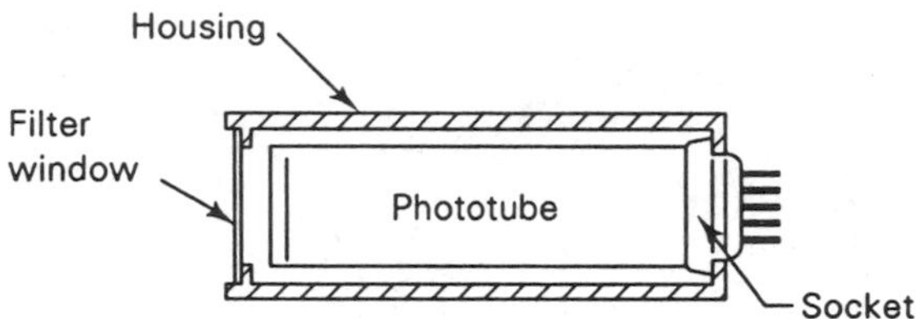

Figure 12–10 Housing and PM tube form the entire sensor.

PHOTOVOLTAIC CELLS

A *photovoltaic* (also called *photogalvanic*) *cell* is a device in which an electrical potential difference is generated—and thus a current made to flow in an external circuit—by shining light onto its surface. The common "solar cell" is an example of a photovoltaic cell.

Figure 12–11 shows three forms of photovoltaic cells. In Fig. 12–11(a) a metal disk (copper, gold, or platinum) is coated with a layer of copper oxide, which is in turn covered with a semitransparent layer that passes light and collects emitted electrons. The copper oxide cell was invented prior to World War I by Bruno Lange and eventually marketed by Westinghouse under the trade name Photox cell.

A similar photovoltaic cell is made of selenium [Fig. 12–11(b)]. This cell was invented in the 1930s and marketed by Weston Instruments under the trade name Photronic cell. In the selenium cell, a layer of photosensitive selenium is coated onto an iron, steel, or aluminum plate. In both forms of metal photovoltaic cells the thin insulator forms a *barrier* layer. When light illuminates the barrier layer, the impinging light photons are absorbed, and in the process electrons are emitted. The existence of free electrons causes a difference of electrical potential to appear across the barrier layer with the result that the selenium side is negative, while the transparent thin metal film side is positive.

Selenium cells produce an output of 0.2 to 0.6 V dc, and 0.45 V dc under 2000 foot-candles (fc) of illumination. Photovoltaic cells designed for power applications produce between 20 and 90 mw of dc power per square inch of photoactive surface exposed to light. The selenium cell covers a spectrum of 300 to 700 nm, with its peak near 560 nm. As with other sensors, it is common practice to alter the response of a selenium cell with special filters over the transparent window. In most instruments the selenium cell must be loaded with a resistor. Otherwise, the characteristic curve is highly nonlinear.

Figure 12–11(c) shows the structure of the silicon photovoltaic cell, discovered in 1958 by scientists at Bell Telephone Laboratories. The silicon cell consists of a *PN* junction of *P*- and *N*-type silicon. In the form shown in Fig. 12–11(c), the *N*-type silicon is deposited onto a metallic substrate that

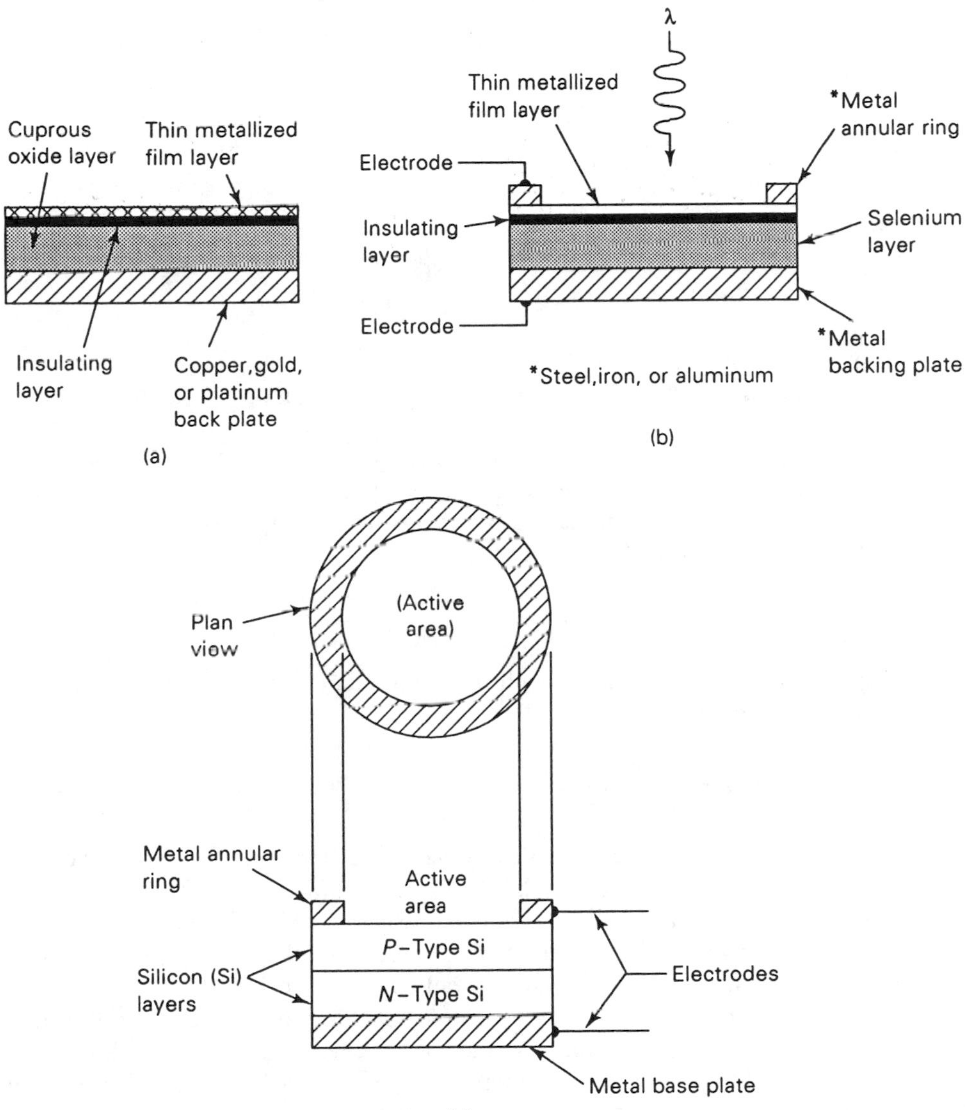

Figure 12–11 (a) Copper oxide form of solid-state photocell; (b) selenium photocell; (c) *PN*-junction silicon photocell.

also forms the negative terminal of the cell. The *P*-type layer is diffused over the *N*-type and forms the surface exposed to light. The positive electrode is an annular ring deposited onto the exposed surface of the *P*-type silicon region. These cells output a potential of 0.27 to 0.6 V under illumination of 2000 fc.

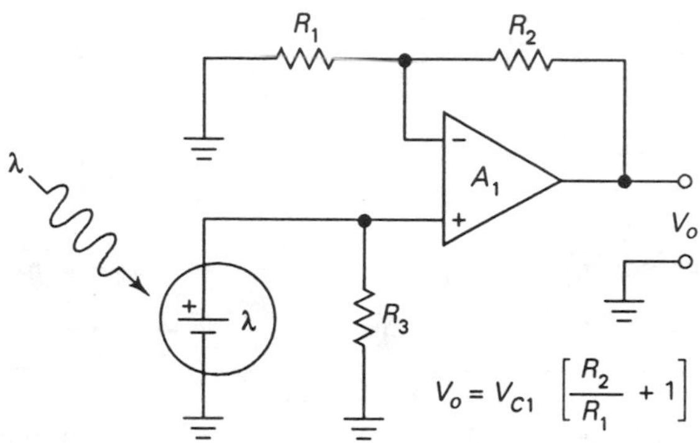

Figure 12–12 Photocell using noninverting follower amplifier for amplification.

Figure 12–12 shows a typical circuit for instrumentation applications of the photovoltaic cell. The cell is connected across the input of a high-impedance amplifier, such as the noninverting op-amp shown. The output voltage is found from the equation

$$V_o = V_{C1}\left(\frac{R_2}{R_1} + 1\right) \tag{12–7}$$

The op-amp provides a high-input-impedance buffer between the cell and the environment.

PHOTOCONDUCTIVE CELLS

A *photoconductive cell* (*photoresistor*) is a device that changes electrical ohmic resistance when light is applied. Figure 12–13(a) shows the usual circuit symbol for a photoresistor: The normal resistor symbol is enclosed within a circle and denoted as a resistor that responds to light by the Greek symbol lambda. An actual photoconductive cell is shown in Fig. 12–13(b).

The active region [Fig. 12–13(c)] of a photoconductive cell is a thin film of silicon, germanium, selenium, a metallic halide, or a metallic sulfide (e.g., cadmium sulfide, CdS). When these types of materials are illuminated with light, free electrons are created as photon energy drives them from a valence band into the conduction band. As in any conductor, free electrons mean that current can flow if an electrical potential is applied. Because increased illumination creates additional free electrons—which are available for conduction—the resistance of the photoconductive cell decreases with increases in illumination level.

When photoresistors are specified, a dark resistance and a light/dark ratio are typically given. Changes from megohms when dark to hundreds of ohms (Fig. 12–14) under maximum illumination are common. Because the intensity of the light affects the resistance, photoresistors are used in photographic light meters, densitometers, colorimeters, and the like.

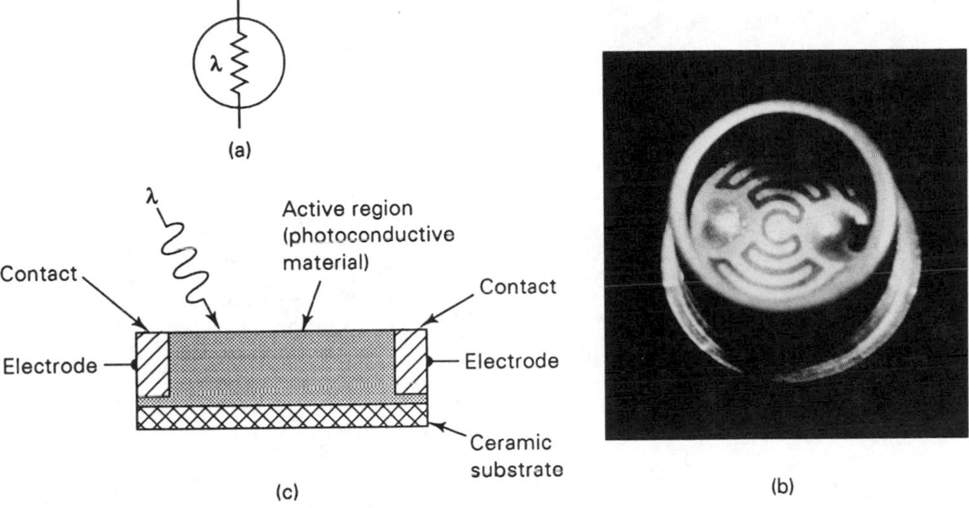

Figure 12–13 (a) Symbol for photoconductive cell (photoresistor); (b) example of photocell; (c) structure of photoconductive cell.

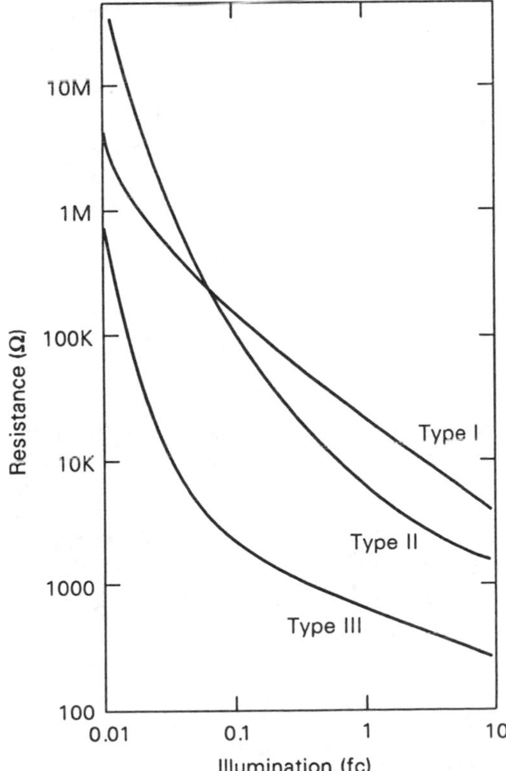

Figure 12–14 Response curves of three classes of photoconductive cells.

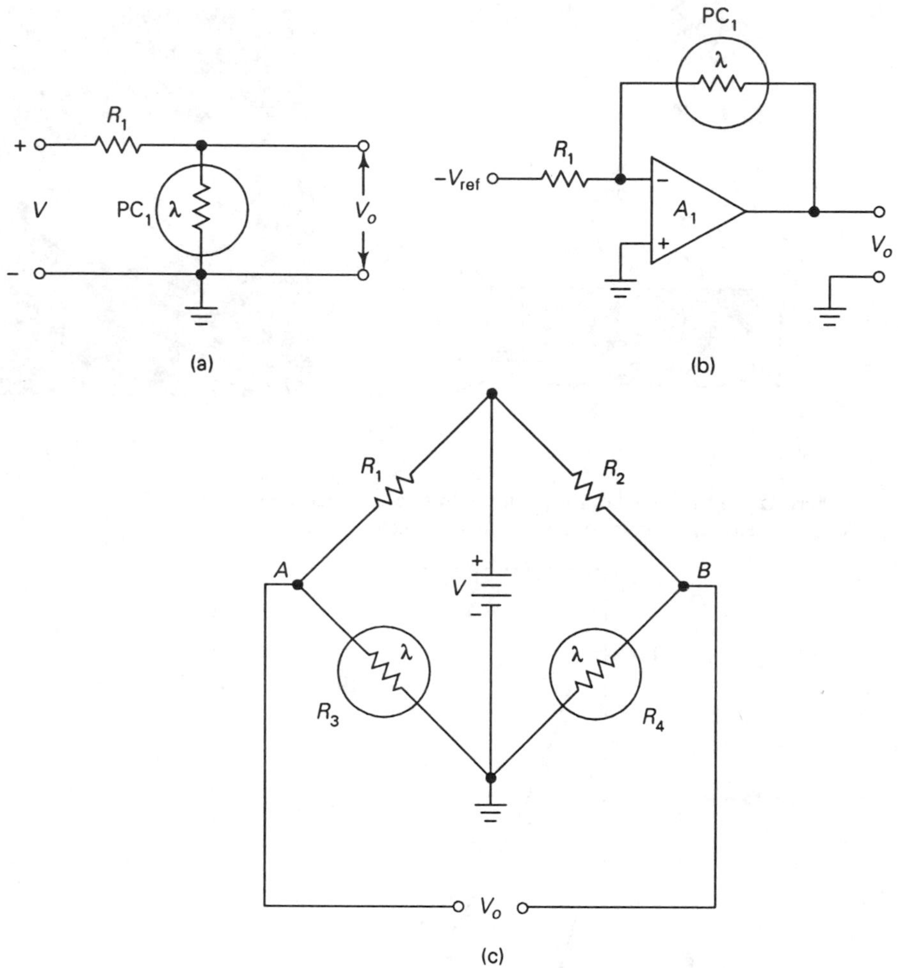

Figure 12–15 (a) Passive form of photoconductive cell (PC) circuit; (b) PC cell used in inverting follower circuit; (c) PC sensors used in Wheatstone bridge circuit.

Figure 12–15 shows three circuits in which photoresistors can be used. The half-bridge circuit is shown in Fig. 12–15(a). In this circuit the photoresistor is connected across the output of a voltage divider made up of R_1 and PC_1. The output voltage is given by

$$V_o = \frac{V \times PC_1}{R_1 + PC_1} \tag{12–8}$$

where V_o is the output potential
V is the applied excitation potential
R_1 and PC_1 are in ohms

A problem with this circuit is that the output potential does not drop to zero, but always has an offset value.

A second way to use the photoresistor is shown in Fig. 12–15(b). Here the photoresistor is the feedback resistor in an op-amp inverting follower circuit. The output voltage V_o is found from

$$V_o = -\frac{(-V_{ref})(PC_1)}{R_1} \tag{12–9}$$

The circuit of Fig. 12–15(b) provides a low-impedance output, but as with the other half-bridge circuit [Fig. 12–15(a)], the output voltage does not drop to zero. An additional problem is that photoresistors the dynamic range of the op-amp may not match the dark/light ratio of the photoresistor at practical values of $-V_{ref}$ potential.

The last photoresistor configuration is the Wheatstone bridge shown in Fig. 12–15(c). This circuit allows the output voltage to be zero under the right circumstances and is the circuit favored by most designers. If a low-impedance output is required or additional amplification is needed, then a differential dc amplifier can be connected across output potential V_o.

The Wheatstone bridge can be considered as two half-bridges in parallel. The output voltage is equal to the difference between the respective half-bridge output voltages, that is, the voltages at points A and B. The voltages at these points are found from Eq. (12–8), so the output voltage from the bridge is equal to

$$V_o = V \times \left(\frac{R_3}{R_1 + R_3} - \frac{R_4}{R_2 + R_4}\right) \tag{12–10}$$

PHOTODIODES AND PHOTOTRANSISTORS

Perhaps the most modern light sensor is the PN junction in the form of special photodiodes. But before discussing this sensor, we briefly review conduction in simple semiconductor devices.

Semiconductors are made into conductors by doping the material with impurities that contribute charge carriers. In N-type semiconductor material the impurities form covalent bonds with the semiconductor atoms and thereby create an excess of free electrons, which are negative charge carriers. In P-type semiconductor materials the impurities also form covalent bonds with the semiconductor atoms, but in this case there are insufficient numbers of electrons. As a result, there are places in the lattice of bonds where an electron should be, but isn't. These points are called *holes* and serve as positive charge carriers. Keep in mind that holes are not real entities, but rather they are places where an electron should be, but isn't. The dynamic of semiconductor materials is a constant attempt to reestablish electroneutrality. Thus, electrons and holes are constantly attempting to recombine.

The hole acts like a positive charge carrier with the same mass as an electron, but a positive charge. The "flow" of positive charge carriers (holes) is actually the flow of electrons in the opposite direction. This concept is illustrated in Fig. 12–16(a). Consider an electron at point *A* and a hole at point *B*. When the electron and hole recombine, the electron must move from point *A* to point *B*. This action can be viewed as an electron moving from *A* to *B*, or a hole moving from *B* to *A*.

A *PN*-junction diode [Fig. 12–16(b)] consists of *N*-type and *P*-type material joined together. If a positive potential is applied to the *P*-type side, and a negative potential to the *N*-type side, the respective charge carriers are repelled from the ends of the structure toward the junction. Under this condition a large amount of electron-hole recombination can occur, which produces a current flow across the junction. But when the potentials are reversed, as shown in Fig. 12–16(b), charge carriers are attracted away from the junction toward the end electrodes. This action forms a *depletion zone* between the *N*- and *P*-type materials in which few charge carriers (ideally zero) exist. Recombination of those few charge carriers that do exist in the depletion zone creates a minute *leakage current* across the *PN* junction. Ideally, the leakage current is zero, but in practice tends to be on the order of 10^{-7} A.

A certain *junction potential* (V_γ) also exists across the *PN* junction. In germanium-based semiconductors V_γ is on the order of 200 to 300 mV, while in silicon-based semiconductors it is 600 to 700 mV.

When the *PN* junction of certain types of diodes is illuminated, the level of reverse leakage current available increases. These diodes are built with an extremely thin *N*-type region and are configured so that light can illuminate the region. The photon energy causes electrons to strip away from their

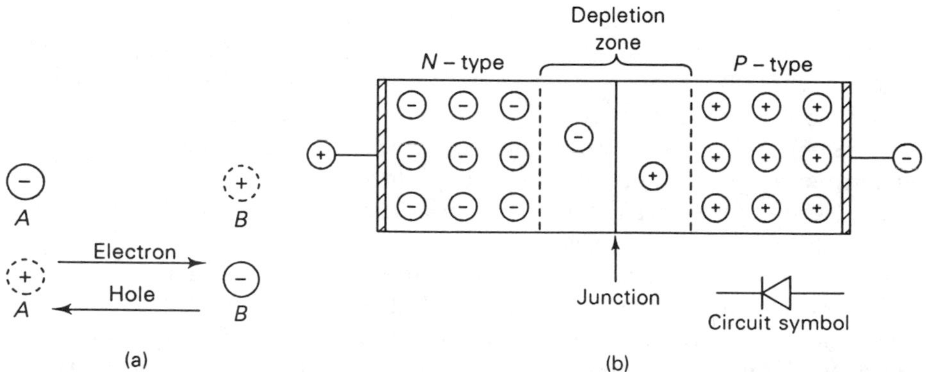

Figure 12–16 (a) Hole conduction is electron conduction in reverse. (b) *PN*-junction diode.

respective atoms to become negative charge carriers. Their former spots in the lattice become holes. If a reverse bias voltage is applied [as in Fig. 12–16(b)], then a very large electrical field is created in the region of the junction. This field prevents electrons and holes from recombining, but rather drives them toward the opposite polarity end of the electric field. Thus is the leakage current I_L increased by light illuminating the junction.

Figure 12–17(a) shows the basic circuit used for *PN*-junction-diode light sensors. The diode is normally reverse biased, with a current-limiting resistance R in series. In practical instrumentation applications either the leakage current I_L or the voltage drop V_o across resistance R is measured to find the light level. This voltage drop is the product of the current and the resistance (I_LR).

The same basic principle also creates a class of bipolar *NPN* or *PNP* transistors called *phototransistors* [Fig. 12–17(b)]. In these devices collector-to-emitter current flows when the base region is illuminated. These devices are the heart of optoisolator and optocoupler integrated circuits and are also used in various linear instrumentation sensor applications.

Figures 12–17(c) and 12–17(d) show two ways in which photodiodes and phototransistors are used. The current flow through the device under light conditions is the transducible event, so the device must be connected into a circuit that makes use of that property. The inverting follower op-amp circuit [Fig. 12–17(c)] serves that function. In all such cases the output voltage V_o is equal to the product of the input current I_L and the feedback resistance R_1:

$$V_o = -I_LR_1 \qquad (12\text{–}11)$$

In the case of Fig. 12–17(c) a zero control is added, which can be added to the other circuit as well.

The noninverting follower version of this circuit is shown in Fig. 12–17(d). In this circuit the phototransistor or diode is connected to the noninverting input of the op-amp. A voltage drop V_1 is created by transistor output current I_L flowing in resistor R_3. This voltage becomes the input signal of the op-amp and is proportional to the light level applied to the transistor base region. The op-amp input voltage is found from

$$V_1 = I_LR_3 \qquad (12\text{–}12)$$

and output voltage V_o from

$$V_o = V_1\left(\frac{R_2}{R_1} + 1\right) \qquad (12\text{–}13)$$

or

$$V_o = I_LR_1\left(\frac{R_2}{R_1} + 1\right) \qquad (12\text{–}14)$$

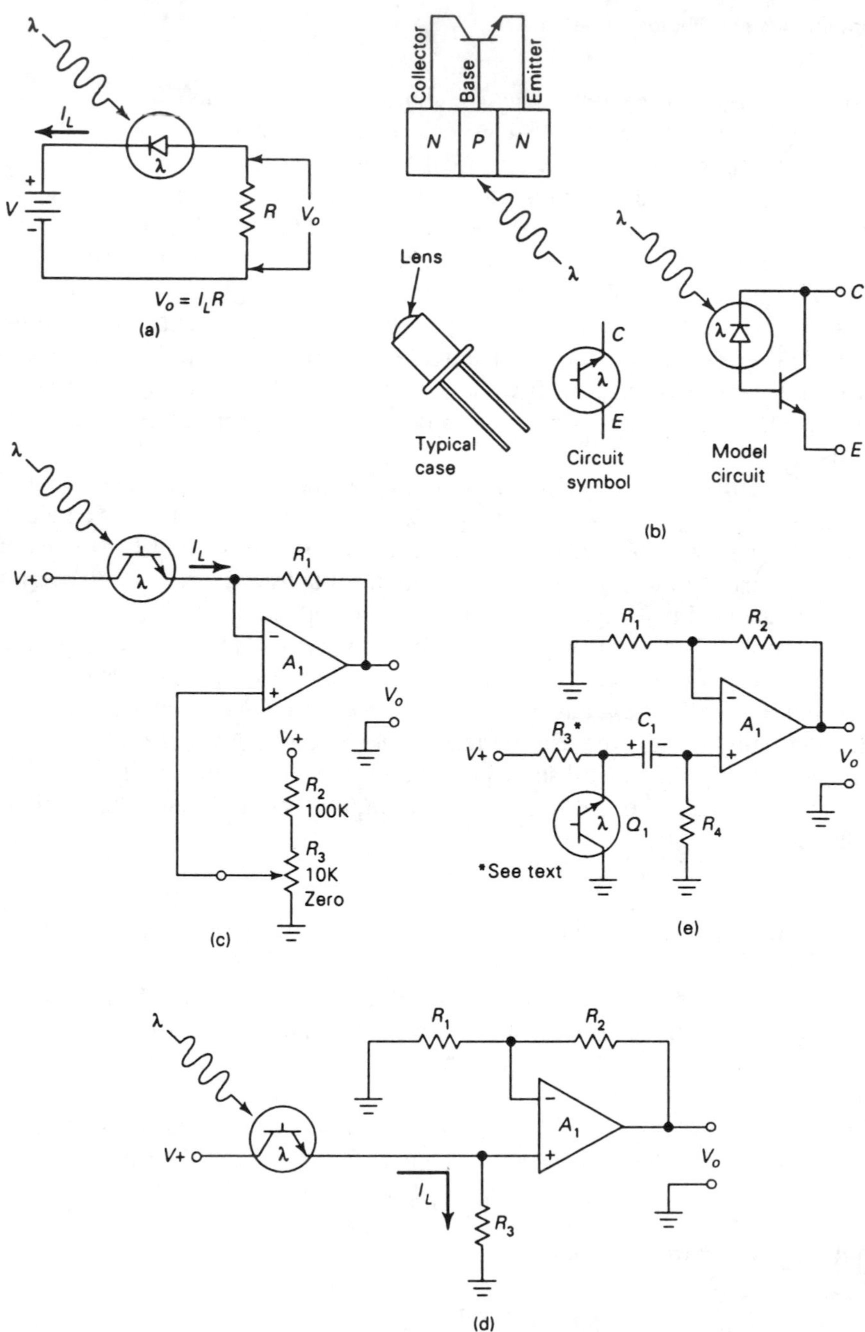

Figure 12–17 (a) *PN*-junction-diode circuit requires a voltage source for reverse bias to generate a leakage current that is proportional to the applied light level. (b) *NPN* bipolar phototransistor (structure and package styles); (c) operational amplifier circuit for phototransistors; (d) a noninverting follower (produces dc output level); (e) ac-coupled version of the noninverting circuit.

Figure 12–17(e) shows a circuit used when the light signal is either varying or modulated. An example of the former is seen in medical electronics in pulse or heart rate meters. These devices, called *photoplethysmographs*, use a red light source transmitting through the tissue of a thumb, finger, or ear lobe with a phototransistor sensing the light on the other side of the path. When blood pulses through the tissue, it changes the optical density to red light by a small amount, and this change can be used for taking the pulse waveform. Modulated light sources are used in some alarms and other applications to help the circuit distinguish between light from a pulsed emitter and ambient light. It also helps in burglar alarm circuits where the felon tries to defeat the alarm by shining a light into the sensor. If the source emitter is pulsed, then shining a constant light, or a light pulsed at the wrong frequency, will not help at all.

The circuit of Fig. 12–17(e) consists of a phototransistor Q_1 driving an ac-coupled noninverting follower op-amp circuit. The sensitivity of Q_1 is determined in some measure by the resistor in series with the collector and $V+$ supply (i.e., R_3). A high value, 100K ohms to 1M ohms, will increase sensitivity, while a low value, 10K ohms to 100K ohms, will increase operating speed.

The collector of Q_1 has a large dc offset, with variation due to the superimposition of stimulus light levels on the dc component. The capacitor C_1 strips off the dc component, allowing only the variations to pass on to the amplifier. The low-frequency response of this circuit is dependent on the capacitor value and the resistance of R_4:

$$F_{-3\text{dB}} = \frac{1}{2\pi R_4 C_1} \tag{12–15}$$

ELECTRO-OPTICS AND INSTRUMENTATION

Electro-optical devices have long been used in scientific, medical, and engineering instruments and will continue to be so for a long time to come. In this section we take a look at several different electro-optical techniques.

PHOTOCOLORIMETRY

One of the most basic techniques is both the oldest and most commonly used: *photometry*, also known as *photocolorimetry* or *absorption spectrophotometry*. These electro-optical methods are used to measure oxygen (O_2) content of blood, carbon dioxide (CO_2) content of air, water vapor content in a gas, electrolyte—that is, sodium (Na) and potassium (K)—levels of blood, and a host of other similar measurements. The basis for all these measurements is the absorption of different wavelengths of light when they are passed through a substance, or the emission of different wavelengths when the substance is burned. For example, when IR "light" passes through a gas containing CO_2, certain IR wavelengths are heavily absorbed. The amount of carbon dioxide present in the mixture can be determined from the relative absorption of these wavelengths.

Photocolorimetry is a comparison measurement technique in which IR, visible, or UV light transmission over two paths is compared. Figure 12–18(a) shows the basic circuit of the most elementary form of visible light colorimeter, a form of Wheatstone bridge circuit. The sensors are photoresistors, such as cadmium sulfide (CdS) cells. Although the circuit is very basic, this is the actual circuit used in a number of widely used instruments. The circuit uses a pair of photoresistor cells (R_2 and R_4) as the light sensors. Potentiometer R_5 in Fig. 12–18(a) is used as a bridge balance control. It is adjusted for zero output ($V_o = 0$) when the same light shines equally on both photoresistors. The output voltage from the bridge V_o is zero when the two legs of the bridge are balanced. In other words, V_o is zero when R_1/R_2 =

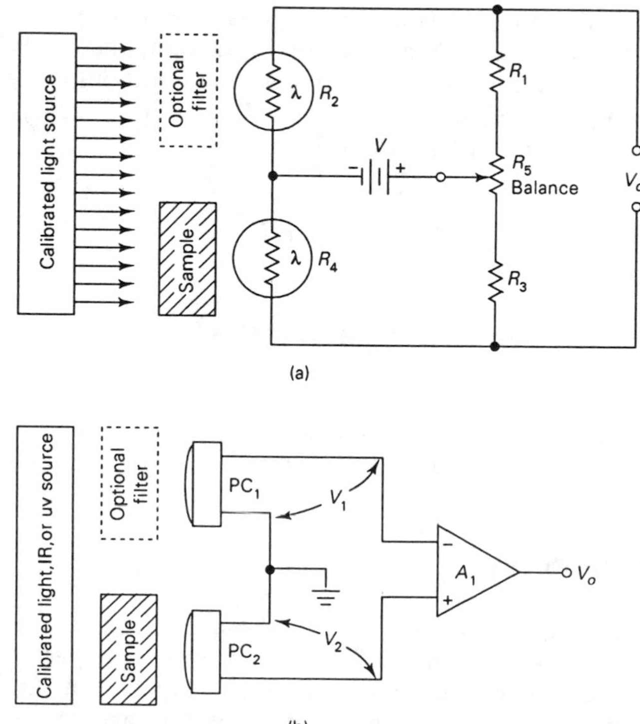

Figure 12–18 (a) Photocolorimeter uses a Wheatstone bridge circuit; (b) colorimeter based on photovoltaic cells.

R_3/R_4. It is not necessary for the resistor elements to be equal (although that is often the case), only that their resistance ratios be equal. Thus, a 500K/50K ratio for R_1/R_2 will produce zero output voltage when R_3/R_4 = 100K/10K.

The photoresistors are arranged so that light from a calibrated source illuminates both of them equally and fully, except when an intervening filter or sample is present in one or both pathways. Thus, the bridge can be nulled to zero using potentiometer R_5 under this zero condition. In most instruments that are based on this principle, a translucent sample is placed between the light source and one of the photocells. The amount of light transmission allowed by the sample is a measure of its optical density and is thus a transducible property. Let's look at a couple of different types of instruments to see how this principle is applied.

A variation on this design is shown in Fig. 12–18(b). This circuit uses either silicon solar cells or silicon phototransistors (a dc power supply is required for the latter) as the sensors. These sensors have a wider spectral response than CdS cells and will work down into the IR region. The two matched sensors (PC_1 and PC_2) are used to drive a differential amplifier A_1. The output of the amplifier V_o is the difference between the output of PC_1 (V_1) and PC_2 (V_2). This voltage V_o is proportional to the difference in the relative light (or IR) levels applied to the two sensors.

Blood O_2 level. A classical (but still widely used) method for measuring blood oxygen level is based on a colorimetric principle: The "redness" of blood is a measure of its oxygenation. Oxygenated arterial blood (HbO_2) is redder than deoxygenated venous blood (Hb). Figure 12–19 shows the relative optical spectral absorption of blood according to its oxygen content. At a light wavelength of approximately 800 nm the two types of blood show the same absorption. This point is called the *isoabsorption point* or, more commonly, the *isobestic point*. A comparison of the light absorption properties of a sample of blood with the absorption at 800 nm indicates the oxygen content.

An instrument such as that diagramed in Fig. 12–18(b) can be built to read blood O_2 content. An 800 nm standard filter cell is introduced between the light source and one sensor, and a blood sample is placed in a standardized tube between the light source and the other sensor. The degree of blood O_2 saturation in the sample is thus reflected by the difference in the bridge reading between the sample path and filter path. A noninvasive method (i.e., does not require a blood sample to be drawn) is discussed later in the section in which both the IR and visible red portions of the spectrum are considered.

Respiratory CO_2 level. The exhaled air from humans is approximately 2 percent to 5 percent CO_2, while the percentage of CO_2 in normal room air is negligible. A popular form of "end tidal CO_2 meter" is based

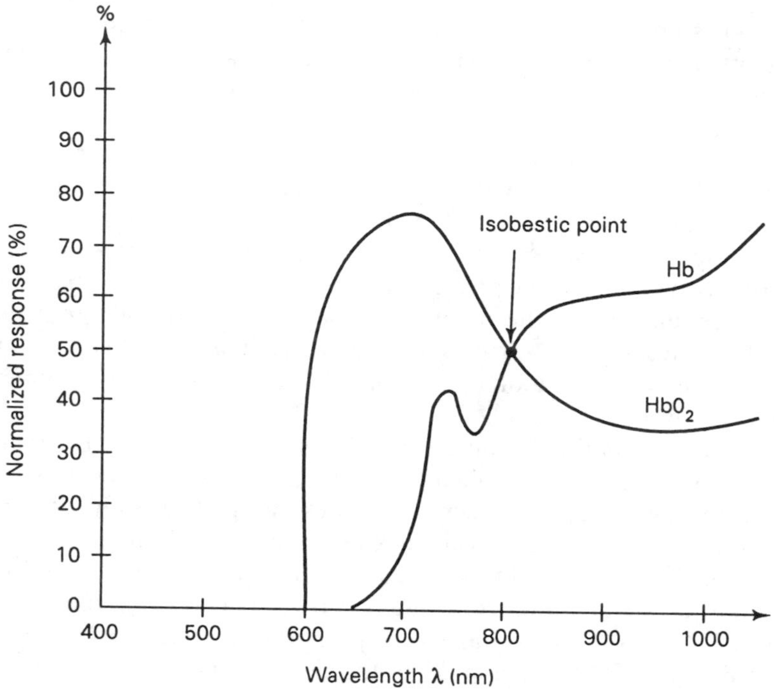

Figure 12–19 Spectra of light absorption by blood are the basis for a blood oxygen meter sensor.

on the fact that CO_2 absorbs IR waves at several discrete wavelengths (Fig. 12–20). The "light source" in many of these instruments is actually just a Cal-Rod identical to the one that heats your coffee pot! The photosensors are selected for good IR response, so they tend to be silicon cells. In this type of instrument, room air is passed through a glass cuvette placed between one sensor and the heat source, while patient expiratory air is passed through the same type of cuvette placed between the heat source and the other sensor. The difference in IR transmission across the two paths is a function of the percentage of CO_2 in the sample circuit.

Water vapor also absorbs IR energy (see Fig. 12–20), but at different wavelengths than CO_2. It is not easy to fine-tune the IR emitter to produce only those wavelengths that are absorbed by CO_2, but it is possible to filter out the wavelengths absorbed by water by placing a filter between the source and the sample.

The associated electronics for a circuit such as that in Fig. 12–18(b) (not shown) will allow zero and maximum span (i.e., gain) adjustment. The zero point is adjusted with room air in both cuvettes, whereas the maximum scale (usually 5 percent CO_2) is adjusted with the sample cuvette purged of room air and replaced with a calibration gas (usually 5 percent CO_2, 95 percent

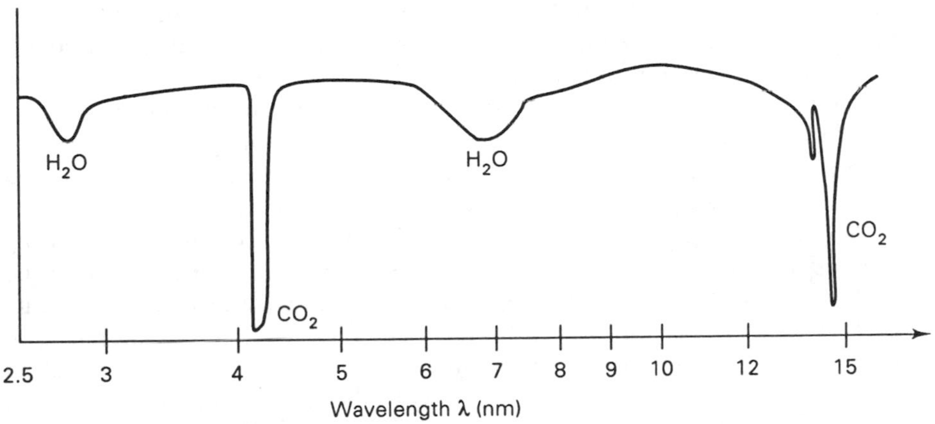

Figure 12–20 Absorption spectra for water and carbon dioxide.

N_2). This calibration gas must be obtained from a local scientific laboratory supplies outlet and be specified for use as a calibration gas; otherwise, the quantities may be only approximate.

Blood electrolytes. Blood chemistry tests often check for levels of sodium (Na) and potassium (K). An instrument commonly used for these measurements is the flame photometer (Fig. 12–21). This instrument replaces the light source with a flame produced by burning a gas in a carburetor. The sample is injected into the carburetor and burned along with the gas/air mixture. The intensities of the colors emitted on burning are proportional to the concentrations of Na and K ions in the sample. A special gas is used

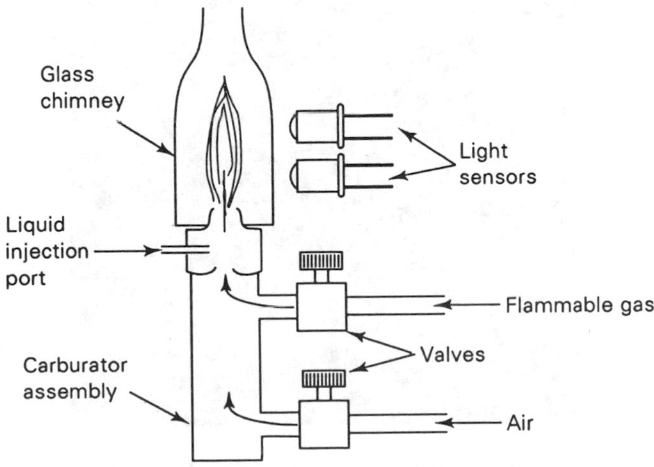

Figure 12–21 Flame photometer carburetor.

to burn cleanly with a blue flame when no sample is present. (A yellow flame indicates that unwanted sodium is present.) In medical applications a specified volume of the patient's blood is mixed with a precise, predetermined amount of a lithium calibrating solution. The solution is well mixed and injected into the carburetor. The concentrations of the two elements are determined by comparing the intensities of the colors emitted by the excited Na and K ions with the intensity of the calibration color.

It is important to follow proper maintenance procedures when using flame photometers to ensure accurate results. In most cases, errors are the fault of the operator, not the instrument. The carburetor and surrounding glass structures must be cleaned frequently, or the buildup of material from past tests will bias the results of the present test. In addition, carbonization of the associated glass windows will obscure the flame and may create both a lack of sensitivity and an erroneous reading (especially if the windows are not carboned uniformly).

PHOTOPLETHYSMOGRAPHY (PPG)

We previously discussed the use of a colorimeter to measure blood oxygen levels. That method required the sample to be compared with a standard filter cell in a Wheatstone bridge circuit similar to those depicted in Figs. 12–18(a) and 12–18(b). Another form of blood oxygen sensor that is non-invasive is the *photoplethysmograph* (PPG), a device that uses a light source and sensor to produce a waveform that closely resembles the blood pressure pulse (Fig. 12–22).

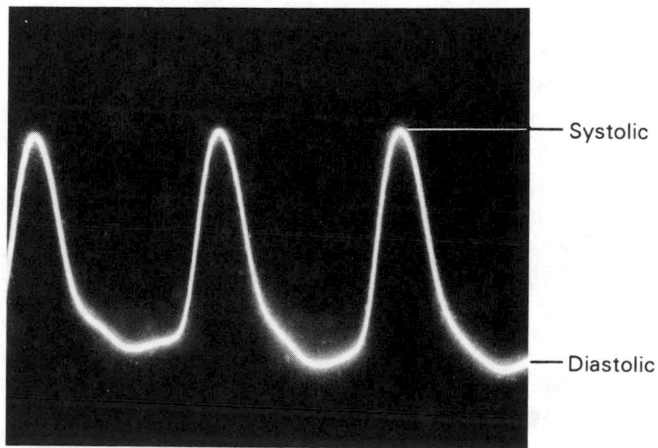

Figure 12–22 Human arterial pulse measured by a photoplethysmograph (PPG) sensor.

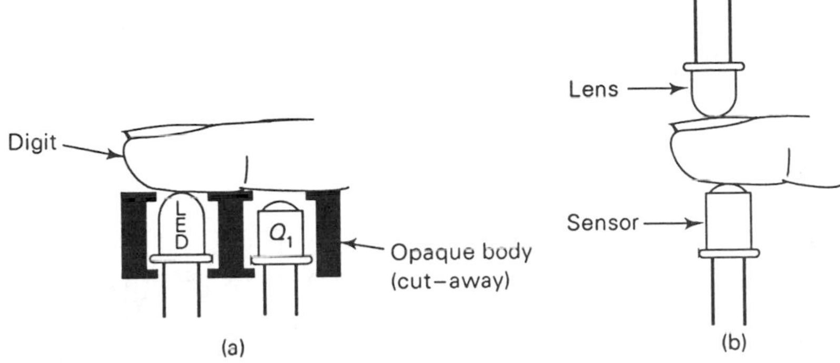

Figure 12–23 (a) Reflective PPG; (b) direct PPG.

Figure 12–23 shows the two basic forms of PPGs: *reflective* and *direct*. Both have as their principle of operation the *pulsing* of blood through tissue as the heart beats. As a result, the optical density of the tissue in which the blood flows changes as the pulse passes. The PPG is the form of sensor attached to exercisers who use a heart rate meter. The PPG can be clipped to an earlobe or placed over a finger or thumb to acquire the pulse signal. The reflective PPG shown in Fig. 12–23(a) is based on the scattering of light from the source (usually a red LED today, but small white "grain of wheat" incandescent lamps were once widely used) by bone in the finger or thumb. The direct type of PPG (Fig. 12–23(b)) transmits the light through the tissue to a sensor on the other side. Of these two methods, the direct form is easier to implement because the reflective form requires a very high degree of isolation between the source and sensor as well as good aiming of the two.

The circuit for a basic PPG sensor is shown in Fig. 12–24. This circuit uses a high-intensity red LED as the emitter and a high-gain silicon photo-Darlington transistor (e.g., *Digi-Key* L14R1GE[1]) as the sensor. The tissue (e.g., a finger or thumb) is placed between the emitter and the sensor. The signal from the collector of the phototransistor Q_1 is a low-frequency pulse, so a relatively high value coupling capacitor is needed. The amplifier is an operational amplifier connected in the noninverting follower configuration. The op-amp selected must have a high input impedance, so the CA-3140 BiMOS device or one of the BiFET op-amps is preferred. The voltage gain A_v of the amplifier is given by

$$A_v = \frac{R_5}{R_4} + 1 \tag{12–16}$$

The output of the op-amp is a voltage-pulse waveform with an amplitude between 200 mv and 1 V, but it will contain a fair amount of 60 hZ and high-frequency noise on the baseline. The capacitor C_2 shunting the op-amp feedback resistor R_5 will help cut some of the noise by limiting the frequency

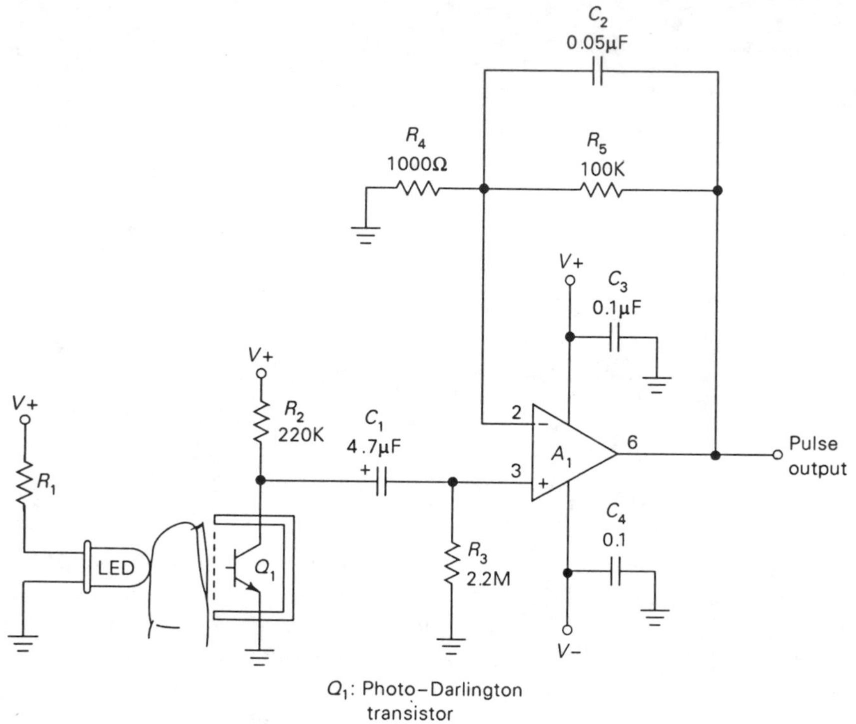

Figure 12–24 Practical PPG preamplifier circuit.

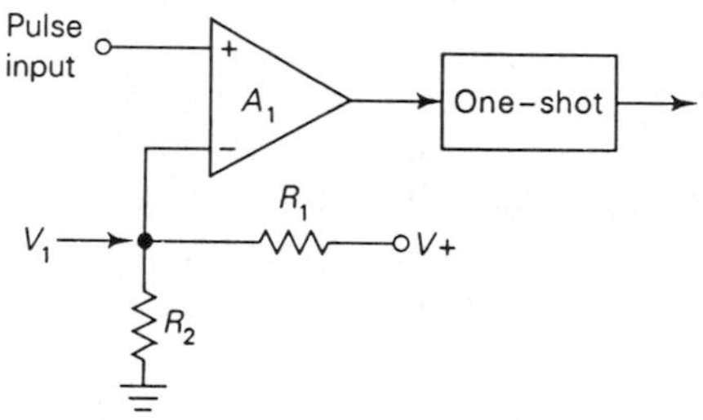

Figure 12–25 Comparator circuit to clean up the signal from a PPG.

response of the circuit. If additional amplitude is required, then the circuit of Fig. 12–24 is followed with another amplifier.

The PPG preamplifier can be used to drive a pulsometer circuit. A voltage comparator A_1 is used to clean up the output signal (Fig. 12–25). The comparator is used to produce an abrupt output level shift when the pulse voltage rises above V_1, the voltage at the output of voltage divider R_1/R_2. Voltage V_1 is set so that noise signals do not cross the threshold. This level can then be used to trigger a monostable multivibrator ("one-shot") circuit, if desired. A count of the one-shot output pulses—for example, in an integrating voltmeter—will yield a dc voltage that is proportional to heart rate.

Construction of the pulsometer sensor is shown schematically in Fig. 12–26(a), and in a photo in Fig. 12–26(b). The sensor housing is a large rubber boot normally used for insulating jumbo-size alligator clips. These can be purchased at most complete electronics parts stores. A size is selected

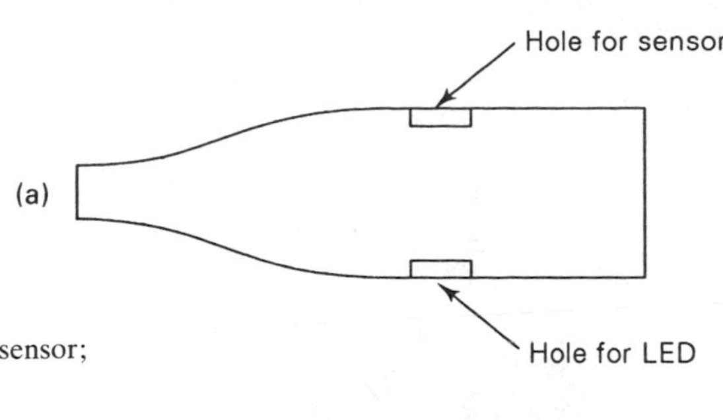

Figure 12–26 (a) "Boot" for PPG sensor; (b) actual PPG boot/sensor.

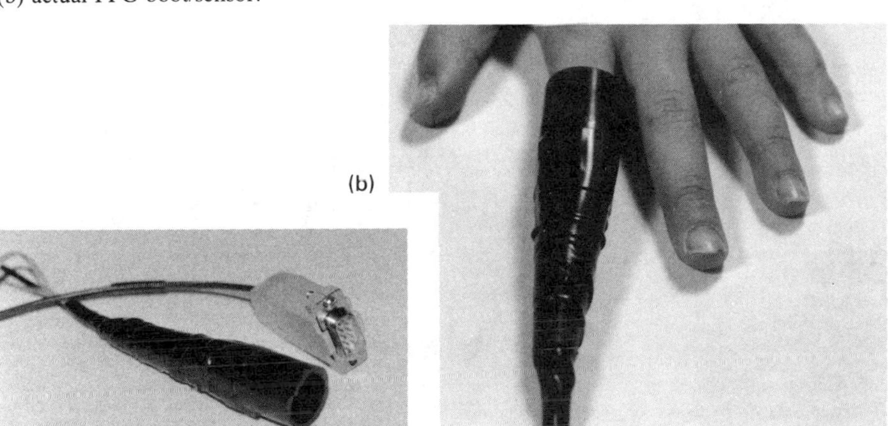

that fits over the thumb, the index finger, or the middle finger. Two holes are cut into the rubber boot to accommodate the sensor and LED emitter. Once these are wired into a circuit such as in Fig. 12–24, the whole assembly is covered with black insulating tape. It is very important to prevent ambient light from affecting the sensor transistor.

BLOOD OXYGEN MEASUREMENT PPG

One of the critical problems in surgery is ensuring that the patient receives an adequate oxygen supply. If the ratio of oxygen to anesthesia gas is too low, or if the flow of oxygen ceases altogether, then a catastrophe can result — the patient either dies or is reduced to a vegetative state. Both tragedies have been caused by human error and mechanical failure of the equipment. However, if the anesthesiologist has some means for monitoring the O_2 content of blood, even if only from a trending (rather than absolute) measurement, then action can be taken instantly when trouble occurs. One popular blood oximetry method is based on the PPG principle.

Figure 12–27(a) shows the basic PPG sensors required for this measurement, while Figs. 12–27(b) and 12–27(c) show actual medical products. A single wide-spectrum sensor is placed opposite a pair of LEDs. One LED emits visible red light near 660 nm, and the other emits IR light near 920 nm. The *isobestic point*, also called *isoabsorption point* (recall the HB and HbO_2 spectrum of Fig. 12–19) is a reference point between the oxygenated and deoxygenated blood O_2 saturation levels. The instrument produces an output

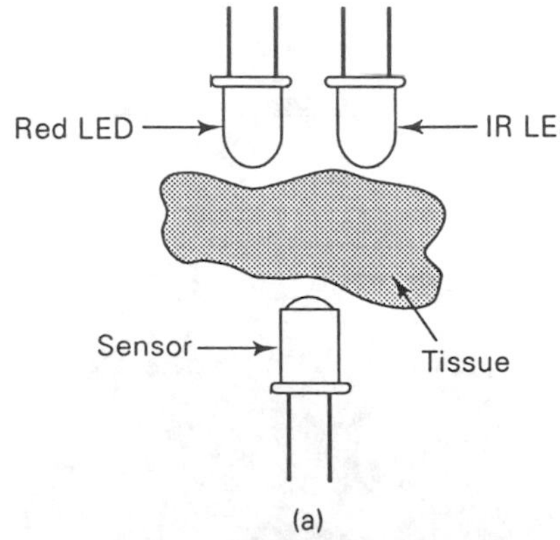

Figure 12–27 (a) Dual-sensor PPG that measures blood oxygen; (b) two forms of commercial blood oximeter sensors; (c) placement of oximeter sensor on finger.

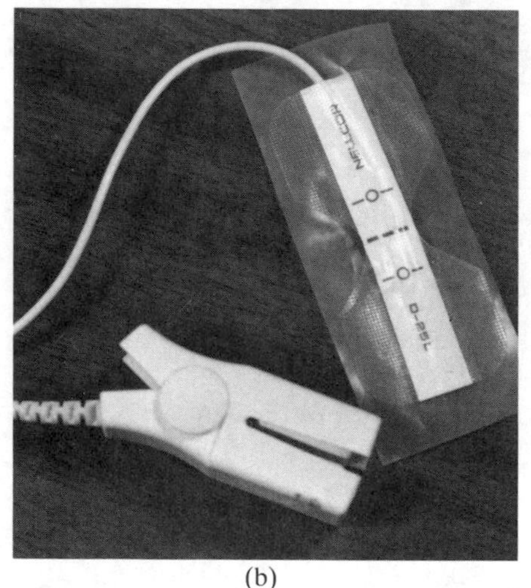

(b)

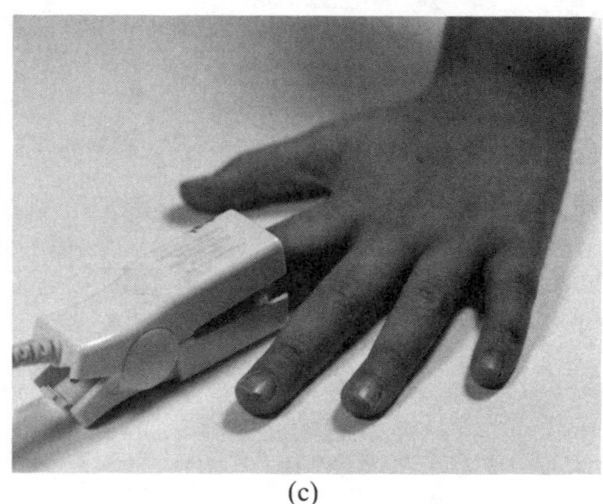

(c)

that measures blood O_2 level by comparing the relative absorption of a sample at the two absorption bands.

Two forms of sensors are available. The permanent form uses an alligator-clip-like device that can be placed over a finger of the patient's hand. Alternatively, disposable units are wrapped around the digit or placed over some other tissue on the body.

REFERENCE

1. Digi-Key Corporation, P.O. Box 677, Thief River Falls, Minn., 56701-0677; 1-800-344-4539.

Biological Impedance Measurements

13

Impedance is defined as the total opposition to the flow of alternating current (ac) and comprises three components: resistance R, inductive reactance X_L, and capacitive reactance X_c. All conductive materials have impedance, including living tissues. The impedances of various organisms are known to vary with the seasons, diurnally, and with assorted stimuli. In humans, physicians and physiologists have used impedance as an indirect, noninvasive measure of respiration rate, pulse rate, endocrine activity, cardiac activity, and a number of other physical factors.

In this chapter the term impedance is defined, some of the applications are explored, and practical circuits for measuring impedance are discussed. All cases involve either direct current (dc) or sinusoidal alternating current (ac), as appropriate. Waveforms other than the sine wave are composed of a fundamental sine wave and a collection of harmonic sine and cosine waves that are unique to the specific waveshape in question. As a result, nonsinusoidal waveforms are very difficult to analyze, so such tasks are left as an exercise for the reader.

WHAT IS IMPEDANCE?

Simply defined, impedance is the total opposition to the flow of alternating current from both resistive and reactive components in the measuring path. The resistive component should be well understood, for it is the same as the electrical resistance discussed in Chapter 3. Briefly restated, resistance is opposition to the flow of dc and (in part) ac caused by the material through which current is flowing. All material possesses resistance at temperatures above absolute zero ($\approx -273.16°C$). Such resistance is inversely proportional to the cross-sectional area A of the material between electrodes and directly proportional to the length L of the current path and a property of the material called *resistivity* ρ, that is,

$$R = \rho \frac{L}{A} \tag{13–1}$$

where ρ is the resistivity in ohm · centimeters
L is the length of the conductor path in centimeters
A is the cross-sectional area in square centimeters

The resistivity factor ρ varies somewhat with species, and considerably for different types of tissue. Blood, for example has been measured at approximately 150 ohm · cm, urine at 30 ohm · cm, skeletal muscle at 300 to 1600 ohm · cm, lung tissue at 1275 ohm · cm, and fatty tissue at 2500 ohm · cm.[1]

For many purposes the simple resistance is sufficient. Indeed, many of the "impedance" measurements made in physiology laboratories are actually ac resistance measurements. There are, however, some problems with the simple case in which only resistance is measured. First, although tissue is electrolytic in nature, it is not a simple electrolyte. While it is convenient to model tissue as a simple electrolytic solution (as indeed we did in Chapter 8 when discussing biopotentials electrodes), that model fails in some situations. The problem with resistance measurements in living organisms is that tissue tends to contain suspensions of cells and large molecules in addition to dissolved salts of potassium and sodium.[2] A second problem is that alternating currents do not flow uniformly throughout a conductor, and these effects are a function of frequency. Although ac flow patterns in complex tissues are dauntingly difficult to describe, the analogous phenomenon in cylindrical metal conductors (e.g., copper wire) is considerably easier. In such conductors the ac tends to flow near the surface of the conductor to a depth that is a function of frequency. The center portion of such a conductor does not conduct ac, so the ac resistance of a conductor is higher than the dc resistance of the same conductor.

The second component of impedance is *reactance*, which is of two types: *inductive reactance* X_L and *capacitive reactance* X_c. Reactance, which like resistance is measured in ohms, was discussed in depth in Chapter 3.

The conventional way to graphically show the relationship between resistance and the reactances is to use a Cartesian coordinate system [Fig. 13–1(a)] in which the east direction represents the resistance and is the point of reference for all angle measurements, inductive reactance $+X_L$ is plotted along the north axis, and capacitive reactance $-X_c$ is plotted along the south axis.

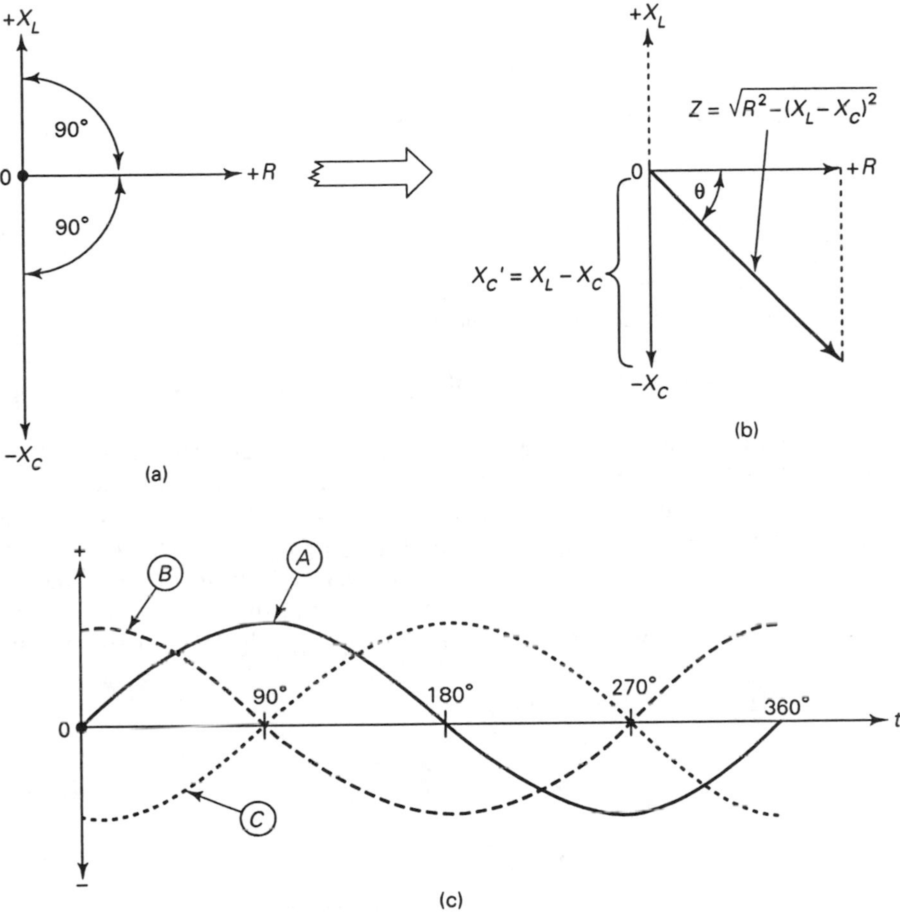

Figure 13–1 Vector relationships showing impedance: (a) *LCR* circuit; (b) *RC* circuit; (c) phase relationships in *RLC* circuits.

Both reactances are along the vertical axis in Fig. 13–1(a), so they can be algebraically added to form a resultant, as shown in Fig. 13–1(b). The resultant is at a certain *phase angle* θ, which derives from the fact that pure inductive reactance produces a current-lagging-voltage angle of +90°, whereas a pure capacitive reactance produces a current-leading-voltage in which the phase angle is −90°. These angles are shown graphically in Fig. 13–1(a) and in the form of sine-wave diagrams in Fig. 13–1(c). The reference curve in Fig. 13–1C is *A*, which represents both the voltage and current in a purely resistive circuit, and the voltage in a reactive circuit. In the inductive circuit the current (*B*) will lag 90° behind the voltage, whereas in a capacitive circuit the current (*C*) will lead the voltage by the same amount.

When a resistance is added to the circuit, the phase angle changes from

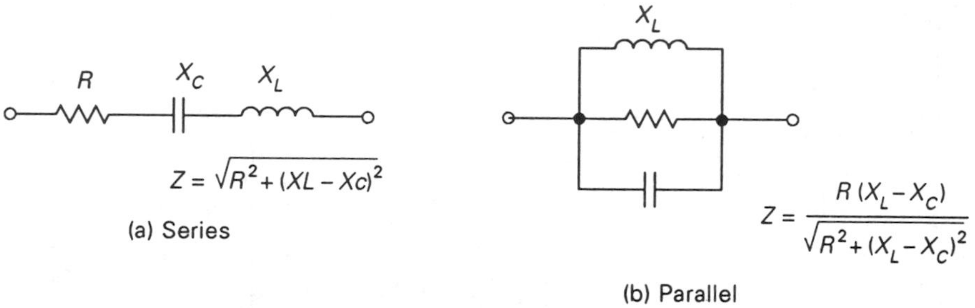

Figure 13–2 (a) Series-resonant circuit; (b) parallel-resonant circuit.

±90° to something less than 90°, as shown by the resultant vector Z in Fig. 13–1(b).

To find the impedance requires that both the *magnitude* and the *phase angle* of the impedance vector Z be found. Because reactances cause a phase shift in the angle between voltage and current, and resistance does not, the reactance terms cannot be simply combined with the resistance terms. Rather, the root-of-the-sum-of-the-squares method must be used to find the magnitude of the impedance vector. Further, the impedances generated by the same components in series and parallel circuits (Fig. 13–2) differ somewhat:

Series circuits [Fig. 13–2(a)]:

$$|Z| = \sqrt{R^2 + (X_L - X_c)^2} \tag{13–2}$$

Parallel circuits [Fig. 13–2(b)]:

$$|Z| = \frac{R(X_L - X_c)}{\sqrt{(R)^2 + (X_L - X_c)^2}} \tag{13–3}$$

It is a common, but very erroneous, practice to drop the parallel bars around the Z-terms in Eqs. (13–2) and (13–3). So, when you see Z in an equation, do not automatically assume that it is the complex form that is meant. In the discussions that follow the series case is used in order to keep the discussion simple.

SERIES-PARALLEL EQUIVALENT CIRCUITS

Impedance models are typically based on series impedance, even though parallel impedances are often found in practice. It is possible to mathematically convert a parallel impedance into an equivalent series impedance by using the following transformation equations (Fig. 13–3). The object is

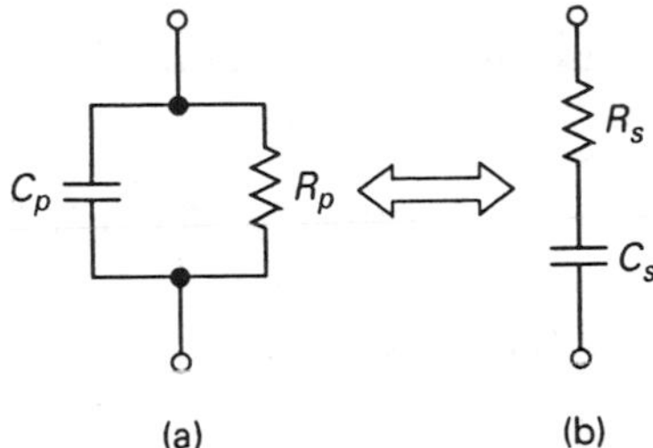

Figure 13–3 (a) Parallel *RC* network; (b) series *RC* network.

to calculate an equivalent series resistance R_s from a known parallel resistance R_p, and a series capacitance C_s from a parallel capacitance C_p:[3]

$$R_s = \frac{R_p}{1 + (2\pi f C_p R_p)^2} \tag{13–4}$$

and

$$C_s = C_p\left[1 + \frac{1}{(2\pi f R_p C_p)^2}\right] \tag{13–5}$$

To perform the opposite calculation, that is, to transform series impedance component values into component values for a parallel impedance, the following equations are used:

$$R_p = R_s + \frac{(X_s)^2}{R_s} \tag{13–6}$$

$$R_p = \frac{1 + (2\pi f C_s R_s)^2}{4\pi^2 f^2 (C_s)^2 R_s} \tag{13–7}$$

and

$$C_p = \frac{C_s}{1 + (2\pi f C_s R_s)^2} \tag{13–8}$$

$$X_p = X_s + \frac{(R_s)^2}{X_s} \tag{13–9}$$

The phase angle θ is a function of the reactance magnitude and the resistance:

$$\theta = \tan^{-1}\left(\frac{X}{R}\right) \tag{13–10}$$

where θ is the phase angle in degrees
X is the algebraic sum of X_L and X_c

The proper way to specify impedance is to state the magnitude of $|Z|$

and the phase angle. For example, in a situation where the impedance magnitude is 1200 ohms, and the reactive component is 200 ohms, the phase angle is 80.5°, so the impedance would be specified as "1200 ∠80.5."

Another way to denote impedance is to recognize that it is mathematically a complex number. Impedances can be represented by complex numbers because the effect of inductance and capacitance is to phase-shift the voltage and current 90° with respect to each other. Thus, impedance can be represented by the *j*-operator ($\sqrt{-1}$)* in the form

$$Z = R \pm jX \tag{13–11}$$

or, when representing either inductive or capacitive reactances,

$$Z = R + jX_L \tag{13–12}$$

and

$$Z = R - jX_c \tag{13–13}$$

or, in combined X_L and X_c circuits,

$$Z = R + j(X_L - X_c) \tag{13–14}$$

In physiological situations the inductive reactance term tends to be very small relative to the capacitive reactance ($X_L << X_c$), so it is usually ignored in physiological impedance measurement schemes. Thus, for most physiological applications, the impedance magnitude equations reduce to Eq. (13–13) and a simplified version of Eq. (13–4):

$$|Z| = \sqrt{(R)^2 - (X_c)^2} \tag{13–15}$$

The measurement of physiological impedances can be difficult because the forms taken by the tissue structures are neither simply geometric nor homogeneous in makeup. The capacitance of tissue can be generalized, however, by a model that takes into consideration the area A of the measuring electrodes and the distance *d* between them:[4]

$$C = \frac{\varepsilon \varepsilon_0 A}{d} \tag{13–16}$$

where ε is the dielectric constant of the tissue relative to free space
ε_0 is a factor related to units rationalization
A is the electrode area
d is the electrode separation

*Mathematicians and physicists may object that the term $\sqrt{-1}$ should be represented by *i*. The term *j* is used in electrical engineering because *i* represents the instantaneous value of an alternating current.

IMPEDANCE MEASUREMENT

The measurement of impedance requires an ac oscillator and involves passing the ac through the tissue being investigated. The choice of ac frequency is important, as is the level of current I_{ac} used.

The frequency is important because the reactive components of impedance are frequency-sensitive. As a result, differing impedances are seen at different frequencies. For example, when ordinary Ag–AgCl electrodes ($d \simeq 1$ cm) are used, the impedance of skin as a function of frequency is approximately as shown in Fig. 13–4. Obviously, when measuring skin impedances, as in making GSR measurements, of a low frequency, typically less than 100 Hz is necessary. The selection of frequency for galvanic skin resistance (GSR) is further constrained by the fact that the ac power lines operate at 60 Hz in the United States and Canada, and at 50 Hz in Europe, so if frequencies near 50 to 60 Hz are selected, they would interfere with the measurement. As a result, most GSR measurements are made using frequencies in the range around 100 Hz or <25 Hz.

When looking at deeper organs, the investigator might want to consider higher frequencies (>20 kHz) to reduce the effect of skin impedance on the final measurement.

Another reason for selecting higher frequencies is that cells and other

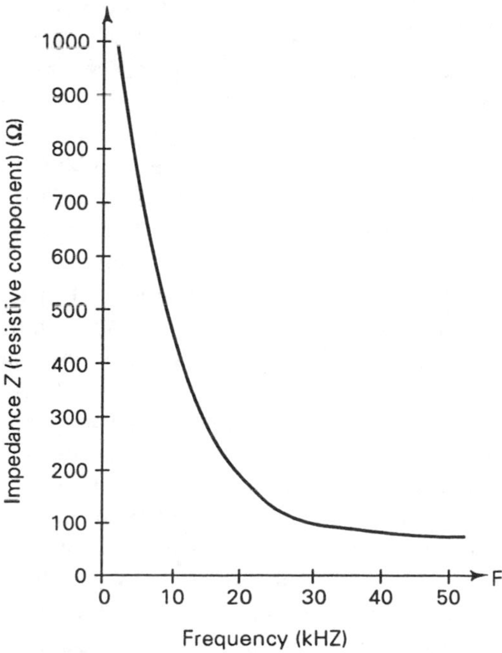

Figure 13–4 Electrode as a function of frequency.

structures react to the current with a specific strength-duration (S-D) characteristic. The S-D is sometimes specified in terms of the *chronaxie* of the tissue, which is defined as "the duration of a stimulus having twice the intensity of a stimulus that would be just sufficient (threshold) to produce stimulation if allowed to remain on for an infinite time."[5] In other words, it is the stimulus duration at twice the threshold for a steady-state stimulus. Geddes and Baker report chronaxie values from 0.2 ms for nerve tissue to 100 ms for smooth muscle. An implication of the chronaxie is that the period of the ac signal used to excite tissue in impedance measurements must be less than the chronaxie value for that tissue. The frequency f of an ac signal is the reciprocal of its period T, so from the relation $f = 1/T$, the minimum frequency should be 5000 Hz for nerve tissue and 10 Hz for smooth muscle. As a practical matter, the selected frequency is usually much larger than the minimum value.

The current level is critical for two reasons: (1) Higher currents stimulate cells and cause measurement artifact, and (2) higher currents can cause discomfort in the subject. The threshold of perception for dc and low ac currents is on the order of 1 mA, while for 25 kHz ac it is an order of magnitude higher. In general, the best rule is to make the current as small as possible, with 100 μA being a commonly used rms value.

One method for measuring impedance is the ac Wheatstone bridge circuit shown in Fig. 13–5. Four impedances are used in this circuit. Two arms of the bridge are represented as fixed resistors (R_1 and R_2), one arm is a standard impedance Z_{std}, and the fourth arm is the unknown impedance Z_{unk}. Either series or parallel standard impedances can be used, and some ac bridges make it possible to use either one by properly setting a switch. It is typical, but not necessary, for R_1 and R_2 to be equal resistances.

The bridge is excited from an ac source by voltage V. The bridge is constructed so that the values of R and C in the standard impedance can be read from a dial. In most instruments, the C component is measured in terms of capacitive reactance at the excitation frequency, unless that frequency is variable. In the latter case it is common practice to use units of reactance (ohms) at some standardized frequency such as 1 kHz or 10 kHz to normalize the capacitor dial and then multiply or divide the dial reading by the actual frequency.

There are two ways to use the bridge. The *null-condition method* uses an output detector to find when the output voltage drops to zero. A typical output detector is an oscilloscope or an ac voltmeter. In this type of measurement the electrodes are connected to the unknown impedance, and both the resistor and capacitor on the standard impedance side of the bridge are adjusted to find the deepest null. These adjustments are somewhat interactive in most bridges, so they must be made repeatedly until the deepest null possible is obtained. The values of C and R can then be read from the respective dials.

In the *off-null method* the R and C components of the standard imped-

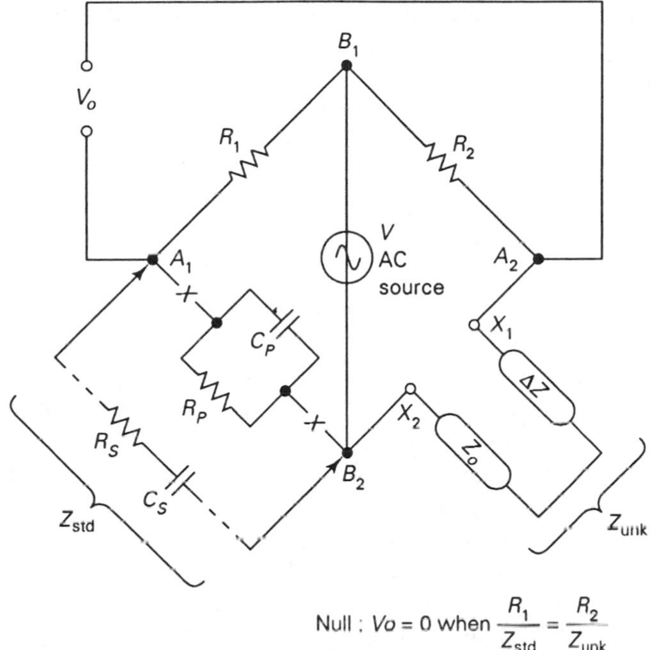

Figure 13–5 Wheatstone bridge used to measure impedance.

ance are set to a standard value (the selection of settings in any particular case depends on the magnitudes of the expected unknowns). If the magnitude and phase angle of the output voltage V_o are measured under a standard condition, for example, when a resistance is connected across the unknown impedance terminals (X_1–X_2), and the same parameters are remeasured when the unknown is present, then it is possible to determine the values of the unknown impedance. A number of instruments on the market do this task automatically.

Another impedance measurement scheme, the *bipolar electrode method*, is shown in Fig. 13–6. In this case the impedance is modeled in terms of a

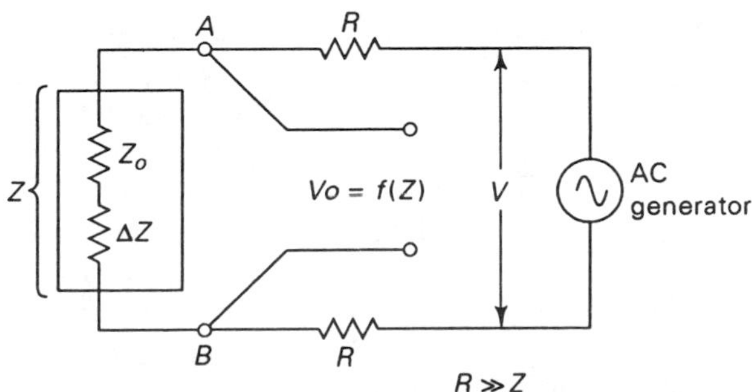

Figure 13–6 Four-wire impedance measurement.

fixed component Z_0 and a change in impedance ΔZ. The impedance varies with physiological activity around the value $Z_0 \pm \Delta Z$. For all cases the output voltage is given by

$$V_o = \frac{V(Z_0 + \Delta Z)}{2R + Z_0 + \Delta Z} \tag{13–17}$$

or, for the usual case in physiological measurements where $Z_0 >> \Delta Z$ and $R >> Z_0$,

$$V_o = V\left(\frac{Z_0}{2R} + \frac{\Delta Z}{2R}\right) \tag{13–18}$$

The two terms in Eq. (13–18) represent the fixed or resting impedance Z_0 and the change of impedance ΔZ. The fixed term $VZ_0/2R$ is very much larger than the changing term $V\Delta Z/2R$, so the former must be stripped from the output by using either a cancellation null circuit or a simple capacitor coupling scheme (the preferred method).

In the case where $R \gg Z_0$, the circuit acts as a pseudo-constant-current source because the total resistance seen by the ac generator $(R + R + Z_0 + \Delta Z)$ is very large compared with $Z_0 \pm \Delta Z$. Thus, relatively large changes in ΔZ represent only small changes in the overall resistance loading the ac source. The current is therefore approximately constant.

If a dynamic constant-current source is used, as in Fig. 13–7, then the constant current can be set as needed, and it will remain constant regardless of wide fluctuations in Z_0.

A *tetrapolar electrode method* of measuring physiological impedance is shown in Fig. 13–8. In this configuration four electrodes (current electrodes IE_1, IE_2, and measurement electrodes ME_1, ME_2) are spaced around the area of measurement. An ac voltage V is applied to the two outer electrodes (IE_1 and IE_2), causing a current to flow through the tissue being measured. This current flows back and forth between IE_1 to IE_2, causing an ac voltage drop to appear between ME_1 and ME_2. This voltage drop is output voltage V_o and is proportional to the impedance between ME_1 and ME_2.

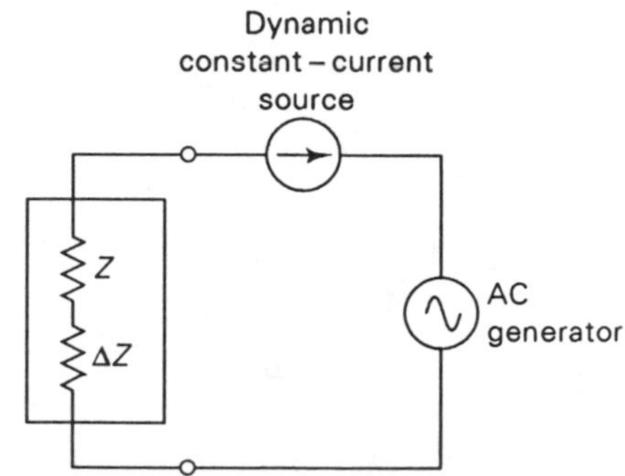

Figure 13–7 Constant-current measurement.

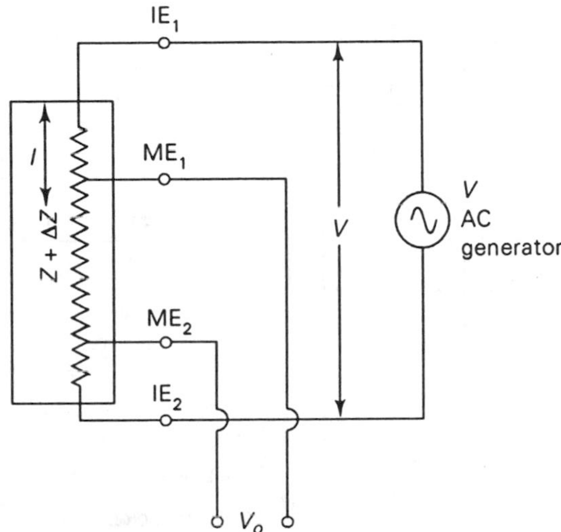

Figure 13–8 Four-electrode impedance measurement.

Another tetrapolar scheme, used in impedance encephalography, is shown in Fig. 13–9. In this variation on the method, a voltage signal V is applied through transformer T_1. Another voltage signal, phase-shifted by standard impedance Z_{std} comprising C and R, is applied to the patient through T_2. The output signal V_o is taken from the reference end of T_2 and the fourth electrode on the patient's head.

A *hexelectrode* scheme is shown in Fig. 13–10. In this method a series of six electrodes are arranged in two groups of three electrodes each. The center electrode in each triplet is the signal electrode (S_1 and S_2), while the

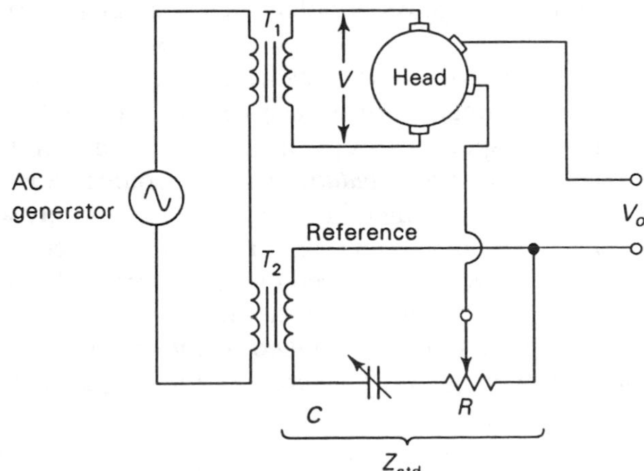

Figure 13–9 AC impedance measurement.

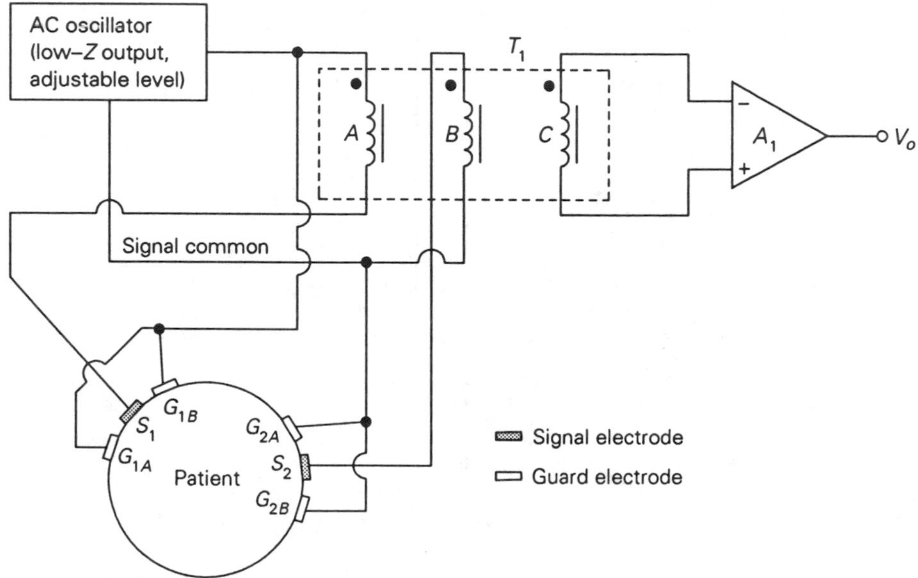

Figure 13–10 Multielectrode impedance measurement at subject's scalp.

outer two are guard electrodes (G_{1A}, G_{1B}, G_{2A}, and G_{2B}). The signal is applied to these electrodes by way of a special three-winding transformer. Although ordinary three-winding transformers have been used successfully, a better solution is to use a *trifilar-wound toroidal transformer*. These transformers are wound on a doughnut-shaped form made of ferrite, powdered iron, or ferroceramic (depending on the operating frequency). The wires making up the windings are wound over the form together so that wires for windings A, B, and C are always parallel to each other at all points on the form.

The ac source used for the impedance measurement can be an audio generator or oscillator circuit. Most such devices are *single-ended*, that is they are unipolar with respect to ground or common. Some of the impedance measurement schemes require a *balanced output* source (e.g., Fig. 13–8), whereas in others it is recommended because of patient safety and measurement artifact considerations. Figure 13–11 shows two circuits that convert single-ended signal sources into balanced output sources. The circuit in Fig. 13–11(a) uses an op-amp connected in the noninverting follower with gain configuration. The output voltage applied to the isolation resistors R is a function of the input voltage V_{in} delivered by the single-ended source, according to the equation

$$V_o' = V_{\text{in}}\left[\left(1 + \frac{R_2}{R_1}\right)\left(\frac{N_p}{N_s}\right)\right] \tag{13–19}$$

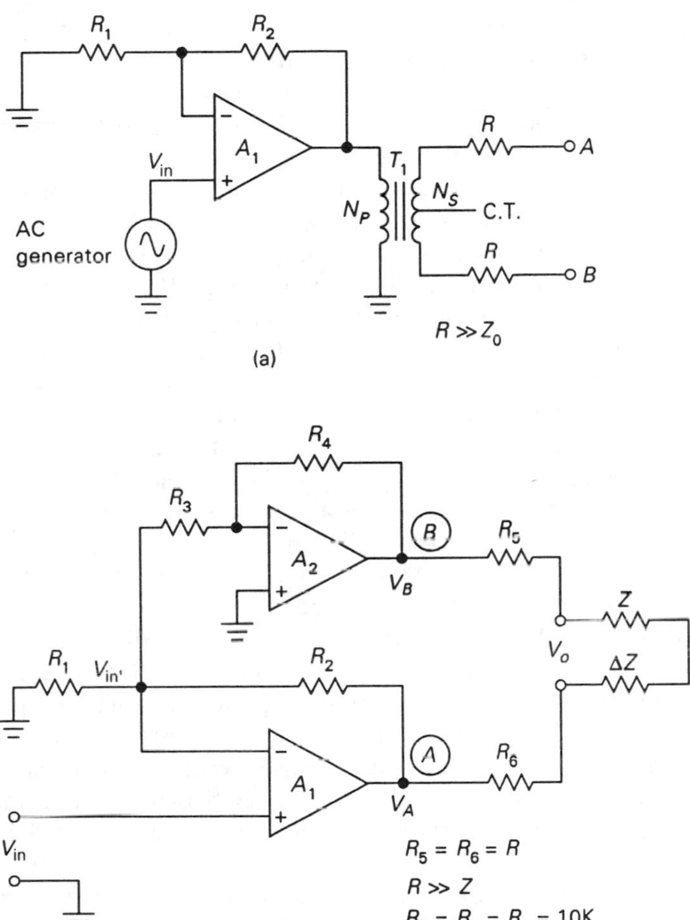

Figure 13–11 (a) Generator circuit for ac impedance measurements using a center-tapped transformer; (b) op-amp version.

where V_{in} is the input voltage applied to the noninverting input
V_o is the output voltage applied to the isolation resistors across the transformer secondary
R_1 and R_2 are the resistances in the amplifier feedback network
N_p is the number of turns in the transformer primary winding
N_s is the number of turns in the transformer secondary winding

The transformer selected for T_1 must be a low-capacitance type and must be capable of operating at the frequency used to make the impedance measurement. Because most low-cost transformers are audio models and have a poor frequency response at the frequencies above 20 kHz where physiological impedances are typically measured, it is uncommon to find these transformers used—despite their ready availability.

The alternative circuit shown in Fig. 13–11(b) uses two op-amps to form a balanced output, without using a transformer. The input amplifier A_1 is connected in the noninverting follower configuration. The output of A_1 (V_A) appears at point A and is twice the applied voltage V_{in}.

The input of A_2 is taken from the inverting input of A_1. Because of one of the properties of the ideal op-amp, this voltage (V'_{in}) is equal to V_{in}. Because A_1 has a gain of 2, A_2 must also be designed to provide a gain of 2. This is accomplished by making $R_1 = R_2 = R_3$, and making R_4 equal to twice R_3. In practical circuits a common value for R_1 through R_3 is 10K ohms, so R_4 would be 20K ohms. The voltage at point B (V_B) has the same magnitude as V_A but is inverted 180°. Thus, the algebraic sum of these is the voltage appearing across $A-B$ and is balanced.

A simple circuit such as that in Fig. 13–12 can be used to discern the magnitude and phase angle of the impedance being measured. The excitation signal is applied to the patient in the manner discussed previously. The resultant output voltage V_o of the impedance-measuring electrodes becomes the input signal to a differential amplifier A_1. The output of this amplifier is filtered to remove artifact and then passed to two circuits: a *magnitude detector* (such as a full-wave precise rectifier)[6] and a *phase-sensitive detector* (PSD). The PSD circuit discerns the phase of angle of the impedance signal by comparing it to a reference phase signal ϕ provided by the ac generator. As long as the propagation time through A_1 and the filter is small compared with the period of the ac excitation signal, the measurement provided by the PSD will be an accurate reflection of the impedance angle contribution.

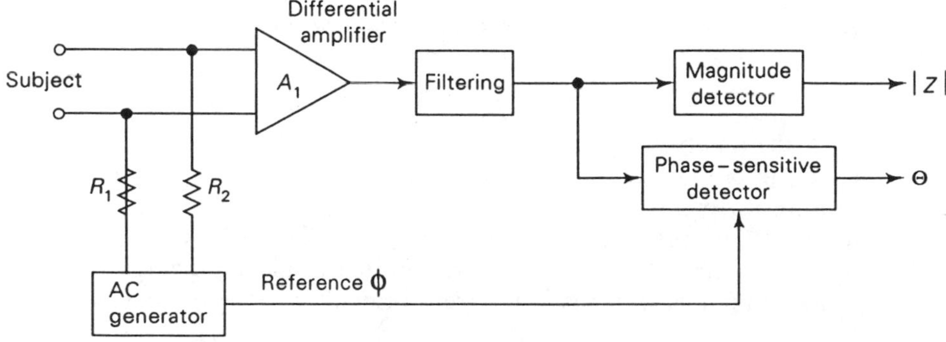

Figure 13–12 Impedance measurement using the previous generators.

SOME APPLICATIONS

In this concluding section we examine two simple applications of impedance or resistance measurement in physiology: galvanic skin resistance and impedance pneumotachometry.

Galvanic Skin Resistance (GSR)

Galvanic skin resistance (*GSR*) is a generic term referring to either of two phenomena or to both: *electrical skin resistance* (*ESR*) and *galvanic skin responses* (also *GSR*). The ESR phenomenon is due to the operation of sweat glands in the skin. Sweat glands secrete a saltwaterlike fluid, so they markedly decrease the resistance of the tissue in which they reside when sweat activity is high.

The galvanic skin response comprises two components: the exosomatic *Fere effect* and the endosomatic *Tarchanoff effect*. The Fere effect is a change in resistance (δR in Fig. 13–13) due to external stimuli. Some research indicates that a decrease in the Fere resistance indicates arousal, whereas relaxation is accompanied by an increase in Fere resistance. Unfortunately, the Fere resistance is not narrowly enough defined for exosomatic stimuli to be reliable as an arousal indication, the spurious claims of the lie detector industry notwithstanding. The Tarchanoff effect is the generation of a minute potential (≤ 1 mV) between an area that is rich in sweat glands (hence also rich in ions) and an area that is devoid of sweat glands (see Fig. 13–13).

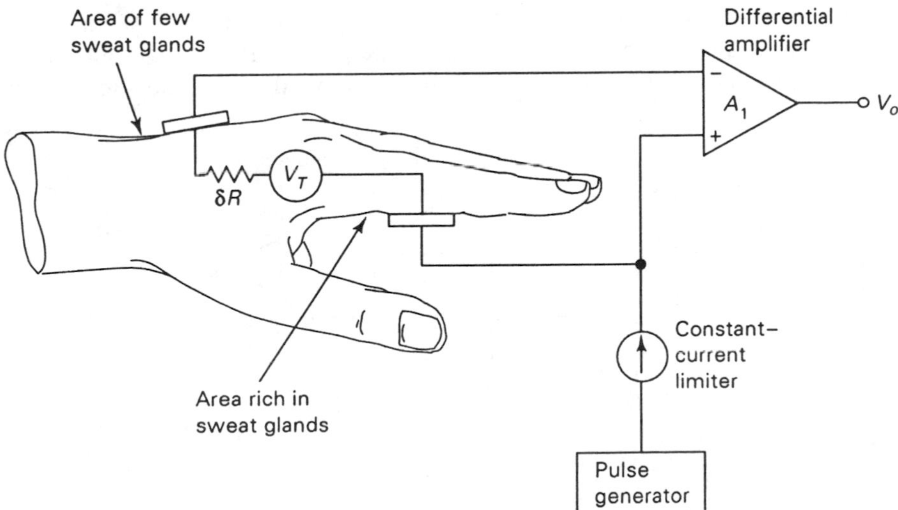

Figure 13–13 Galvanic skin resistance measurement.

This potential can theoretically be measured, but it is very tiny and rides on a large electrode potential. It is therefore not often measured.[7]

The GSR measurement is made by either of two methods. With one of the schemes shown earlier in this chapter, the GSR can be measured using an ac excitation frequency on the order of 2 to 30 Hz. Another method (shown in Fig. 13–13) uses a pulse generator driving a constant-current limiter (CCL), such as a FET diode. The pulse generator and CCL should be designed to provide a 4 ms, 10 μA peak pulse, with a pulse repetition rate of 2.5 Hz or so. Expected values range from 20K ohms to 200K ohms in healthy people, and as much as 1M ohm in cases of certain disease functions. (The GSR is not, however, diagnostic.)

Because of the exosomatic Fere response, galvanic skin resistance can be interpreted to represent psychological state, or at least changes of emotional state, under controlled circumstances. Interpretation of such events remains an obscure art, however. The so-called lie detector industry has promulgated GSR and other physiological changes as indications of deceptiveness. According to the theory, telling a lie causes certain small changes in respiration, blood pressure, and GSR. Although those claims have long been discredited by credible scientists, security personnel continue to sell their services to clients.[8]

Impedance Pneumotachometry

Figure 13–14 shows the configuration for a thoracic impedance pneumotachometer, that is, a device that measures respiration rate using thoracic impedance. When a person breathes, the lungs expand and thereby change the dielectric constant of the lung tissue by a small amount. This change is reflected as a ΔZ component of the chest impedance. If a low-amplitude 50 to 250 kHz ac constant excitation current I_e is applied across the chest, as in

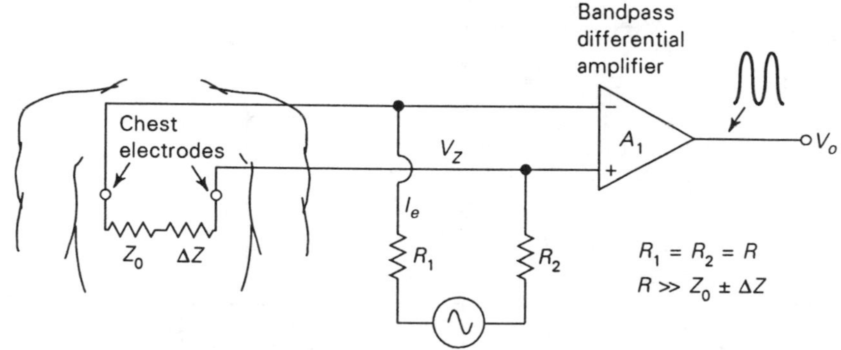

Figure 13–14 Thoracic impedance measurement to acquire data about respiratory system.

Fig. 13–14, then the voltage drop appearing across the chest electrodes is proportional to the thoracic impedance Z_0. As the person breathes, the impedance changes to $Z_0 \pm \Delta Z$, so the voltage V_z across the electrodes changes in step with the breathing. This produces a series of amplitude changes that resembles the respiratory waveform. Unfortunately, the only datum implicit in the impedance waveform is the breathing rate, not the volume, flow rate, or lung compliance.

In some clinical medical instruments, especially those intended for neonatal intensive care use, the impedance pneumotach electrodes are the same as the chest electrocardiograph (ECG) electrodes. Although the excitation signal is on the same order of magnitude as the ECG signal, the ECG is bandwidth-limited to <100 Hz, and the pneumotach signal is 50 to 250 kHz, depending on the design of the excitation circuit. It is therefore relatively easy to separate the two components through simple filtering.

REFERENCES AND NOTES

1. Geddes, L. A., and L. E. Baker. *Med. Biol. Eng.* 5: (1971) 271–293. Reproduced in Geddes, L. A., and L. E. Baker. *Principles of Applied Biomedical Instrumentation.* New York: John Wiley, 1968.
2. Ibid., *Principles.*
3. Ibid.
4. Lifshitz, Kenneth. "Electrical-Impedance Cephalography (Rheoencephalography)." Chapter 2 in *Biomedical Engineering Systems.* New York: McGraw-Hill, 1970.
5. Geddes and Baker, *Principles.*
6. Carr, Joseph J. *Integrated Electronics.* San Diego, Calif.: Harcourt, Brace, Jovanovich, Publishers, 1990.
7. Strong, Peter. *Biophysical Measurements.* A Tektronix Measurement Concept Series Book. Beaverton, Oreg.: Tektronix, Inc. 1973, pp. 195ff.
8. Lykken, David Thoreson. *A Tremor in the Blood: Uses and Abuses of the Lie Detector.* New York: McGraw-Hill, 1981. Lykken's treatment is thorough and fair, and in the end he finds "lie detectors" lacking in scientific evidence and indeed finds substantial evidence against them.

14 Chemical Sensors

In this chapter we discuss several different forms of sensors: humidity, gas/vapors, and pH. These electrodes are used in a variety of applications from medicine, to industry, to scientific research. Some are even used to run automobiles today, especially since computerized controls are now routinely used to control unwanted emissions.

HUMIDITY SENSORS

Dry air is a gas consisting of approximately 78 percent nitrogen (N_2) and 21 percent oxygen (O_2); the remaining 1 percent encompasses "all others." When water vaporizes, it becomes gaslike and enters into the air. *Humidity* is a measure of the water vapor content of air. Dry air has zero humidity, while air that holds all the water that it possibly can is said to be *saturated*.

Absolute humidity (AH) is measured in terms of water mass per unit volume of air (e.g., kilograms per cubic meter, kg/m^3) and gives the amount of water in the air. The humidity most often quoted in weather forecasts is the *relative humidity* (RH), which is specified in terms of water parts per million parts of air, or as a percentage. By definition, relative humidity is defined as the ratio of the absolute humidity of the air to the saturated absolute humidity at the same temperature, or

$$\text{RH} = \frac{\text{mass } H_2O/m^3}{\text{mass } H_2O/m^3 \text{ saturated}} \tag{14–1}$$

Figure 14–1 shows that humidity is a nonlinear function of air temperature. For any given temperature and relative humidity a maximum water vapor content is possible. If any more water tries to evaporate, the *dew point* is reached, and condensation (rain or fog) takes place.

The dew point can be detected by using a pinpoint light source reflected from a chilled mirror surface into a photocell. As long as there is no moisture on the mirror surface, the light level into the photocell is high. But when the mirror fogs (which occurs at the dew point), the reflected light is diffused, and the photocell therefore sees a reduced light level. The sudden drop in light level indicates the dew point.

Absolute humidity is sometimes measured with a *chemical hygrometer*. This device consists of a series of U-shaped tubes filled with a dry material that can absorb water. The mass of the dry tubes is measured prior to the test. A known volume of air is passed through the tubes so that the water vapor can be absorbed into the material. When the known volume of air

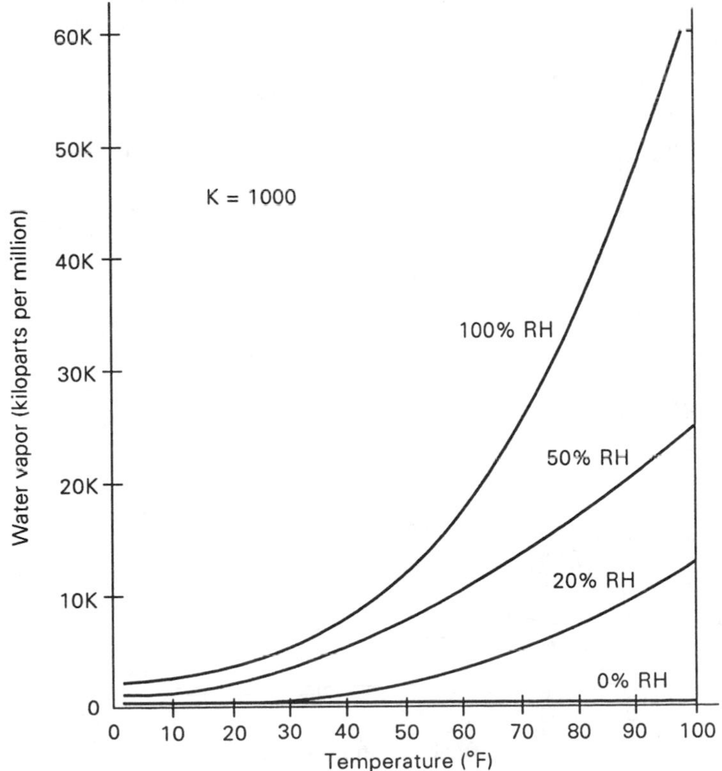

Figure 14–1 Graph of water vapor vs. temperature for three different relative humidities.

has cleared the tubes, they are weighed again. The difference in mass is the water vapor content of the air. Because a known volume of air was used, it is possible to calculate the mass per unit of volume.

Relative humidity is sometimes measured with a device called a *sling psychrometer*. This instrument contains two thermometers, one of which is directly open to the air, while the other is immersed in a wet cloth, wick, or sponge. The psychrometer is rotated in the air being measured for several minutes, enough to permit a difference in the two temperatures to register. The wet bulb will show a lower ("suppressed") temperature that is a function of the relative humidity. Figure 14–2 shows the relationship among the suppressed temperature, the air temperature, and the relative humidity. The RH can be determined from the intersection between the two temperatures.

Figure 14–3 shows two relative humidity sensors that are based on the psychrometer principle. The sensor shown in Fig. 14–3(a) is the Shibaura HS-5. It consists of two thermistors (thermal resistors), one inside a sealed (dry) chamber and the other open to the air. Both thermistors are operated close to the self-heating point so that relatively small temperature changes create large resistance changes. When operated in a Wheatstone bridge

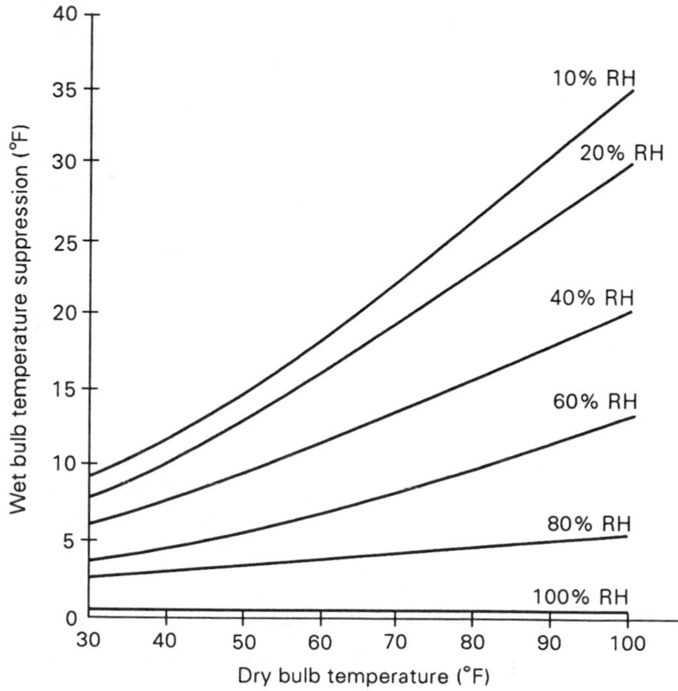

Figure 14–2 Graph of wet bulb vs. dry bulb temperatures for different relative humidities.

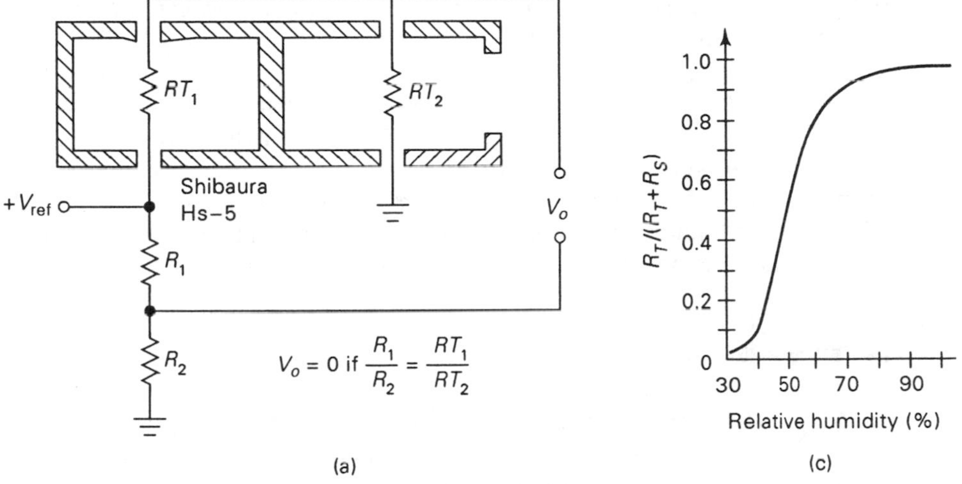

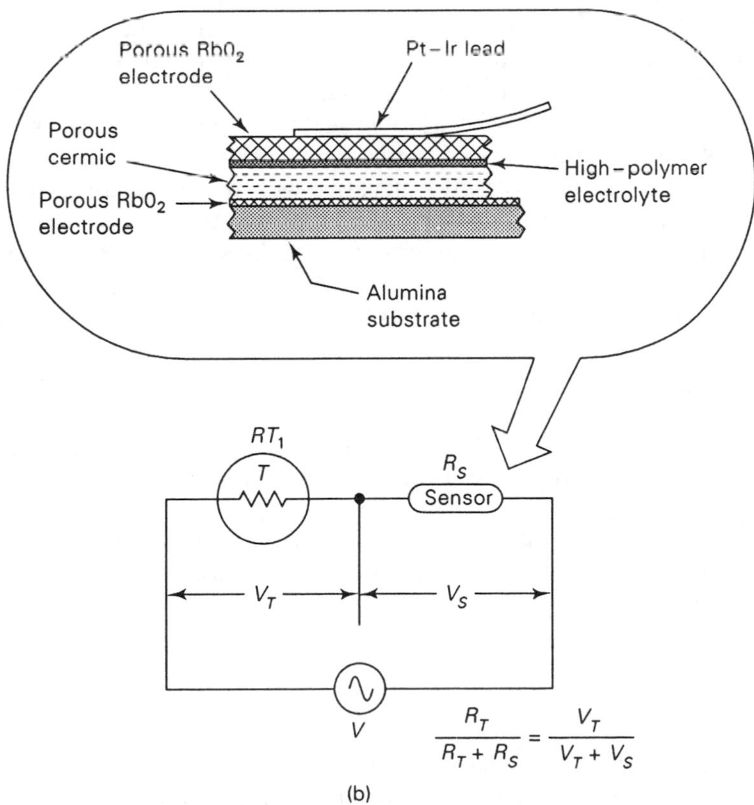

Figure 14–3 (a) Shibaura resistive relative humidity sensor; (b) structure of the sensor element; (c) calibration curve.

configuration, the HS-5 will produce a voltage (V_o) that can be used to determine the relative humidity.

Figure 14–3(b) shows the Figaro Engineering NH-series humidity sensor. It consists of a single assembly containing a thermistor RT_1 and a special absorptive sensor resistor element R_s. The sensor element consists of a high-polymer electrolyte deposited onto a porous ceramic bed mounted on an alumina substrate. A porous rubidium oxide (RbO_2) electrode covers the polymer. The humidity of the air causes the resistance of the ceramic/polymer to change in an exponential manner [Fig. 14–3(c)].

OXYGEN SENSORS

The oxygen (O_2) content of gases and liquids is measured using any of several different sensor methods. Some older sensors used the paramagnetic properties of oxygen, but those instruments were fragile and touchy to operate. Later sensors relied on the operation of oxygen on platinum (Pt) and silver–silver (Ag–AgCl) chloride electrodes. Still later methods used a kind of fuel cell in which the voltage generated was proportional to the oxygen level.

The *Clarke electrode* represents a class of oxygen-sensing electrodes that measure the partial pressure of O_2 in solution. These electrodes operate by bringing an electrode made of a noble metal, Pt being most common, into contact with the solution. The noble-metal electrode is given an electrical potential bias of about 1 V and is negative with respect to a reference electrode, usually calomel or Ag–AgCl.

Figure 14–4 shows a typical configuration for a Clarke electrode in which both a Pt noble metal electrode and a Ag–AgCl reference electrode are immersed in a potassium chloride (KCl) electrolytic solution. The two electrodes and KCl solution are housed in a glass probe. The Pt electrode consists of a small plug of the expensive noble metal, while the reference electrode consists of a length of Ag–AgCl wire.

The open end of the glass probe is covered by a very thin (~25 μm) membrane of polyethylene or Teflon®. This membrane is said to be *semi-permeable* to oxygen, that is, it will pass oxygen but not other molecules. Oxygen passes through the membrane and diffuses through the 0.1 *M* KCl electrolytic solution to the interface with the 8 to 10 μm Pt plug. An electrical current is set up at the interface with a value of

$$I = \frac{4\mathcal{F} A \alpha_m D_{im} P_{O_2}}{\Delta x_1} \tag{14–2}$$

where $\mathcal{F}$ is Faraday's constant (9.65×10^7 coulombs)

A is the electrode area in square meters

α_m is the oxygen solubility constant for the membrane

D_{im} is the diffusion coefficient for the membrane in square kilometers

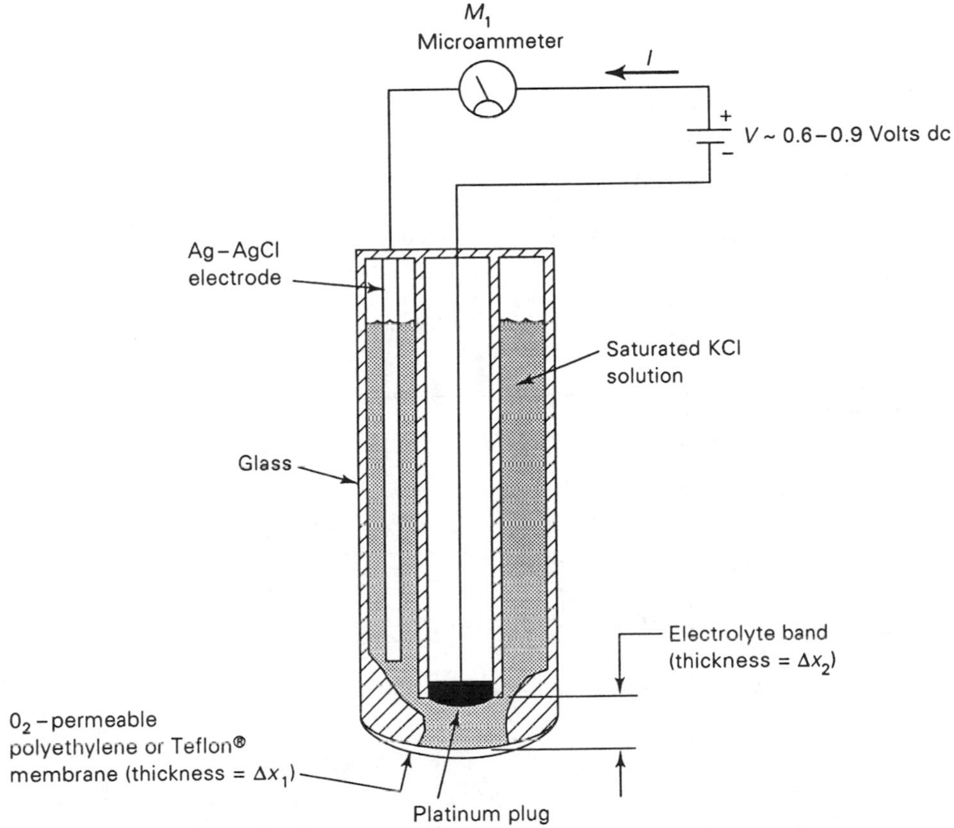

Figure 14–4 Clarke oxygen electrode.

per second
P_{O_2} is the partial pressure of oxygen in newtons per square meter
Δx_1 is the thickness of the membrane in meters

When the electrode is saturated, that is, at a potential greater than about -900 mV, the current will be on the order of 10^{-12} A/torr, or 7.5×10^{-12} A/kPa.

The time constant t of the electrode is the time required for the current to rise to 95 percent of its final value and is given approximately by

$$t_{95\%} = \frac{\Delta X_m^2}{D_m} \times e^{-1} \text{ s} \tag{14–3}$$

Depending on the thickness and diffusion coefficient of the electrode, the time constants of commercial Clarke electrodes are on the order of 5 to 90 s, with 25 s being a commonly encountered value.

A different approach to oxygen sensor design is shown in Fig. 14–5(a).

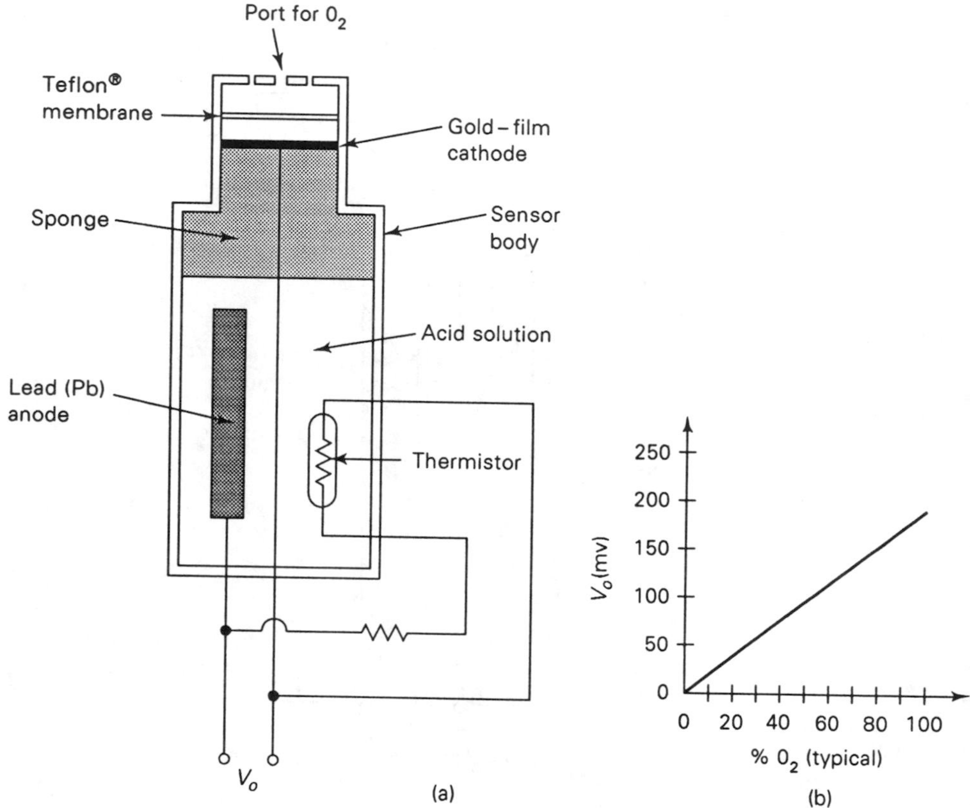

Figure 14–5 (a) Galvanic-cell oxygen electrode; (b) typical calibration curve.

This type of sensor is a *lead-acid galvanic cell* that develops a potential difference that is a function of the oxygen level. A lead (Pb) anode is immersed in an acidic solution inside the sensor body. At the sensing end of the device, a thin gold-film cathode is in contact with a sponge that separates it from the acidic solution, even though the acidic solution is absorbed into the sponge. Oxygen (O_2) diffuses through a Teflon® membrane that is semipermeable to O_2, to be reduced at the cathode surface. The current flowing between the electrodes is proportional to the oxygen concentration, so the potential difference appearing across the electrodes is also proportional to oxygen concentration.

A typical sensitivity curve for a galvanic-cell oxygen sensor is shown in Fig. 14–5(b). The characteristic approximates a straight line between 0 and 100 percent oxygen concentration. For such a response a simple millivoltmeter will suffice for the output display. The meter is calibrated in percentage oxygen but actually displays voltage.

ELECTRODES FOR REDUCING GASES/VAPORS

A number of sensor-based instruments on the market measure the concentrations of reducing gases or vapors in the air. Examples include breath-alcohols analyzers used by police departments, carbon monoxide (CO) analyzers used in performing emission control measurements on vehicles, and methane detectors used to protect against explosions and other dangers from natural gas. All these applications have three things in common: They are relatively low cost, they are operated by ordinary people rather than scientists and engineers, and they are manufactured using similar technology.

The Figaro TGS gas sensors are based on a technology that uses powdered tin dioxide (SnO_2) sintered onto a semiconductor substrate [Fig. 14–6(a)].[1] In normal operation the sensor element is heated to approximately 400°C. Oxygen is adsorbed onto the surface of the SnO_2, where the oxygen molecules accept electrons. These electrons create a relatively high electrical potential barrier that is difficult for free electrons to cross. As a result, the electrical resistance is high [Fig. 14–6(b)] and is a function of the partial pressure of oxygen (pO_2). When a reducing gas or vapor (e.g., CO, methane, methanol) is present, it is adsorbed onto the surface and reacts with the oxygen, thereby reducing the resistance of the device. Figure 14–6(c) shows the ratio of the actual sensor resistance R_s to a standard resistance R_0 for several different elements. The standard resistance R_0 is the value of R_s in an atmosphere of 1000 parts per million (ppm) methane gas.

A typical circuit for the TGS sensors is shown in Fig. 14–6(d). The heater voltage V_H heats the sensor element to the required temperature, while the operating voltage V_c provides excitation to the sensor element. A load resistance R_L is used to convert current flowing in the sensor to an output voltage V_o. The values of V_c and V_H vary from one sensor to another, but are typically in the range of 0.5 to 12 V.

pH ELECTRODES

The pH electrode is used to measure the acidity or alkalinity of solutions. A solution is acidic when the hydroxyl ion activity is less than the hydronium ion activity; it is alkaline (basic) when the hydroxyl ion activity is greater than the hydronium ion activity.[2] An acid is a solution having an excess of hydrogen ions, although any solution that gives up protons is an acid.[3] An operational definition of pH measurement compares the pH of the unknown solution (pH_x) to the pH of a standard buffer solution (pH_s):

$$pH_x = pH_s + \frac{E}{2.302RTF^{-1}} \tag{14–4}$$

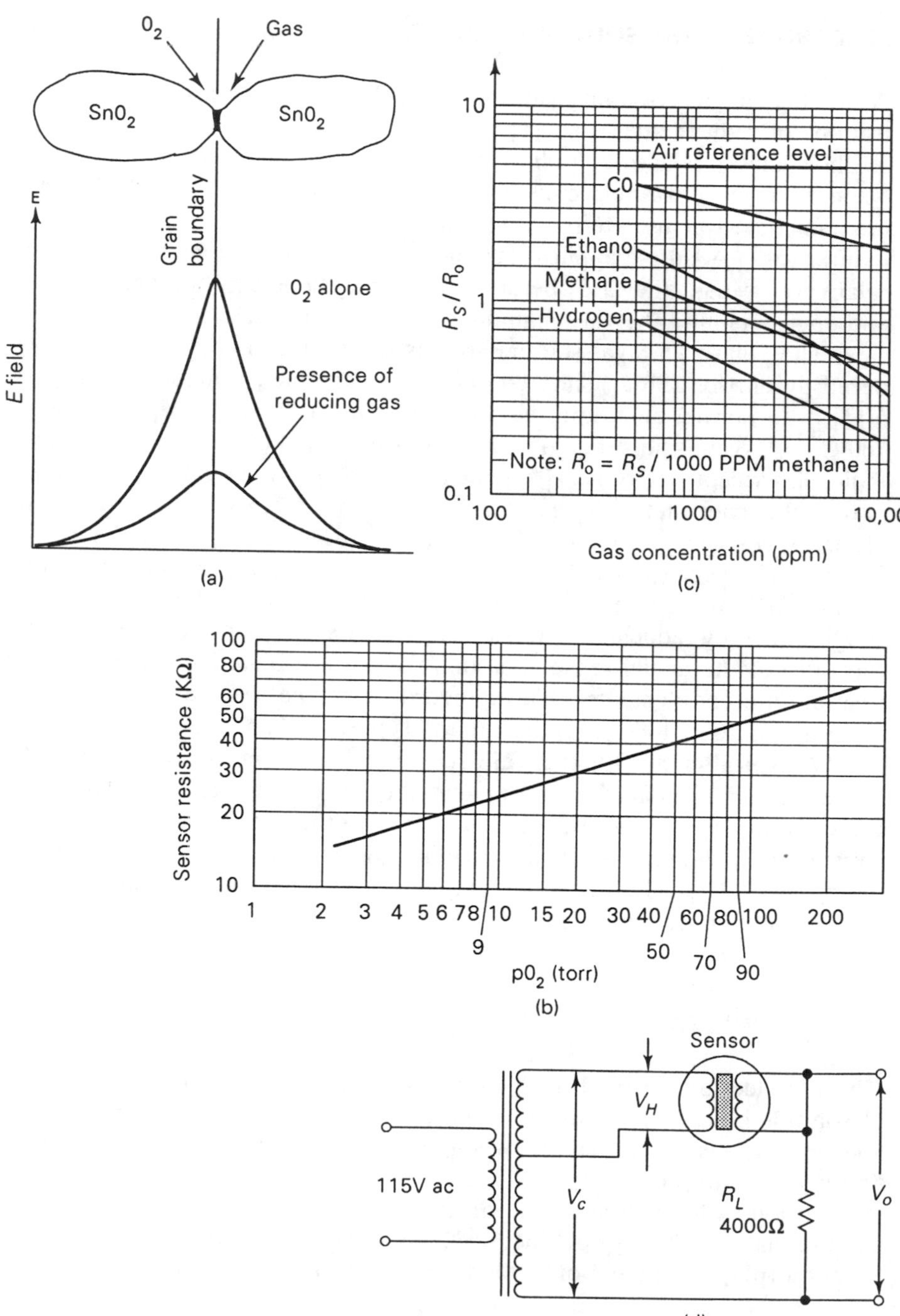

Figure 14–6 (a) Figaro gas/vapor sensor uses sintered SnO_2; (b) Sensor resistance vs. partial pressure of oxygen; (c) R_s/R_0 ratios for various gases and vapors; (d) typical circuit diagram.

where E is the electrical potential produced by the action of the unknown solution pH on a standard electrode
R is the gas constant (8.314×10^3 J · K^{-1}/kmol)
F is the Faraday constant (9.65×10^7 coulombs)
T is the ambient temperature in kelvins

The pH scale extends from 0 to 14 within error boundaries of ±0.4 pH. The pH of ultrapure water, which is neither acid nor alkaline, is 7.0; pH values lower than 7 indicate an acidic solution, while those higher than 7 indicate an alkaline solution.

A typical pH electrode sensor pair is shown in Fig. 14–7. The pH of the unknown solution is measured with the *pH electrode*, while the *reference electrode* measures the pH of a standard solution.

The pH electrode consists of a Ag–AgCl metallic electrode immersed in a chloride buffer and insulated by a glass sleeve (the glass sleeve may have a metal outer shield against electromagnetic interference). The tip of the Ag–AgCl electrode is inside a chloride buffer–filled glass bulb that is a very thin permeable membrane that allows the hydrogen ions to diffuse into the buffer. An electrical half-cell potential is formed at the interface with the Ag–AgCl electrode.

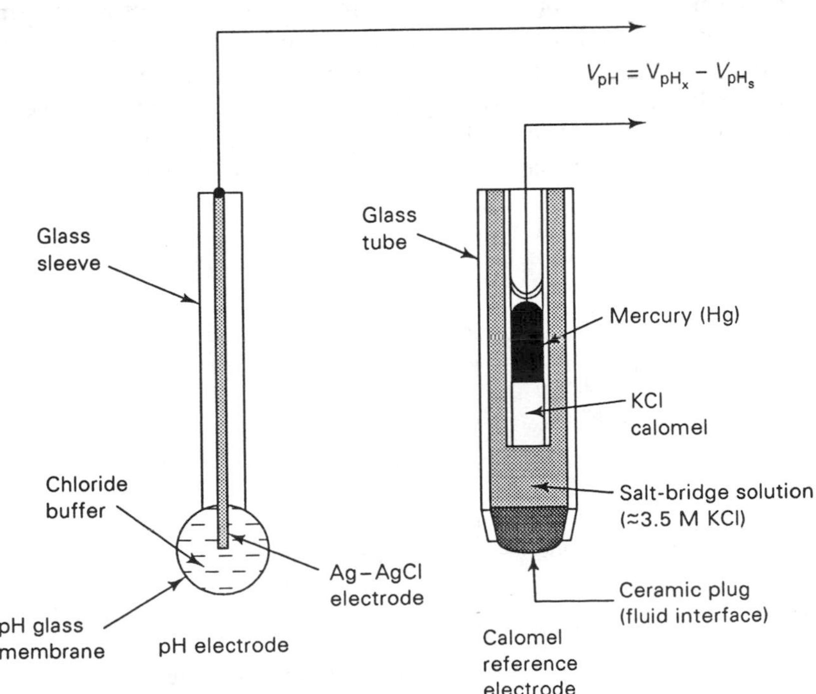

Figure 14–7 pH electrode and reference electrode construction.

The reference electrode consists of a glass tube filled with a salt-bridge solution (usually 3.5 *M* KCl). The actual electrode surface is a KCl-calomel plug backed with a mercury (Hg) electrode. This electrode also produces a half-cell potential. The pH measurement is made by measuring the differential voltage between the pH electrode and the reference electrode.

A millivoltmeter can be used as the pH meter instrument, although some means must be provided to compensate for temperature and for differences in the actual voltage vs. pH characteristic of the particular electrode in use. It is common practice to calibrate pH electrodes using pure water for pH = 7.0, and standard solutions for at least two acid and base points.

CARBON DIOXIDE (CO_2) SENSORS

Carbon dioxide (CO_2) can be measured in at least two different ways. In liquid solutions, the partial pressure of CO_2 can be measured with a selective pH electrode, whereas in air or other gases measurements are made with instruments that utilize the absorption of infrared by CO_2.

Figure 14–8 shows the basic form of a sensor for measuring CO_2 dissolved in a liquid. An ordinary pH electrode is used in this measurement, but the pH glass bulb is protected from the liquid by a semipermeable membrane that will pass CO_2 but not other compounds. The electrode measures the pH of a buffer solution. When CO_2 molecules diffuse through the membrane, they alter the pH of the buffer solution, causing a change of pH that is proportional to the concentration of CO_2 in the unknown liquid.

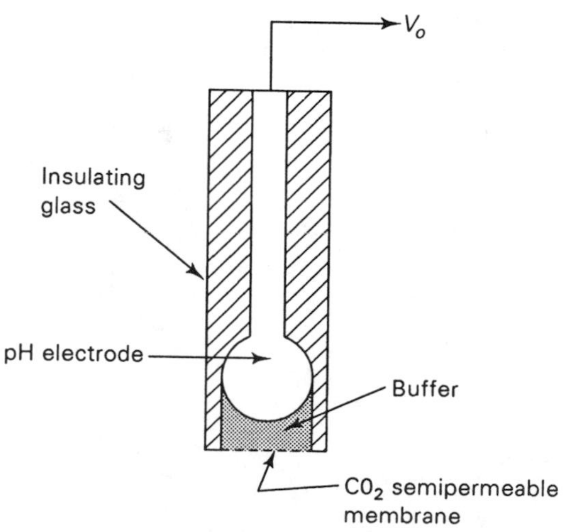

Figure 14–8 CO_2 electrode uses a pH electrode and a CO_2–sensitive membrane.

An instrument for measuring the percentage of CO_2 in air or another gas is shown in Fig. 14–9. It is a photocolorimeter that uses an IR sensor and an IR source. Carbon dioxide absorbs IR at three wavelengths, but the 4.26 μm wavelength is easy to detect using lead selenide (PbSe) photoresistors (PC_1 and PC_2). The circuit is a Wheatstone bridge, so the output voltage is zero when the following relationship is satisfied:

$$\frac{R_{PC1}}{R_1} = \frac{R_{PC2}}{R_2} \tag{14–5}$$

The instrument in Fig. 14–9 uses two PbSe sensors: One (PC_2) looks at the IR source through a channel that is either empty or contains a pipette carrying a standard sample of the gas known to be free of CO_2, while the other (PC_1) looks through a pipette containing the unknown gas. If both paths are equal, with no CO_2 (or other IR-absorbing material), then the IR level at each sensor is equal, so the resistances of the two sensors are equal. Under these conditions $V_o = 0$. But if the CO_2 content of the unknown gas is nonzero, then the IR passing through that sensor is partially absorbed, so less reaches PC_1. An imbalance is therefore created, and V_o is nonzero by an amount proportional to the concentration of CO_2 in the unknown gas.

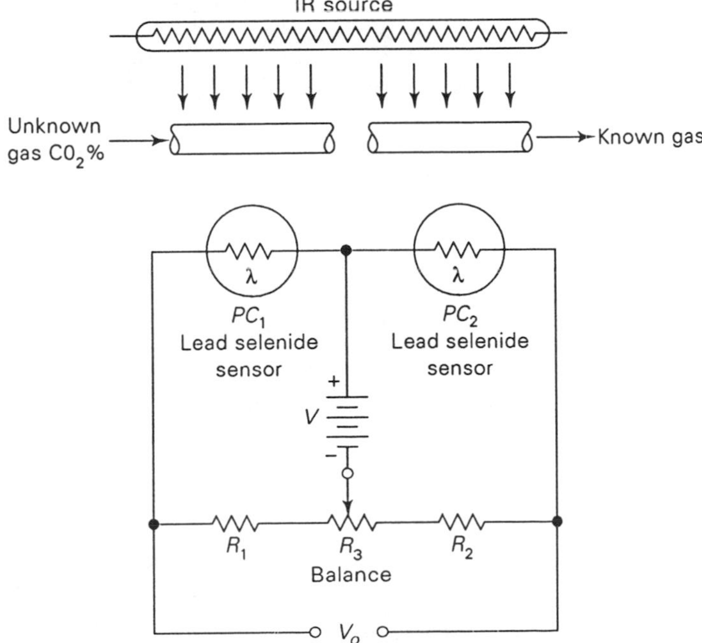

Figure 14–9 Photocolorimeter for measuring CO_2 in air.

REFERENCES AND NOTES

1. Figaro Products and TGS sensors catalogs 1991. Figaro USA, Inc., P. O. Box 357, Wilmette, Illinois 60091.
2. Cobbold, Richard S. C. *Transducers for Biomedical Measurements*. New York: John Wiley, 1973.
3. *Handbook of Chemistry and Physics, 56th ed.* Cleveland, Ohio: CRC Press, Inc., 1976.

Index

Q, R

U

V

W, X